FOCUS ON BIODIVERSITY RESEARCH

FOCUS ON BIODIVERSITY RESEARCH

JAN SCHWARTZ
EDITOR

Nova Science Publishers, Inc.
New York

For permission to use material from this book please contact us:
Telephone 631-231-7269; Fax 631-231-8175
Web Site: http://www.novapublishers.com

LIBRARY OF CONGRESS CATALOGING-IN-PUBLICATION DATA

Focus on biodiversity research / Sulehka Bhati, (editor).
p. cm.
Includes bibliographical references and index.
ISBN 13 978-1-60021-372-4
ISBN 10 1-60021-372-3
1. Biodiversity. I. Bhati, Sulehka.
QH541.15.B56F63 2006
578.7--dc22 2006023094

Published by Nova Science Publishers, Inc. ✢ New York

CONTENTS

PREFACE

Biological diversity, or "biodiversity," refers to the variety of all life on earth, and the complex relationships among living things, and between living things and their environment. Biodiversity includes genetic variety, species diversity, and variability in communities, ecosystems and landscapes. Biodiversity sustains the environments in which we live and on which our lives and those of every other living creature on Earth depend. Thanks to biodiversity, we are able to obtain such necessary goods as food, clothing, medicine, and fuel. Equally important are the ecosystem services that biodiversity provides, such as clean air and drinkable water. Conservation scientists have identified a number of universal threats to biodiversity: habitat loss and degradation, invasive species, pollution, overpopulation, overexploitation and consumption, and global climate change. This new book examines critical issues in this field from researchers around the globe.

Chapter 1 - Based on the present species loss, perhaps half of all existing plant, animal, and microbe species on earth will become extinct by the end of the 21st century. This rate of extinction is alarming because of the vital role that most species play in a sustainable environment and economy. The prime causes of extinction are human population growth and associated loss of habitat, plus introduced alien species. Most of the 10 to 50 million species on earth exist in the 78% of the terrestrial ecosystem that includes managed agriculture, forestry, and human settlements. In a rich hectare of soil more than 20,000 species may be present and just the biomass of species of fungi, bacteria, earthworms, and arthropods may weigh 9,000 kg. Insect pollinators are essential for one-third of U.S. food production with a value of $40 billion per year. Thus, conserving biodiversity is vital to our environment and economy. One win-win conservation approach is to maintain sustainable agriculture and forestry systems because most plant, animal, and microbe species exist in these favorable ecosystems.

Chapter 2 - Global climate change has been widely recognized as a major threat to biodiversity. A significant number of empirical studies provide evidence of species decline or extinction due to climatic change. By applying different modeling frameworks, the scientific community has ambitiously tried to provide politicians, stakeholders, and the public with predictions of the response of biodiversity to climate change. For example, a recent analysis based on a climate envelope model hypothesized that 15-37% of species in various regions of the world may go extinct within the next 50 years. However, despite the general appeal of such quantitative predictions, empirical distribution models have been criticized as being highly uncertain since they do not reflect local species interactions, genetic responses, or spatial processes. In contrast, bottom-up approaches such as mechanistic simulation models often consider species' life-history processes in a spatially explicit, stochastic context. However, mechanistic models rarely consider multi-species interactions and predictions may be invalid for other populations within the species' range. On the other hand, analytical food web models often do consider the dynamics of such multi-species interactions. However, the majority of such models is non-spatial and based on the assumption of a non-linear environment. In this chapter, I discuss the constraints and advantages of these three major modeling frameworks (empirical distribution models, bottom-up mechanistic simulation models, and analytical food web models) and synthesize how they have been applied to study the effects of climate change on biodiversity. I conclude that the combined consideration of a stochastic, spatial, and non-linear environment applied within the context of multiple species interactions may be useful in assessing the potential consequences of climate change for biodiversity. However, the current lack of empirical data, the complexity of such potential models, and the propagation of uncertainty remains a great challenge to be overcome in the future.

Chapter 3 - The 1992 United Nations Convention on Biological Diversity urges nations to conserve, to value, and to use biodiversity in a sustainable manner. Commercial developers of products based on biodiversity are urged to share equitably the benefits that may arise as a result of the effort with the steward of the biodiversity. In line with these objectives, a reciprocity measure as part of a plant-based drug discovery and development endeavor using the plant diversity of Cuc Phuong National Park, Vietnam, in the form of a micro-loan program was implemented for poor families of the Muong ethnic minority in the Cuc Phuong Commune located just outside of the park. Families identified by an Economic Development Village Board were provided a no-interest loan to be returned to the Commune People's Committee within 3 years, which the Committee then used to provide a rotating loan to other identified poor families in the community. Under this program, for the period of 2000 to 2006, 45 poor Muong families participating in the program have improved their household economy as a result of increased productivity, primarily, through the purchase of farming animals (cows, water buffaloes) to work their lands. This improvement of household economy has a beneficial effect in helping reduce the human pressure on the forest, and the biodiversity therein, of the Cuc Phuong National Park, aiding in the effort to preserve, value and use the Park's biodiversity for the future.

Chapter 4- Conflicting opinions regarding the relative importance of dispersal versus vicariance in understanding Caribbean biogeography continue to stimulate lively debate, despite recent advances in regional tectonics, geology, palaeogeography and palaeontology. The Greater Antilles are young geographical features, probably mid-Miocene and Hispaniola has a rich terrestrial invertebrate fossil record in the form of Dominican Republic amber inclusions. Spiders are common in Dominican amber and all fossil species belong in extant families. The island is unique in terms of its known spider fauna, in that a similar number of families have described species recorded on Hispaniola from both fossils in amber and from the extant fauna. It is also the region of the world where the amber fauna is most similar to the Recent fauna, and Miocene Hispaniolan spider biodiversity was presumably similar to that at present.

Fossils form empirical data with both phylogenetic and temporal implications and are of paramount importance for understanding historical biogeography. They can play a decisive role in falsifying hypotheses proposed solely on the distributions of extant taxa. It is also possible to generate hypotheses for both fossil and extant faunas that are ultimately falsibiable through the discovery of new Hispaniolan taxa. The amber spider fauna has clear affiliations to the present South American fauna, so how did they colonize the island from the mainland: dispersal or vicariance? The high Miocene biodiversity (including taxa with poor dispersal capabilities) on a geologically young island would seem to add more weight in favour of the vicariance model for explaining on-island spider lineages, although dispersal cannot be excluded for some of the highly dispersive taxa. However, a case-by-case approach is required in order to determine the relative contributions of the different models. The future research direction for understanding the biogeographical patterns will lie in the application of cladistic and phylogenetic biogeographical approaches. It is interesting to note that preliminary obersvations based on pedipalp morphology suggest some fossil taxa are more derived than extant forms. The contribution that the Dominican (and Mexican) amber fossil fauna can make to our understanding of the origins of Hispaniolan biodiversity, and Caribbean biogeography in general, should not be underestimated.

Chapter 5 - Under consideration are some metaphysical backgrounds of the concept of biodiversity (BD) as an immanent property of the evolving biota. Such a consideration makes it possible to deduce an idea of hierarchy as a principal property of BD, as well as to analyze the latter's structure in a general form.

It is stated that the recent "species-oriented" BD concept, though operational, is outdated, as it does not correspond to the current predominating taxonomic concept. Representation of taxic diversity (as a component of overall BD) in the form of hierarchical phylogenetic pattern seems to be more adequate to the recent developments in taxonomy. Its operational interpretation implies that different species could be "weighted" in respect to their position in the taxonomic hierarchy. This presumes a certain system of fixed taxonomic ranks and does not enjoy an idea of "rankless classifications" declared by neo-cladists.

As far as the species notion is concerned, the major challenge of recent taxonomic theory to the BD researches includes pluralistic treatment of the species which put certain limitations on counting and direct comparison of number of species in different groups of organisms. In this respect, the so called "Hull's principle" is reconsidered to stress that various species definitions are not just mutually exclusive but complementary. Some consideration is given to the focal point of the species problem: if any ontological model exists which might ensure there is *the* species as a universal fundamental unit of the structure of the entire biosphere.

It is stressed that one more component of overall BD structure, in addition to ecological and taxic diversity, could be recognized: it is the diversity of biomorphs (life forms). Some questions of their definition and classification are considered, and ideas of urgent need for a special classificatory discipline (biomorphics) dealing especially with that diversity is declared.

Chapter 6 - Spontaneous pattern formation, the appearance of inhomogeneities that are not directly imposed by external forces, can be an important but overlooked factor in the origin and maintenance of genetic diversity in populations, and in parapatric speciation. Using agent-based simulation models with multilocus genetics, we show that such patterns form in spatially distributed populations with local mating neighborhoods. Under these conditions, disruptive selection induces spontaneous pattern formation and coarsening in the spatial distribution of genotypes, and thus increases the persistence of genetic diversity. The lateral motion of hybrid zones between different genotypes of similar fitness is affected by their topological shape. Hence, the shape of these boundaries plays a crucial role in determining the rate of genetic variation and the long time behavior of the population. These patterns also interact with boundary and internal obstacles of habitat structure, so that diversity persists the longest in habitats with the most irregular geographical features. Localized obstacles contribute to the maintenance of genetic diversity even if they do not preclude interbreeding, by inhibiting the movement of hybrid zones between areas dominated by different genotypes. Our results imply that the most effective size of such obstacles for maintaining diversity is about the same as the radius of a local deme. These results are relevant to spatial ecology and conservation biology, in which both genetic diversity and the size and shape of population reserves have great import for the long-term survival of species in the wild. They also imply significant coupling of population level and ecological properties to evolutionary dynamics. The persistence and amount of diversity, the relationship of diversity to evolutionary change, and the survival of types is controlled by the genetically undetermined spatial distribution of types and the structure of the habitat.

Chapter 7 - The elucidation of biological and biochemical mechanisms of natural products with therapeutic action is of great utility for unveiling new chemical structure platforms on which to base the development of new drugs. Various organisms have been valuable sources for natural products with therapeutic activity. Among these, plants occupy a prominent position for the isolation of pharmacologically relevant metabolites. Approximately 25% of prescribed drugs come from plants. Paradoxically, there is a considerable lack of knowledge on plant chemistry compared to the existing diversity, particularly in the tropics. Tuberculosis (TB) is the leading cause of mortality due to a bacterial pathogen, *Mycobacterium tuberculosis*. More than 95% of 1.8 million deaths caused by TB in the year 2000 occurred in developing countries. Malaria, caused by the protozoan parasites of the genus *Plasmodium*, continues to be a major cause of children´s morbidity and mortality throughout the world, particularly in the tropics. Even though antitubercular and antimalarial agents are available to treat these infectious diseases, drug resistance has posed a major challenge to public health programs. Accordingly, new drugs are needed to face resistant strains and improve current drug regimens. Brazil has extensive areas regarded as biodiversity hotspots, including the Atlantic Forest in the coast and the Cerrado in central Brazil. Moreover, the Amazon is probably going to be regarded as a hotspot as its species are described and their chemical diversity is investigated. In order to maximize the chances of finding bioactive components amongst a vast number of plant species, ethnobotanical and chemotaxonomical approaches have been employed. In addition, a description of a few biological targets to screen against natural-compound library is presented because this is an important decision that is critical to the long-term success of a drug discovery program. Here we also intend to present an array of plants from the Brazilian biodiversity and related taxa that have been reported as having antitubercular or antimalarial activity.

Chapter 8 - Understanding the factors that drive spatial patterns in biodiversity is fundamental to developing effective conservation strategies. The relationships between net primary productivity (NPP) and species richness are well documented in the literature, yet the influence of spatial variation in productivity on species turnover has received relatively little attention. The aims of this study were to: i) compare patterns in species turnover among different taxonomic groups occupying spatial gradients that encompassed low variation in NPP and high variation in productivity; and ii) contrast these patterns with those produced by null models. We examined species turnover for birds, mammals and butterflies along replicate transects of approximately 1000 km split into 10 sampling sites. Six transects encompassed spatial gradients with little variation in NPP and six transects encompassed gradients with high variation in NPP. We calculated three representative indices of turnover and plotted these against distance between sampling sites to determine the decline in similarity among species assemblages with distance. Computer simulations were used to generate random (null) species distributions along transects based on simple 'random distribution models'. These models assumed no differences between species and a consistent decline in species assemblage similarity with distance. Species turnover was greatest along transects with high variation in NPP, especially for birds. Significant turnover for mammals and butterflies was also recorded along many low variation transects. Birds and butterflies exhibited the lowest turnover overall – mammals had high turnover in almost all transects. Patterns in turnover varied depending on the index used. The index emphasising continuity in species composition always recorded the greatest turnover compared to indices representing differences in species identities or richness. A null model assuming a linear decline in species assemblage similarity with distance was able to match observed patterns in turnover in many cases. It was most successful in predicting turnover along high variation transects and for mammals. In many cases, species turnover was predicted from the simple assumption of consistent decline in the similarity between sites with increasing distance irrespective of the gradient in environmental variability. However, important differences occurred between taxonomic groups and gradients of low vs high variation. These differences suggest that protecting a few species rich sites in regions of low variability may capture a substantial proportion of regional diversity for highly mobile groups with large range sizes, although this wouldn't be the case for less vagile and more geographically restricted species. Regions with high variation in NPP require a greater number of reserves owing to strong and consistent turnover irrespective of taxonomic group.

Chapter 9 Despite considerable immunological, epidemiological, genetic and pharmacological efforts, various diseases continue to take their toll in death and disability. Although the fear of infectious epidemics has begun to be controlled, contemporary humanity may become the victims of rapidly growing global epidemics of very dangerous non-infectious diseases including atherosclerosis, heart and liver diseases, stroke, diabetes, high blood pressure, obesity, different kinds of cancer, gout, disease of the joints, schizophrenia, senescence, etc. Many questions about the nature of the diseases do not receive the answers they deserve. Moreover many very important questions have not been formulated. Nothing, for example, is known about the factors and mechanisms of epidemic spread of non-infectious diseases or the origin of extraordinary diversity in clinical manifestations and severity of such diseases. This paper is designed to formulate and examine a new paradigm exploring the genesis and role of biodiversity in determination of the spread, pathogenesis, and individual variation in the signs, course and severity of both infectious and non-infectious diseases.

In: Focus on Biodiversity Research
Editor: Jan Schwartz, pp. 1-25

ISBN: 1-60021-372-3

Chapter 1

CONSERVATION OF BIOLOGICAL DIVERSITY IN AGRICULTURAL, FORESTRY, AND MARINE SYSTEMS

David Pimentel, Tsveta Petrova, Marybeth Riley, Jennifer Jacquet, Vanessa Ng, Jake Honigman and Edwardo Valero
College of Agriculture and Life Sciences
Cornell University, Ithaca, NY

ABSTRACT

Based on the present species loss, perhaps half of all existing plant, animal, and microbe species on earth will become extinct by the end of the 21st century. This rate of extinction is alarming because of the vital role that most species play in a sustainable environment and economy. The prime causes of extinction are human population growth and associated loss of habitat, plus introduced alien species. Most of the 10 to 50 million species on earth exist in the 78% of the terrestrial ecosystem that includes managed agriculture, forestry, and human settlements. In a rich hectare of soil more than 20,000 species may be present and just the biomass of species of fungi, bacteria, earthworms, and arthropods may weigh 9,000 kg. Insect pollinators are essential for one-third of U.S. food production with a value of $40 billion per year. Thus, conserving biodiversity is vital to our environment and economy. One win-win conservation approach is to maintain sustainable agriculture and forestry systems because most plant, animal, and microbe species exist in these favorable ecosystems.

INTRODUCTION

The present rate of species loss projects that half of all existing plant, animal, and microbe species on earth will become extinct by the end of this century [Myers, 2003a; Wilson, 2003]. This projected high rate of extinction due to human activities is alarming because many of these organisms are vital to the safe and productive function of ecological systems that sustain our planet and the global economy [Dirzo and Raven, 2003]. Indeed, agricultural productivity and public health depend on the activities of diverse natural biota. From 10 to 50 million species of plants, animals and microbes fulfill the ecological needs on earth [Pimm et al., 1995; Pimm, 2002].

Although efforts to curb the loss of biodiversity have intensified in recent years, we have not been effective in countering the accelerating human population growth and the increasing destruction of natural habitats. The introduction of alien invasive species throughout the world continues to alter and damage natural and managed ecosystems [Pimentel, 2002]. Moreover, additional complementary strategies are needed to protect small organism species, such as insects, bacteria, and fungi, which are essential to the structure and function of natural ecosystems [Price, 1988; Wilson, 1987; Soil Organisms, 2004].

Establishing national parks has been the prime focus of world biological conservation. Often overlooked but equally vital is the protection of biological diversity existing in our vast agricultural, and forest ecosystems, as well as within human settlements [Pimentel et al., 1992; Daily et al., 2001; Ricketts, 2001; Vandermeer, et al., 2002; Homer-Divine et al., 2003]. Together these cover approximately 78% of productive terrestrial ecosystems [AAAS, 2001].

In this article, we examine the necessity of species diversity in maintaining healthy and productive ecosystems as well as specific threats to biodiversity. The goal is to identify ecological strategies and policies that help enhance the conservation of biological diversity in natural, agricultural, forest, marine, and natural ecosystems.

Causes of Reduced Species Diversity

Over the past billion years, the natural processes of adaptation and diversification have tended to increase the number of species. However, especially over the past several centuries, the rapid escalation of human numbers [PRB, 2003] and the movement of humans into wild areas have resulted in a major and continuous decline in species diversity [Wilson, 1988; Novacek, 2001].

The rapid disappearance of animal and plant species is clearly illustrated in Britain. Thomas et al. (2004) reports that butterfly species have declined 71% during ~20 year period, bird species declined 54% over 20 years, and native plants species showed a 28% decline over 40 years.

In developed countries the use of natural resources is 100 to 600times more per capita than that in developing countries [Conservation Issues, 2003; Pimentel and Pimentel, 2003]. Even where progressive laws and land preservation efforts exist, the increasing demands of humans and their industries undermine biodiversity preservation and the value of various ecosystems [Machlis and Tichnell, 1985; Pimentel and Pimentel, 2003].

When the demand for resources necessary for sustaining the increasing human population exceeds what is available, the result is increasing malnutrition and stress. More than 3 billion people are currently malnourished worldwide -- facing shortages of calories, protein, vitamins A, B, D, and E, iron, and iodine [WHO, 2000]. This is the largest number of malnourished people in history.

Deforestation and Destruction of Natural Habitats

Tropical forests occupy only 6.5% of the world's terrestrial ecosystems, but are home to about 70% of the world's species [Myers, N. personal communication, Oxford University, 2004; Pimm, 1991; Pimm, 2002; Pimm and Brown, 2004]. To date, humans have destroyed more than half of the world's tropical forests. For example, two thirds of the forest area in Central America has been converted to agriculture for livestock production [Robbins, 2001; Rainforest, 2003]. About 19 million hectares of forests (about the size of Florida) are destroyed each year worldwide [World Bank, 2002] and approximately 60% of this deforestation is due to the conversion of forests to agricultural lands to maintain fiber and food production [Earthscan, 2002; Deforestation, 2003].

The destruction of even one tree species can have cascading effects resulting in significant losses of species, because up to 1,000 arthropod species, including parasites and predators, may be associated with that single tree species [Erwin, 1983]. Also, approximately 30 associated animal and microbe species die out, for every plant species that becomes extinct [Edwards, 1998].

Equally disturbing to the loss of plant species is the rapid loss of crop genotypes now occurring. Crop genotype diversity is vital to farmers in different climatic and regional conditions, as they encounter different complexes of pest insects, weeds, and plant pathogens [James, 1999].

Soil Erosion and Salinization

Each year an estimated 10 million ha of cropland worldwide are abandoned due to lack of productivity caused by soil erosion [Faeth and Crosson 1994]. Another 10 million ha/yr are critically damaged due to salinization, in large part as a result of irrigation and/or improper drainage methods [Thomas and Middleton, 1993].

Erosion rates on cropland in Asia, Africa, and South America average 30 to 40 t/ha/yr [Taddese, 2001]. In some regions, like Ethiopia, erosion rates on cropland may average as high as 100 t/ha/yr [Taddese, 2001]. While the United States' current rate of soil erosion from cropland is less than in many developing countries, it is still about 10 t/ha/yr, which is 10 times faster than the soil formation rate [NAS, 2003; Uri, 2001]. The economic cost of erosion in the United States is estimated at about $38 billion per year [Uri, 2001], while for the rest of the world economic losses are estimated to be $400 billion per year [Myers, 2003b].

The first component of the soil loss, due to erosion by rainfall and wind, is soil organic matter. This loss directly reduces the biomass of soil, vital to the survival of plants, animals, and microbes. In addition, the reduction of soil depth can hinder or totally destroy the land's productivity [Uri, 2001].

In streams and lakes, the eroded sediment clouds water, reduces sunlight, and kills fish and other biota [Ontario, 1987; Uri, 2001]. Also, the addition of nitrogen, phosphorus, and other nutrients to waterways may increase "undesirable" algae and other aquatic weeds and cause oxygen shortages (hypoxia), making the water inhospitable to fish and other animal species [USDA, 1999]. For example, approximately 40% of U.S. fresh water is deemed unfit for recreational or drinking water uses because of erosion-induced contamination with dangerous microbes, pesticides, and fertilizers [UNESCO, 2001].

Finally, a growing number of acres are sacrificed to urban and economic development. Presently, nearly 16 million ha of U.S. land are devoted to roads and parking lots [Brown, 2001]. Each person requires 0.4 ha of land for urbanization and highways [Anok and Peace, 2003], plus 0.5 ha of cropland, and 1.8 ha for pasture and forests [USDA, 2002]. As the human population expands, so do the number of hectares required for economic development.

Alien Biological Invaders

Rapid human population growth and loss of natural habitats is reducing species numbers, but equally important are biodiversity losses due to the invasions of alien biological species [Pimentel et al., 2000; Sala et al., 2000; Pimentel, 2002]. To date, more than 50,000 alien invasive species have been introduced into United States and some are causing more than $120 billion in damages and control costs each year [Pimentel et al., 2005a]. Harmful competition, predation, and parasitism from invasive species is causing an estimated 40% of all native species extinctions in the U.S. and forcing other native species to become endangered or threatened [Pimentel, 2002].

Human activities are facilitating the introduction of alien species and removing native vegetation in natural ecosystems for crops, livestock, and urbanization, and thereby contributing to the decline in species diversity [Tuxill, 1999; Henry, 2001; Pimentel, 2002].

Chemical Pollutants

Globally, one of the major factors causing species loss is the growing presence of chemicals throughout the world [Environmental Threats, 2003; Do or Die, 2003]. Worldwide, humans use 100,000 chemicals. More than 500,000 kg of synthetic pesticide chemicals are applied annually for U.S. agriculture, public health, and other purposes [Pimentel, 1997]. Even the recommended use of pesticides destroys many beneficial species each year. Pesticide use in U.S. agriculture, for instance, kills more than 72 million adult birds each year, damages a half million colonies of honey bees and is equally hazardous to wild bees [Pimentel, 2005]. Additionally, some pesticides wash into streams and lakes, destroying fish and other aquatic organisms [Pimentel, 2004]. In addition to animals, there are 300,000 human poisonings with 45 deaths in the U.S. and 26 million human poisonings worldwide with 220,000 deaths [Pimentel, 2005].

Marine Ecosystems and Aquaculture

Although studies most often focus on loss of terrestrial biodiversity, the oceans and other aquatic ecosystems are also suffering tremendous species losses. FAO (2003b) estimates that 47% of the world's fishery stocks are fully exploited and 28% are over-exploited. Today, the estimated biomass of large predatory fish is only 10% of pre-industrial levels [Myers and Worm, 2003], and marine foodwebs have significantly changed [Pauly et al., 1998]. Although pollution, climate change, and invasive species threaten marine ecosystems, the fishing industry exerts the greatest pressure, even though fish make up a relatively small part of the world's human diet (0.01% of calories or 16% of animal protein [FAO, 2002a]. In seeking target fish for harvest many nontarget fish are caught and destroyed. For every kilogram of shrimp harvested, 8 to 9 kg of "trash" fish are mangled and discarded as bycatch [Earle, 1995].

Some suggest that aquaculture will replace ocean fishing. Already, aquaculture is supplying nearly 30% of the world's fishery products [FAO, 2003b]. However, overlooked is the fact that the production of aquaculture fish requires the harvesting of ocean fish. For instance, each kilogram of farm-fed salmon requires from 3 to 5 kg of ocean fishmeal [Gomiero et al., 1999; Goldberg et al., 2001]. In this way, aquaculture is putting additional pressure on ocean fisheries rather than reducing pressure.

About two-thirds of fish species have an early stage of development in coastal wetlands [Ramsar, 2000]. Thus, when coastal wetlands are drained or destroyed, fish production is seriously affected.

There are also economic consequences of reduced marine biodiversity. For example, if a fishery collapses, thousands of fishermen may lose their jobs. This happened in 1992 when 40,000 fishermen on the eastern coast of Canada lost their jobs because of the cod fishery collapse, due to overfishing [Harder, 2003].

Global Climate Change

Many birds and insects have already extended their distributions northward or to higher altitudes, mainly attributed to global warming. In response to climate change many species in the 25 biodiversity hotspots will be forced to extinction [Myers, 2003c]. In the coming decades, projected global climate change could seriously damage the world's species [Hansen et al., 1996; WRI, 2002]. Moreover, many species, like trees, may not be able to change their distributions rapidly enough to keep up with the changing climate [Krajick, 2001, 2004; WRI, 2002]. The World Conservation Union projects global warming may threaten 37% of world's species by 2050 [Thomas et al., 2004].

Marine ecosystems are also at great risk due to climate change. For instance, even mild increases in ocean temperatures cause major epidemics of coral bleaching. Such increases in 1998 resulted in a loss of one-sixth of the world's coral colonies [Goreau et al., 2000; Earthscan, 2002; Dennis, 2002].

Biological Diversity

Although about 90% of the global food supply today comes from 15 plant and 8 animal species [Pimentel and Pimentel, 1996], throughout history people have used as many as 20,000 plant species for food, out of the more than 80,000 species that could be utilized [Vietmyer, 1995; Tuxill, 1999]. Humans obtain 99% of their food and all of their wood products by harvesting them from 70% of terrestrial temperate and tropical ecosystems [Pimentel et al., 1999; AAAS, 2001]. Of the earth's terrestrial area, approximately 11% is devoted to cropland, 37% to pasture land, 30% to forests, and about 5% to urbanization and highways [UNESCO, 2002; AAAS, 2001; Wiebe 1997]. The remaining 17% consists of unproductive areas, including mountains and deserts. Overall, most species are located on land area that is maintained for agriculture, forestry, and human settlements [Western and Pearl, 1989; Pimentel et al., 1992; FAO, 2004]. Therefore, major efforts should be made to conserve the many species that now exist in these extensive managed, terrestrial environments [Paoletti, 1999a].

Current data suggest that from 10 to 50 million species exist on earth [Pimm, et al., 1995]. Most of what is known about biodiversity pertains to large plants and animals, such as flowering plants and vertebrates. The extent of diversity of small organisms like bacteria, fungi, insects, mites, and other minute organisms remains relatively obscure. The United States is home for an estimated 750,000 species, of which small organisms -- such as arthropods and microbes comprise 95% [Pimentel et al., 1992; Dorworth, 2002].

In temperate crop ecosystems, the numbers of arthropod species range from 600 to 1000 species per hectare, while an estimated 20,000 bacteria species may be present in a favorable soil habitat (Table 1). Worldwide, arthropods make up the majority (~90%) of multi-cellular species. To illustrate, in a tropical forest in Uganda, on 80 trees of just two tree species, a total of 1,352 beetle species were identified. In Borneo on 10 trees, a total of 2,800 arthropod species were reported (Table 2). Arthropods and microbes, such as bacteria and fungi, contribute large amounts of biomass and large numbers of species to soil, crop, and forest ecosystems (Tables 1, 2, and 3).

Table 1. Plant, animal, and microbe species and biomass potential in a favorable soil ecosystem with ample soil organic matter and moisture

No. Species/ha		*Sources*	*Kg/ha*	*Sources*
Bacteria	20,000	a,b	3.000	a
Fungi	50	a	3.000	c
Algae	5	h	100	c,d
Protozoa	60	a	100	c
Nematodes	30	a	50	c
Earthworms	15	a	3.000	a,f
Mites	114	e	10	a
Collembola	70	a	3	c
Enchytraeids	22	a	70	a
Termites	60	a	30	a
Ants	40	a	100	a
Isopoda	4	a	1	I
Beetles	46	g	70	a
Diptera	10	a	400	a
Arachnida	. 62	g	400	a
Total				

(a) Lavelle and Spain, 2001; (b) Wayne et al., 1987; (c) Metting, 1993; (d) Alexander, 1977; (e) Osler and Beatie, 2001; (f) Edwards and Bohlen, 1996; (g) Rushton et al., 1989; (h) Masyuk, 2002; (i) Thimmayya, 1998

Table 2. Arthropod biodiversity in various ecosystems

Ecosystem	Arthropod		
	Location	**Species**	**Source**
Alfalfa (per ha)	New York	600	a
Corn (monoculture)	Minnesota	600	b
Cotton (monoculture)	Arkansas	600	c
Pasture (per ha)	Britain	1.000	d
Two tree species (80 trees)	Uganda (beetles)	1,352	e
Forest tropical (10 trees)	Borneo	2,800	f

(a) Pimentel and Wheeler, 1973; (b) Warters, 1969; (c) Whitcomb and Godfrey, 1991; (d) MacFadyen, 1961; (e) Wagner, 2003; (f) Stock, 1988

Marine ecosystems also have an abundance of species. For instance, 1 liter of seawater may contain from 100 to 1,000 species of bacteria [Fred Dobbs, Old Dominion University, personal communication, 2004]. On coral reefs, for example, it is estimated that only 10% of the species have been described; over one million species are thought to inhabit these ecosystems making reefs rivals of rainforests in terms of diversity [Thorne-Miller, 1999].

Preservation of Biological Diversity: Large and Small Organisms

Clearly, plants, fish, birds, and mammals are invaluable contributors to ecosystem health [Krajick. 2001]. An estimated 275,000 species of plants have been identified, and perhaps as many as 100,000 more plant species have yet to be discovered [IUCN, 2002a]. Despite their general resilience, the survival of many plants is now in peril: for every 10 species of plants and animals that are listed as endangered, approximately 6 of these species are plants [IUCN, 2002a]. According to Walter and Gillett (1998), at least 1 of every 8 known plant species is threatened with extinction.

The fate of larger organisms too remains a concern: 1 in every 4 mammals and 1 in every 8 birds is facing a high risk of extinction in the near future [IUCN, 2004]. However, the importance of small organisms that dominate the basic structure and function of ecosystems cannot be overstated [Terborgh, 1988; USGS, 2003]. Small organisms, such as insects, are useful indicators for the overall "health" of an ecosystem and its capacity to provide vital services to humans [Paoletti, 1999b; ESA, 2003]. Insects and other "little things", like bacteria and fungi, perform crucial functions that sustain ecosystems in ways that are still scarcely understood, including pollinating plants and degrading wastes [Wilson, 1987; Price, 1988; Soil Organisms, 2003; FAO, 2003a)] Because small organisms may be more specialized and more closely associated with a plant species than larger animals, they are likely to be more susceptible to environmental changes [Dourojeanni, 1990; IUCN, 2002b]. For example, it is estimated that for each tropical plant species facing extinction, approximately 20 species of arthropods feeding on a particular plant may also be forced to extinction [Erwin, 1983].

In general, ecosystems require a sound relationship among the various species that make up the system. The elimination or addition of even one species to a relatively balanced ecosystem can have profound, cascading, and largely unpredictable effects [Fritts and Rodda, 1998].

Biodiversity and Economic and Environmental Benefits

U.S. agriculture and forestry depend upon most of the estimated 750,000 species of natural plant, animal, and microbe species for production and sustainability [Pimentel et al., 1997]. Plant, animal, and microbe species provide the basic food, fiber, and shelter to support U.S. agriculture and forestry, and contribute more than $15 trillion dollars annually to the U.S. economy [USBC, 2002].

A most vital activity carried out primarily by invertebrates and microbes is recycling wastes produced by agriculture, forestry, and human activities. A conservative estimate of the annual benefits of these processes in the U.S. alone is $62 billion annually [Pimentel et al., 1997].

Moreover, approximately two-thirds of the world's flowering plants depend on insect and other pollinators for reproduction and survival [Native Pollinator, 2003]. Specific pollinators are sometimes vital to a particular species of plant [LaSalle and Gould, 1993; Comba and Corbet, 1998]. This cross-pollination by bees is essential to about one-third of the crops grown in the U.S. and has a value of about $40 billion per year [USBC, 2002]. Some seed eating birds and mammals, like rodents, are essential in the dispersal of some plant seeds as well [Reid and Miller, 1989].

An estimated $10 billion is spent annually in the U.S. for pesticides to control crop pests. But the parasites and predators that exist in natural ecosystems provide an estimated $40 billion per year in benefits for pest control. Without the existence of natural enemies, crop losses by pests in agriculture would increase 10% to 20%. Then the amount of pesticides and costs of chemical pest control would escalate [Pimentel et al., 1997].

Fish, other wildlife and plant materials harvested from the wild in the United States alone have an estimated annual value of $45 billion [USBC, 2002]. For instance, the livelihood of more than 30 million fishers and fish farmers worldwide (most of which live in developing countries) comes from fisheries [FAO, 2002b]. The U.S. alone has a $25 billion fishing industry [USBC, 2002]. In addition, approximately 25% of all pharmaceuticals manufactured in the U.S., valued at $20 billion, are obtained directly or indirectly from plant materials [Tuxill, 1999].

Sustainable and productive agriculture and forestry systems cannot function successfully without the vital activities contributed by a wide diversity of natural plants, animals, and microbes.

PLANT, ANIMAL, AND MICROBE BIOMASS AND DIVERSITY

Biological diversity in an ecosystem is related to the amount of living and nonliving organic matter present [Elton, 1927; Wright, 1983; Sugden and Rands, 1990; Mishra and Dhar, 2004].

Table 3. Biomass of various organisms per hectare in a temperate-region pasture

Organism	Biomass (kg fresh weight)
Plants	20,000 a
Fungi	4,000 b
Bacteria	3,000 b
Arthropods	1,000 e
Annelids	1,320 b
Protozoa	380 b
Algae	100 c
Nematodes	50 c
Mammals	1.2 d
Birds	0.3 d

(a) Estimated; (b) Richards, 1974; (c) Metting, 1993; Alexander, 1977; (d) Walter, 1985; Xerces Society, 2001

In addition to plants, the data in Table 3 indicate that fungi, bacteria, annelids, and arthropods contribute the bulk of the non-plant biomass in a pasture ecosystem. The fungi alone comprise about 4,000 kg/ha (wet), bacteria about 3,000 kg/ha (wet), earthworms 1,300 kg/ha (wet), arthropods 1,000 kg/ha (wet). In contrast, mammals and birds contribute only 1.2 and 0.3 kg/ha (wet) biomass, respectively. Because the abundance of biomass is most often positively correlated with biodiversity, efforts to increase biomass in agricultural and forestry ecosystems are an important factor in the preservation of the wealth in biodiversity [Elton, 1927; Wright, 1983; Sugden and Rands, 1990; Mishra and Dhar, 2004].

STRATEGIES FOR CONSERVING BIOLOGICAL DIVERSITY

Because agriculture, forestry, and human settlements occupy about 78% of the terrestrial environment, a large portion of the world's biological diversity coexists with humans in these ecosystems [Western, 1994; Heywood, 1999]. Therefore, major efforts should be made to conserve the many species that now exist in these extensive, managed, terrestrial environments. Conservation programs based on sound ecological principles will assist agriculture and forestry production in becoming more sustainable, while at the same time maintaining biological diversity [Heywood, 1999; NAS, 2003]. Species diversity benefits from abundant biomass, habitat diversity, stable ecosystems, abundant soil nutrients, high-quality soils, effective biogeochemical cycling, abundant water, and healthy marine systems [Westman, 1990].

Abundant Biomass

Except for green plants that capture solar energy for themselves and certain bacteria that use inorganic material as an energy source, all other organisms rely on plant biomass as their primary or secondary energy source. Crop and forest residues are biomass resources that are vital to agricultural and forest production. They not only protect the soil from erosion and conserve water, but, when recycled, also contribute large quantities (2,000 to 15,000 kg/ha [dry]) of nutrients and organic matter to the soil [ERAB, 1981; NAS, 2003]. Suggestions that crop residues be harvested for fuel and other purposes have proven catastrophic [Fenster, 2003; Pimentel and Wen, 2004]. In China and India, the removal of crop residues has increased the rates of soil erosion and rapid water runoff approximately 10-fold and reduced soil quality and fertility [Fenster, 2003; Pimentel and Wen, 2004]. In addition to reducing soil erosion and water runoff, cover crops are also advantageous for agricultural production because they reduce soil erosion, compaction, suppress weeds, conserve soil nutrients and moisture, and increase soil organic matter [Pimentel et al., 2005b]. Furthermore, cover crops can increase vegetative biomass and diversity in crop ecosystems because they provide additional shelter and refuges for many species.

Plant Species Diversity

Approximately half of the plant species in the United States exist in managed ecosystems. Of the estimated 17,000 plant species in the U.S. [Morin, 1995], approximately 6,000 are crop species and 2,000 are weed species [Pimentel et al., 2000; Hanelt, 2001].

Increased plant diversity, with associated species diversity, can be encouraged in some managed ecosystems. Multi-species crop systems support a diverse group of natural biota that increase productivity. At the same time, farmers benefit from the effective use of soil nutrients and reduced water runoff. Examples of such cropping systems are found in Java, where small farmers cultivate more than 600 crop species in their gardens, making for overall species diversity comparable to subtropical forests [Dover and Talbot, 1987]. In Guatemala, about 279 species were reported in the tropical-humid gardens [IPGRI, 2004]. Similarly, nearly 80% of the farmers in West Africa and Latin America intercrop their gardens, raising upwards of 100 different crop species on their small plantings [Thrupp, 1998]. By increasing the number of plant species on their farms, farmers were able to increase food production with a high degree of diversity.

Intercropping

When leguminous crops, like clover, are grown between crop rows, such as corn, they serve as an intercrop or living mulch. Not only do legumes fix nitrogen in the soil , but they also conserve soil and water resources, and at the same time increase the associated biomass and animal and plant diversity present in the ecosystem [Sigvald and Yuen, 2001].

Strips of different crops are especially helpful when planted across the slope of agricultural fields. Such strips not only help control soil erosion and water runoff, but also increase the diversity of vegetation and thus increase the availability of beneficial parasites and predators for biological control [Francis et al., 1986; Fortin, et al., 1994; Ramert, 2002]. With appropriate combinations of strip crops grown *in rotation*, various pests can be controlled with little or no pesticides. Such pest control occurs, for example, when corn, soybeans, and wheat are grown in rotation in strip patterns [Pimentel et al., 1993; Pimentel et al., 2005b].

Shelter Belts and Hedgerows

Shelter belts and hedgerows planted along the edges of cropland and pasture land also contribute to biological diversity because, like intercrops, they reduce soil erosion and moisture loss as well as increase the biomass and structural and habitat diversity present in managed ecosystems [Elton, 1927; NAS, 1988; HMSO, 1995]. Organic hedgerows are superior to hedgerows associated with conventional agriculture. For example, in Denmark organic hedgerows were comprised of 27% more plant species than conventional hedgerows [Aude et al., 2003]. Furthermore, shelter belts and hedgerows frequently provide refuges for beneficial parasites and predators, like ground beetles, that help control pest insects and weeds, thereby reducing the need for pesticides [Paoletti et al., 1989; Whalon, 2002]. In addition, shelter belts help reduce erosion and moisture loss from crops by buffering winds and are especially beneficial in areas with low rainfall and high winds [Kedziora et al., 1989; Lu and Lu, 2003].

Biomass and Soil and Water Conservation

High quality soils maximize plant biomass productivity and help increase biodiversity. In general, quality soils are rich in nutrients; high organic matter (5% to 10% of soil by weight); store soil moisture (about 20% by weight); are well drained and relatively deep (>15 cm); and have abundant soil biota [Doran and Parkin, 1994].

Abundant vegetative cover, including nonliving plant residues, prevent soil erosion and rapid water runoff [Hayes, 1996]. Organic matter not only harbors large numbers of species but, equally important, sustains the productivity of the soil by improving water-holding capacity, providing a source of nutrients, improving soil tilth, and increasing the number and diversity of soil biota (Table 1). Because soil organic matter is the first to suffer the effects of erosion, soil conservation is vital to maximize biomass productivity and biodiversity.

A strong association exists between precipitation, plant diversity, and productivity. Because all plants and animals require water to sustain themselves, sufficient water is vital for maintaining maximum productivity and biodiversity [Neveln, 2003]. Plants require large amounts of water for photosynthesis. For example, a corn crop producing 18,000 kg/ha of biomass during the growing season requires about 9 million liters of water per hectare [Pimentel et al., 2004a].

Many technologies can be employed to conserve water and soil resources (crop rotations, strip cropping, contour planting, terracing, ridge planting, no-till, grass strips, vegetative cover, drip irrigation, intercropping, and shelter belts) [Troeh et al, 1999]. The adaptability of each technique depends on the particular characteristics of the crop or forest ecosystem

[Troeh et al., 1999]. In general, the presence of abundant biomass also conserves water by slowing rapid water runoff and increases the water holding capacity of the soil.

Livestock Manure

Livestock manure, when properly used, is a valuable resource that increases the biomass and biodiversity in agricultural systems. For example, when manure (100 t/ha wet) was added to agricultural land in Hungary, the biomass of soil microbes increased 10 fold [Olah-Zsupos and Helmeczi, 1987].

Habitat Diversity

Increasing the diversity of physical habitats increases the diversity of associated plants and other organisms present in the ecosystem [Allee et al., 1949; Fletcher, 1995]. For example, when the habitat area was increased 10-fold, the number of bird species increased 1.6- to 2.5-fold [Avian Ecology, 2003].

Arnold (1983) reported that only 5 bird species were present in a pure farming ecosystem surrounded solely by farmland. The bird species increased to 12 when there was a short hedge, 17 species when there was a tall hedge, and 19 species when a strip of woodland was present.

Corridors between habitats are essential for many large predators, like coyotes and mountain lions, that actively move between suitable habitats [Rodriguez et al., 2003].

Agroforestry

Agroforestry is an ecologically based, natural resource management system that integrates trees into cropland and rangeland systems [Leakey, 1997]. Agroforestry increases biomass and conserves soils and water resources by preventing erosion [Kidd and Pimentel, 1992]. Further, crop losses due to pests are often reduced because of increased plant diversity [Schroth et al., 2000]. In addition to all these benefits of agroforestry, biological diversity is conserved and in some cases enhanced [CGIAR, 2003; Griffith, 2000].

For example, in tropical Central America, conventional corn plantings produce approximately 2,000 kg/ha of dry biomass, whereas in an agroforestry system with a leguminous tree, the corn biomass was approximately doubled to about 3,800 kg/ha [Kidd and Pimentel, 1992]. At the same time, 4,500 kg/ha of leguminous tree biomass was produced. Thus, in the agroforestry system, the total biomass produced was increased more than 4 times over that of the conventional system.

Similarly, in Indonesia, for example, agroforestry increased plant diversity above that in conventional farming with some farmer gardens having 50% to 80% of the plant diversity found in natural forests [Leakey, 1999; Nobel and Dirzo, 1997]. When the forests in the Tamaulipas region of Mexico were managed as agroforesty systems, they contained more than 300 plant species [Perfecto et al., 1996].

Mixed Forests

Mixed forests produce approximately 20% more biomass than a homogenous stand of trees [Ewel, 1986; Moore, 2002]. The benefits are attributed to the differing nutritional need of the trees in the forest. In addition, mixed forests improve biological diversity because of the multiple arthropods and microbes associated with each tree species.

Moreover, in commercial forestry, as well as in natural forests, tree diversity increases biomass production by diminishing pest attack on tree hosts [Ewel, 1986; Allen, 2003]. For example, the attack of the white-pine weevil on white pines and the Douglas-fir tussock moth on Douglas fir are significantly more severe in areas with single species forest than in areas with high tree species diversity [Allen, 2003].

Careful selective cutting of forests, however, can maintain high biological diversity and a healthy productive forest ecosystem [Hansen et al., 1991]. Large-scale clear-cutting of forests should be avoided because it not only reduces biomass and biological diversity, but also removes nutrients from the soil, which eventually reduces the productivity of the entire ecosystem. Both biomass production and biological diversity decline as a result. Planting trees along streams is another helpful strategy to increase biodiversity as well as reduce erosion and conserve nutrients [Streams for the Future, 2004].

In addition, important agroecosystems are also found below ground in mature forests. When forests are cleared, vital mycorrhizal fungi and other micro-flora and fauna are reduced or exterminated [Tallis, 2002].

Pasture Management

A pasture management strategy that maintains maximum biomass, while preventing overgrazing, is the most productive strategy for livestock and ecosystem biodiversity [Clark et al., 1986; McIntyre, 2001]. In addition to providing livestock with forage and vegetative cover, pasture productivity prevents soil erosion and rapid water runoff. Parol (1986, 2003) reports that increasing the plant species diversity in pastures can increase the productivity of the pastures up to 10%.

To prevent overgrazing, the pasture should have the appropriate number of animal units per hectare and should employ a sound pasture rotation system [Beetz, 2001; Rotational Grazing, 2004]. For example, in the northeastern region of the United States maximum production of livestock was achieved when pastures were grazed for several weeks and then rested for several weeks to allow vegetative growth [Yohn and Rayburn, 2004].

Pesticide Reduction

Pesticides severely reduce biological diversity by destroying a wide array of both harmful and beneficial species in agricultural ecosystems. In this process, they change the normal structure and function of the ecosystem. Concern for the negative effects of pesticides on natural biota and public health has prompted some nations to reduce pesticide use. For example, Sweden has reduced pesticide use by 68% and Indonesia by 65% without reducing crop yields. In the case of Indonesia, crop yields actually increased 12% [PCC, 2002; Oka, 1991].

By employing appropriate biological controls and other environmental practices in agriculture, pesticide use can be reduced, and in some cases eliminated, while maintaining or increasing crop yields [Pimentel et al., 1993; NAS, 2000].

Conserving Fisheries

While many national and international fishery agreements do exist, they have not been enforced, as evidenced by the declining fish catches in recent years. Pauly and Watson (2003) recommend stricter regulations, including banning certain types of fishing gear. Furthermore, aquaculture that currently produces carnivorous fish, such as salmon, requires large quantities of fish meal [Goldberg et al., 2001]. Alternatively, to be sustainable, aquaculture might have to rely increasingly upon herbivorous fish, such as catfish and talapia [Gomiero et al., 1999; Swing, 2003].

Protected Parks

The maintenance of protected parks and wildlife refuges that occupy about 12% of the terrestrial ecosystem [Chape et al., 2003]. However, these parks are not protected from outside assault. For instance, about one third of the tropical parks are already subject to encroachment by landless individuals who live in poverty [Myers, 2002]. Many of these poor people who have an income of less than $1 per day are forced to attempt to find food or produce food in parks. As for protecting and effectively managing national parks, wildlife refuges, and other protected areas, it is reported that less than one quarter of the declared areas in 10 key forested countries were well managed, many had no satisfactory management at all [Heywood, A. H., personal communication, University of Reading (UK)]. Further support of this concern comes from World Wildlife Fund study entitled, "How Effective are Protected Areas?" [World Wildlife Fund, 1999].

The further concern is that most parks are too small to insure the conservation of the majority of species they contain. For instance, the succulent Karoo biome in South Africa covers a relatively small area, however, this biome holds more than 6,000 plant species of which 40% are endemics, in addition to many endemic animals [Rodriguez et al., 2003]. In Mexico on the El Eden Ecological reserve, only 73 species of slime molds are present, compared with 244 species in all of Mexico [Gomez-Pompa, 2004]. Similarly, in Kenya, about 7% of its land is in protected national parks, however, 75% of the wildlife lives outside parks and within human systems [Western and Pearl, 1989; Muriuki, 2003]. Including arthropods and microbes, more than 90% of species live outside of protected parks [Rodriguez et al., 2001].

Marine reserves, which comprise less than 1% of all marine ecosystems [WRI, 2002], have been widely promoted as conservation and fishery management tools. The benefits of marine reserves are indisputable, for within and around marine parks, fish populations increase dramatically and adjacent fisheries are improved up to 90% [Roberts et al., 2001]. In addition, marine protected areas can provide substantial tourism revenues. For example,

Australia's Great Barrier Reef Marine Park yields over $1 billion per year in revenue for the local economy while costing a mere $20 million to manage [Hinrichsen, 1998].

Global Climate Change

With only 4% of the world's population, the United States is responsible for more release of carbon dioxide than any other nation in the world (about 25% of the total releases) [PRB, 2003]. Reducing the rate of release of carbon dioxide and other greenhouse gases and slowing global warming will require a major effort by Americans and other people of the world. Hopefully, the U. S. will become the leader instead of an opponent of international climate policy. The United States could save as much as $430 billion per year on energy costs while reducing carbon dioxide emissions 30% below 2004 levels in 10 years [Pimentel et al., 2004b].

CONCLUSION

The present rate of species loss suggests that half of all species on earth may be lost at the end of the 21st century. Millions of species of plants, animals, and microbes carry out vital functions in the biosphere, especially for agriculture, forestry, and aquatic systems. The prime threats to biodiversity result from rapid human population growth, and include habitat loss, urbanization, chemical use, introduced alien species, pollution, and global warming.

Conservationists are dedicated to protecting biodiversity and implementing sound conservation policies. Unfortunately, most conservation policies are established by economic planners, agriculturists, foresters, and corporations, and do not come from conservationists themselves [Myers, 2002]. In the light of species loss and growing pressures on biodiversity worldwide, it appears that the only way that biodiversity can be saved is by saving the total biosphere [Myers, 2002].

One win-win approach is to strive for sustainable agriculture and forestry systems because most plant, animal, and microbe species exist in these ecosystems that cover 78% of the terrestrial ecosystem [Pimentel et al., 1992; Daily et al., 2001; Ricketts, 2001; Vandermeer, et al., 2002; Homer-Divine et al., 2003]. Also, agriculture and forestry ecosystems are the most favorable systems in terms of moisture, soil, nutrients, and temperatures. Maintaining biological diversity is essential for sustainable and productive agriculture and forestry systems. Biological diversity can best be protected by maintaining abundant biomass and habitat diversity; conserving soil, water, and nutrient resources; reducing water, soil, and air pollution; and reducing global warming.

The public, as well as political leaders must give high priority to protecting biodiversity and the total biosphere. We recommend that the U.S. and other nations take the following policies to enhance the conservation of biodiversity:

- Encourage and implement ecologically sound and sustainable management practices for agriculture, forestry, and fishery systems.

- Implement policies to prevent the introduction of alien invasive species in the U.S. and other nations.
- Implement various international agreements, including the Convention on Biological Diversity, Framework Convention on Climate change, and the Convention on the Law of the sea [J.A.McNeely, personal communication, Chief Scientist, IUCN, The World Conservation Union, Gland, Switzerland, 2004].
- Reduce water, air, and soil pollution that threaten species survival.
- Conserve and reduce fossil energy consumption to reduce green house gases and global climate change.
- Set aside more ocean as enforced marine protected areas.

REFERENCES

AAAS. (2001). World Land Use. American Association for the Advancement of Science. *http://www.ourplanet.com/aaas/pages/population01.html.* (10/02/03).

Alexander, M. (1977). *Introduction to Soil Microbiology*. New York: John Wiley and Sons.

Allee, W. C., Emerson, A. E., Park, O., Park, T. and Schmidt, K. P. (1949). *Principles of Animal Ecology*. Philadelphia: Saunders.

Allen, D.C. (2003). Biological diversity – is variety the spice of life? *http://www.dec.state.ny.us/website/dif/privland/forprot/health/nyfo/biodiv.pdf.* (10/18/03).

Anok and Peace. (2003). Overpopulation. Anok and Peace. Cols, Ohio. http://noleaders.net/anok/features/nopop/ ((10/02/03).

Arnold, G.W. (1983). The influence of ditch and hedgerow structure, length of hedgerows, and area of woodland and garden on bird numbers on farmland. *Journal of Applied Ecology, 20*: 731-750.

Aude, E., Tybirk, K., and Pedersen, M. B. (2003). Vegetation diversity of conventional and organic hedgerows in Denmark. *Agriculture Ecosystems and Environment, 99*: 135-147.

Avian Ecology. (2003). Avian ecology. *http:://www.ornithology.com/lectures/Avian%20 Ecology.html. (10/17/03).*

Beetz, Camann A. (2001). Sustainable Pasture Management. *Livestock Systems Guide*. October 2001. Fayetteville, AR: University of Arkansas.

Brown, L. R. (2001). *Paving the Planet: Cars and Crops Competing for Land.* Washington, D.C.: Earth Policy Institute.

CGIAR. (2003). Agroforestry – Combining nature and livelihoods. Consultative Group on International Agricultural Research. *http://www.futureharvest.org/earth/agroforestry .shtml ((11/03/03).*

Chape, S., Blyth, S., Fish, L., Fox, P., and Spaulding, M. (2003). United Nations List of Protected Areas. World Conservation Monitoring Centre.

Clark, D.A., Lambert, M.G., and Grant, D.A. (1986). Influence of fertilizer and grazing management on North Island moist hill country. 5. Animal production. *New Zealand Journal of Agricultural Research.*

Comba, L., and Corbet, S. (1998). Living larders for bumblebees. *Plantlife Magazine, Summer Issue*. 3 pages.

Conservation Issues. (2003). Impacts on Biodiversity. Population growth, overconsumption and technology. http://www.redpath-museum.mcqill.ca/Qbp/3.Conservation/impacts.htm. (11/16/03).

Daily, G.C., Ehrlich, P.R., and Sanchez-Azonfeifa, G.A.. (2001). Countryside biogeography: utilization of human dominated habitats by the avifauna of southern Costa Rica. *Ecological Applications* 11: 1-13.

Deforestation. (2003). Deforestation. http://www.bsrsi.msu.edu/rfrc/deforestation.html. (10/04/03).

Dennis, C. (2002). Reef under threat from "bleaching outbreak". *Nature, 415*: 947.

Dirzo, R., and Raven, P.H. (2003). Global state of biodiversity and loss. *Annual Review of Environmental Resources*. 28: 137-167.

Do or Die. (2003). Biodiversity and Its Loss – What does it all really mean? *Do or Die Issue 8*:109-120. (available http://www.eco-action.org/dod/no8/biodiversity.html)

Doran, J. W. and Parkin, T. B. (1994). Defining and assessing soil quality. In J.W. Doran, D. F. Bezdicek, D.C. Coleman, and B.A. Stewart (Eds.), *Defining Soil Quality for a Sustainable Environment* (pp3-21). Special Publication. No. 35. Madison, WI: Soil Science Society of America.

Dorworth, D. (2002). *The New Politics: Developer vs. Ecologists*. Ketchum, Idaho: Idaho Mountain Express and Guide. September 25 – October 1, 2002. 5 pages.

Dourojeanni, M. J. (1990). Entomology and biodiversity conservation in Latin America. *American Entomologist, 36(2)* :88-93.

Dover, N. and Talbot, L. (1987). *To Feed the Earth: Agroecology for Sustainable Development.* Washington, DC: World Resources Institute.

Earle, S.A. (1995). *Sea Change: A Message of the Oceans*. New York: Putman's Sons.

Earthscan. (2002). *Global Environment Outlook 3: Past, Present and future Perspectives.* United Nations Environmental Programme. London: Earthscan Publications.

Edwards, R. (1998). Save Our Pathogens. *New Scientist*. Pages 5, 22. August 1998.

Edwards, C. A. and Bohlen, P. J. (1996). *Biology and Ecology of Earthworms*. London: Chapman and Hall.

Elton, C. S. (1927). *Animal Ecology*. London, Sidgwick and Jackson, LTD.

Environmental Threats. (2003). Why We Need the 1% Alliance. Top 10 Environmental Threats. http://www.onepercentfortheplanet.org/threats.html. (11/16/03).

ERAB. (1981). *Biomass Energy*. Washington, DC: Energy Research Advisory Board, U.S. Department of Energy.

Erwin, T. (1983). Beetles and other insects of tropical forest canopies at Manaus, Brazil, sampled by insecticide fogging tropical rainforest. In: *Tropical Rainforest: Ecology and Management.* Sutton, S. (ed). pp 59-75. Oxford, England: Blackwell Scientific.

ESA. (2003). *ESA Position Statement on Insects and Biodiversity.* Lanham, MD: Entomological Society of America. 4 pages.

Ewel, J. J. (1986). Designing agricultural ecosystems for the humid tropics. *Annual Review of Ecology and Systematics, 17*: 245-271.

Faeth, P., and Crosson, P. (1994). Building the case for sustainable agriculture. *Environment. 36(1)*: 16-20.

FAO. (2002a). *Food Balance Sheets. A Handbook, Food Balance Sheets.* Rome: Food and Agricultural Organization of the United Nations.

FAO. (2002b). *World Agriculture: Towards 2015/2030.* Summary report prepared mainly by Paul Harrison. Rome: United Nations, Food and Agriculture *Organization.*

FAO. (2003a). *Why should soil biodiversity be managed and conserved?* Land and Water Development Division. Rome: Food and Agriculture Organization. United Nations Organization. 9 pages.

FAO. (2003b). *Overview of Fish Production, Consumption and Trade: Based on 2001 Data.* Food and Agriculture Organization. United Nations Organization. Fisheries Information, Data and Statistics Unit, May 2003.

FAO. (2004). *Biological Diversity in Food and Agriculture.* Food and Agricultural Organization of the United Nations. http://www/fao/org/biodiversity/index.asp. (1/10/04).

Fenster, C. R. (2003). *Protect Soil with Vegetative Residues.* Published by Cooperative Extension, Institute of Agriculture and Natural Resources. Lincoln, NB: University of Nebraska.

Fletcher, S. R. (1995). *Biological Diversity: Issues Related to the Convention on Biodiversity.* Washington, DC: National Council for Science and the Environment. 9 pages.

Fortin, M. C., Culley, J., and Edwards, M. (1994). Soil water. Plant growth and yield of strip-cropped corn. *Journal of Production Agriculture, 7*: 63-69.

Francis, C.A., Jones, A., Crookston, K., Wittler, K., and Goodman, S. (1986). Strip cropping corn and grain legumes: A review. *American Journal of Alternative Agriculture, 1*: 159-164.

Goldberg, R., Elliott, M., and Naylor, R. (2001). *Marine Aquaculture in the U.S.: Environmental Impacts and Policy Options.* Pew Oceans Commission Report.

Gomez-Pompa, A. (2004). The role of biodiversity scientists in a troubled world. *Bio Science, 54*: 217-225.

Goreau, T., McClanahan, T., Hayes, R., and Strong, A. E.. (2000). Conservation of coral reefs after the 1998 global bleaching event. *Conservation Biology, 14*: 5-15.

Griffith, D.M. 2000. Agroforestry: a refuge for tropical biodiversity under fire. *Conservation Biology, 14*, 325-326.

Hansen, J., Oppenheimer, M., Pearl, M., and Fuentes, E. (1996). *Panel Discussion: Extirpated Species, Warming Temperatures, and Politics Worry Scientists.* Columbia Forum on the Environment. New York: Columbia University.

Harder, B. (2003). Sea burial for Canada's cod fisheries. *Science News.* Washington; May 17, 2003. *Volume 163* (20): 318.

Hayes, W.A. (1996). *Understanding Wind Erosion and its Control.* Cooperative Extension, Institute of Agriculture and Natural Resources. Lincoln, NB: University of Nebraska. 3 pages.

Henry, D. (2001). *Population Growth Fuels Biodiversity Loss.* SE-QLD Branch, Media Release. Cleveland, QLD: Sustainable Population Australia.

Hinrichsen, D. (1998). *Coastal Waters of the World: Trends, Threats, and Strategies.* Washington, DC: Island Press.

HMSO. (1995). *Biodiversity: The UK Steering Group Report.* Volume 2: Action Plans. London: Her Majesty's Stationary Office.

Homer-Devine, M.C., Daily, G.C., Ehrlich, P.R., and Boggs, C.L. (2003). Countryside biogeography of tropical butterflies. *Conservation Biology.* 17: 168-179.

IPGRI. (2004). *The Home Gardens Case Study.* Submitted by International Plant Genetic Resources Institute to Convention on Biological Diversity/United Nations Environmental

Programme. *http://www.ipgri.cgiar.org/themes/human/Articles/CBDCaseStudy.doc* (1/10/04).

IUCN. (2002a). Plant species in trouble. IUCN, The World Conservation Union. http://www.iucn.org (11/22/03).

IUCN. (2002b). 2002 IUCN Red list of Threatened Species. IUCN, The World Conservation Union. http://www.redlist.org. (10/01/03).

IUCN. (2004). Species Extinction. IUCN, The World Conservation Union. http://iucn.org (02/22/04).

James, C. (1999). Global Status and Distribution of Commercial Transgenic Crops in 1997. *Biotechnology and Development Monitor, No. 35*. June 1999.

Kedziora, A., Olejnik, J., and Kapuchinski, J. (1989). Impact of landscape structure on heat and water balance. *International Association for Ecology. 17*: 1-17.

Kidd, C., and Pimentel, D. (1992). *Integrated Resource Management: Agroforestry for Development.* San Diego, CA, Academic Press.

Krajick, K. (2001). Arctic life on thin ice. *Science, 291*: 424-425.

Krajick, K. (2004). All downhill from here? *Science, 303*:1600-1602.

LaSalle, J., and Gould, I.D. (1993). Hymenoptera: their diversity and their impact on the diversity of other organisms. Pages 1-26. in Lasalle, J. and I.D. Gould. Eds. *Hymenoptera and biodiversity*. Oxon (UK: CAB International.

Lavelle, P., and Spain, A.V. (2001). *Soil Ecology.* Dordrecht: Kluwer Academic Publishers.

Leakey, R.R.B. (1997). Agroforestry for biodiversity in farming systems. In W. Collins and C. Qualset (Eds.), *Biodiversity in Agroecosystems* (pp. 127-145). New York: Lewis Publishers.

Lu, Y., and Lu, J. (2003). *Oasis Forestry in Xinjiang. Agroforesty Systems in China.* International Development Research Centre. Ottawa, Canada. 5 pages.

MacFadyen, A. (1961). Metabolism of soil invertebrates in relation to soil fertility. *Annals of Applied Biology, 49*: 215-218.

Machlis, G.E., and Tichnell, D.L. (1985). *The State of the World's Parks.* Boulder, CO: Westview Press.

Masyuk, N.P. (2002). The genus Chlamydomonas HER. In the flora of Ukraine. *International Journal of Algae, 4(1):* 1-17.

McIntyre, S. (2001). Incorporation of Practical Measures to Assist Conservation of Biodiversity within Sustainable Beef Production in Northern Australia. (CSIRO) Commonwealth Scientific and Industrial Research Organization. *Sustainable Ecosystems.* Australia. 3 pages

Metting, F.B. (1993). *Soil Microbial Ecology.* New York, NY: Marcel Dekker, Inc.

Mishra, U., and Dhar, D.W. (2004). *Biodiversity and Biological Degradation of Soil. National Center for Conservation and Utilization of Blue-Green Algae.* Indian Agricultural Research Institute. New Delhi, India. 8 pages.

Moore, J.A. (2002). Climate Change and Extreme Weather-Related Event Results. Southern Forest Resource Assessment. USDA Forest Service. 4 pages.

Morin, N. (1995). Vascular Plants in the United States. Missouri Botanical Garden. http:://biology.usgs.gov/s+t/nopframe/j084.htm. (10/16/03).

Muriuki, J.N. (2003). Co-operation or Conflict? Managing Scarce Resources of Africa. *A Case of Community-Wildlife Conservation in Kenya.* Kenyatta University, Kenya. pp. 8.

Myers, N. (2002). Biodiversity and biodepletion: the need for paradigm shift. In T. O'Riordan and S. Stoll (Eds.) *Protecting the Protected: Managing Biodiversity for Sustainability* (pp. 46-60). Cambridge, UK: Cambridge University Press.

Myers, N. (2003a). The Last Extinction? *Foreign Policy,* March/April. 2 pp.

Myers, N. (2003b). Earth's Top Environmental Problems. Population Press. http://www/popco.org/press/articles/2003-1-myers.html. (1/10/04).

Myers, R.A. and Worm, B. (2003). Rapid worldwide depletion of predatory fish communities. *Nature, 423*: 6937.

NAS. (1988). *Biodiversity.* Washington, D.C.: National Academy of Sciences.

NAS. (2000). *Importance of Pesticides.* Washington, D.C.: National Academy of Sciences.

NAS. (2003). *Frontiers in Agricultural Research: Food, Health, Environment, and Communities.* Washington D.C.: National Academy of Sciences.

Native Pollinator. (2003). Native Pollinator and Native Plant Demonstration Project. http://www.denis.osd.mil/denix/Public/ES-Programs/Conservation/Legacy/Native_Pollinator_Native_Plant_Demo.pdf. (11/13/03). 46 pages.

Neveln, V. (2003). Organic Mulch: Helpful or Harmful. *Garden and Hearth* 2001-2003. Editors@gardenandhearth.com. 5 pages.

Nobel, I.R., and Dirzo, R. (1997). Forests as human-dominated ecosystems. *Science, 277,* 522-525.

Novacek, M. (2001). *The Biodiversity Crisis: Losing What Counts.* New York: American Museum of Natural History, New Press.

Oka, I.N. (1991). Success and challenges of the Indonesian national integrated pest management program in the rice based cropping system. *Crop Protection, 10,* 163-165.

Olah-Zsupos, A., and Helmeczi, B. (1987). The effect of soil conditioners on soil microorganisms. In J. Szegi (Ed.), *Soil Biology and Conservation of the Biosphere. Vols. 1-2.* (pp 829-837) Proceedings of the 9th International Symposium. Akademiai Kiado, Budapest.

Ontario. (1987). Soil Erosion Causes and Effects. Ontario Ministry of Agriculture and Food. Ottawa, Ontario.

Osler, G.H.R., and Beatie, A.J. (2002). Contribution of orbatid and messtigmatid mites in ecologically based estimates of global species richness. *Austral Ecology 26(1)*, 70-79.

Paoletti, M.G. (1999a). The role of earthworms for assessment of sustainability and as bioindicators. *Agriculture, Ecosystems and Environment.* 74: 137-156.

Paoletti, M.G. (ed). (1999b). *Invertebrate Biodiversity as Bioindicators of sustainable Landscapes. Practical use of Invertebrates to assess Sustainable Landuse.* Amsterdam: Elsevier.

Paoletti, M.G., Favretto, M.R., Ragusa, S., and Strassen, R.Z. (1989). Animal and plant interactions in the agroecosystems: the case of the woodland remnants in northeastern Italy. *Ecological International Bulletin, 17*, 79-91.

Parol, A. (1986). The productivity of cultivated pastures evaluated by dairy cows depending on the botanical composition and fertilization. *Bioproduction of Grasslands*. Tallinn, Estonia, 141-148.

Parol, A. (2003). The Importance of White Clover in Pasture to the Productivity of Dairy Cows. Institute of Grassland Science and Botany, Estonia Agricultural University. Tartu, Estonia. 5 pages.

Pauly, D., and Watson, R. (2003). Counting the last fish. *Scientific American, 289(1),* 42-47.

Pauly, D., Christiensen, V., Dalsgaard, J., Freese, R., and Torres, F. (1998). Fishing down marine food webs. *Science. 279 (5352),* 860-863.

PCC. (2002). News Bites. http://www.pccnaturalmarkets.com/sc/0205/newsbites.html. (1/19/03).

Perfecto, I., Rice, R.A., Greenberg, R., and Van der Voort, M.E. (1996). Shade coffee: a disappearing refuge for biodiversity. *BioScience. 46*, 598-608.

Pimentel, D. (1997). *Techniques for Reducing Pesticides: Environmental and Economic Benefits.* Chichester, UK: John Wiley.

Pimentel, D. (2002). *Biological Invasions: Economic and Environmental Costs of Alien Plant, Animal, and Microbe Species.* Boca Raton, FL: CRC Press.

Pimentel, D. (2005). Environmental and economic costs of the application of pesticides primarily in the United States. *Environment, Development and Sustainability.*(2005).7: 229-252.

Pimentel, D., and Pimentel, M. (1996). *Food, Energy and Society.* Revised edition. Niwot, CO: University Press of Colorado.

Pimentel, D., and Pimentel, M. (2003). World population, food, natural resources and survival. *World Futures, 59,* 145-167.

Pimentel, D., and Wen, D. (2004). China and the world: population, food and resource scarcity. In: *Dare to Dream: Vision of 2050 Agriculture in China,* Tso, T.C. and He Kang (eds), China Agricultural University Press, Beijing, pages 103-116.

Pimentel, D., and Wheeler, A.G. (1973). Species diversity of arthropods in the alfalfa community. *Environmental Entomology, 2(4)*, 659-668.

Pimentel, D., Stachow, U., Takacs, D. A., Brubaker, H. W., Dumas, A. R., Meaney, J. J., O'Neil, J., Onsi, D. E., and Corzilius, D. B. (1992). Conserving biological diversity in agricultural/forestry systems. *BioScience, 42,* 354-362.

Pimentel, D., McLaughlin, L., Zepp, A., Lakitan, B., Kraus, T., Kleinman, P., Vancini, F., Roach, W. J., Graap, E., Keeton, W. S., and Selig, G. (1993). Environmental and economic impacts of reducing agricultural pesticide use. In D. Pimentel (Ed.), *Pesticide Question: Environment, Economics and Ethics* (223-278). New York, Chapman and Hall.

Pimentel, D., Wilson, C., McCullum, C., Huang, R., Dwen, P., Flack, J., Tran, Q., Saltman, T., and Cliff, B. (1997). Economic and environmental benefits of biodiversity. *BioScience 47(11),* 747-758.

Pimentel, D., Bailey, O., Kim, P., Mullaney, E., Calabrese, J., Walman, F., Nelson, F., and Yao, X. (1999). Will the limits of the Earth's resources control human populations? *Environment, Development and Sustainability, 1,* 19-39.

Pimentel, D., Lach, L., Zuniga, R., and Morrison, D. (2000). Environmental and economic costs of nonindigenous species in the United States. *BioScience, 50(1)*, 53-65.

Pimentel, D., Zuniga, R., and Morrison, D. (2005a). Update on the environmental and economic costs associated with alien invasive species in the United States. *Ecological Economics.*52: 273-288.

Pimentel, D., Hepperly, P., Hanson, J., Dodds, D. D., and Siedel, R. (2005b). Environmental, energetic and economic comparisons of organic and conventional farming systems. *Bioscience* 55 (7): 573-582.

Pimentel, D., Berger, B., Filiberto, D., Newton, M., Wolfe, B., Karabinakis, E., Clark, S., Poon, E., Abbett, E., and Nandagopal, S. (2004a). Water Resources: agricultural and environmental issues. *Bioscience* 54 (10): 909-918.

Pimentel, D., Pleasant, A. , Barron, J., Gaudioso, J., Pollock, N., Chae, E., Kim, Y., Lassiter, A., Schiavoni, C., Jackson, A., Lee, M., and Eaton, A. (2004b). U.S. energy conservation and efficiency: benefits and costs *Environment Development and Sustainability*. 6: 279-305.

Pimm, S.L. (1991). The Balance of Nature? :Ecological Issues in the Conservation of Species and Communities. Chicago: University of Chicago Press.

Pimm, S.L. (2002). Food Webs. Chicago: University of Chicago Press.

Pimm, S.L. and J.H. Brown. (2004). Domains of Diversity. Science. 304: 831-833.

Pimm, S.L., G.J. Russell, J.L. Gittlemand, andT.M. Brooks. (1995). The future of biodiversity. Science. 269: 347-350:

PRB. (2003). *World Population Data Sheet.* Washington, DC: Population Reference Bureau.

Price, P.W. (1988). An overview of organismal interactions in ecosystems in evolutionary and ecological time. *Agriculture, Ecosystems and the Environment, 24*, 369-377.

Rainforest. (2003). Rainforest Live. Information pages *http://www.rainforestlive.org.uk/ index.cfm?articleid=214. (10/06/03).*

Rainforest Facts. (2003). *http://www.pbs.org/tal/costa_rica/facts.html* (10/15/03)

Ramert, B. (2002). The use of mixed species cropping to manage pests and diseases – theory and practice. In Powell, J., G. Davies, S. Fowler, M. Hovi, N. Lampkin, M. Lennartsson, S. Padel, B. Pearce, L. Phillips, J. Robertson, E. Stockdale, C. Watson, R. Weller, J. Welsh and M. Wolfe (Eds), *U.K. Organic Research. Proceedings of the Colloquium for Organic Research 2002.* Conference 26-28, March 2002, Aberystwth. (pp. 207-210).

Ramsar. (2000). Background Papers on Wetland Values and Functions. The Ramsar Convention on Wetlands. Gland, Switerland: Ramsar Convention Bureau.

Reid, W.V., and Miller, K.R. (1989). *Keeping Options Alive: the Scientific Basis for Conserving Biodiversity.* Washington, DC: World Resources Institute.

Richards, B.N. (1974). *Introduction to the Soil Ecosystem.* New York: Longman.

Ricketts, T.H. (2001). The matrix matters: effective isolation in fragmented landscapes. *The American Naturalist.* 158: 87-99.

Robbins, J. (2001). *The Food Revolution: How your diet can help your life and our world.* Berkeley, CA: Conari Press.

Roberts, C., Bohnsack, J.A., Gell, F., Hawkins, J., and Goodridge, R. (2001). Effects of marine reserves on adjacent fisheries. *Science, 294*, 1920-1923.

Rodriguez, A.S.L., Andelman, S.J., Bakarr, M.I., Boitani, L., Brooks, T.M., Cowling, R.M., Fishpool, L.D.C. et al. (2003). Global gap analysis: Towards a representative network of protected areas. *Advances in Applied Biochemistry Science Number 5* August 2003), 3-98.

Rotational Grazing. (2004). Protect soil with rotational grazing. http://www.thisland.uiuc.edu/60ways_10.html. (1/12/04).

Rushton, S.P., Luff, M.L., and Eyre, M.D. (1989). Effects of pasture improvement and management on the ground beetle and spider communities of upland grasslands. *Journal of Applied Ecology 26*, 489-503.

Sala, O.E., Chapin, F.S., Armesto, J.J., Berlow, R., Bloomfield, J., Dirzo, R., Huber-Sanwald, E., Huennekke, L.F., Jackson, R.B., Kinzing, A., Leemand, R., Lodge, D., Mooney, H.A.,

Oesterheld, M., Poff, N.L., Sykes, M.T., Walker, B.H., Walker, M., and Wall, D.H. (2000). Global biodiversity scenarios for the year 2100. *Science 287*, 1770-1774.

Schroth, G., Krauss, U., Gasparotto, L., Duarte, A.J.A., and Vohland, K. (2000). Pests and diseases in agroforestry systems of the humid tropics. *Agroforestry Systems, 50* (3), 199-241.

Sigvald, R. and Yuen, J. (2001). *Cereal/Legume Intercropping Systems: Effect on the Associate Crop Pathosystems*. Swedish University of Agricultural Sciences, Department of Ecology and Plant Production Science/Integrated Pest Management, P.O. Box 7043, 75007 Uppsala, Sweden. 2 pages.

Soil Organisms. (2004). Soil Organisms. *http://www.biodiversity.org.uk.ibs/envmath/resoruces/year3/env324/pro*. (1/10/04).

Stock, N.E. (1988). Insect diversity: facts, fiction and speculation. *Biological Journal of Linnaean Society, 35*, 321-337.

Streams for the Future. (2004). Classic Stream Problems; New Stream Solutions. Streams for the Future, Fisheries Division, Missouri Department of Conservation, Jefferson City, Missouri. 7 pp.

Sugden, A.M. and Rands, G.F. (1990). The ecology of temperate cereal fields. *Trends in Ecology and Evolution, 5*, 205-206.

Swing, J.T. (2003). What future for the oceans? *Foreign Affairs, 82,* (5), 139-152.

Taddese, G. (2001). Land degradation: a challenge to Ethiopia. *Environment Management, 27,*(6), 815-824.

Terborgh, J.W. (1988). The big things that run the world: a Sequel to E.O. Wilson. *Conservation Biology, 2,* 402-403.

Thimmayya, A. (1998). The effects of fertilization on soil and litter fauna in oak forest. http://www.courses.mbl.edu/SES/data/project/1998/thimmayya.pdf (3/29/04).

Thomas, C.D., Cameron, A., Green, R.E., Bakkens, M., Beaumont, L., Collingham, Y.C., Erasmus, B.N., De Siqueira, M.F., Grainger, A., Hannah, L., Huges, L., Huntley, B., van Jaarsveld, A.S., Midgley, G.F., Miles, L., Ortega-Huerta, M.A., Peterson, A.T., Phillips, O.L., and Williams, S.E. (2004). Extinction risks from climate change. *Nature, 427,* 145-148.

Thomas, D.S.G., and Middleton, N.J. (1993). Salinization: new perspectives on a major desertification issue. *Journal of Arid Environments, 24,* 95-105.

Thorne-Miller, B., and Cantena, J. (1991). *The Living Ocean: Understanding and Protecting Marine Biodiversity.* Washington, DC: Island Press.

Thrupp, L.A. (1998). *Cultivating Diversity: Agrobiodiversity and Food Security.* Washington, DC: World Resources Institute.

Troeh, F. R., Hobbs, J. A., and Donahue, R. L. (1999). *Soil and Water Conservation.* Englewood Cliffs, NJ: Prentice Hall.

Tuxill, J. (1999). Nature's Cornucopia. Report No. 148. Washington, DC: World Watch Institute

UNESCO. (2001). Protecting Ecosystems. World Water Assessment programme. United Nations Education Scientific and Cultural Organization. Paris.

UNESCO. (2002). Securing the Food Supply. World Water Assessment Programme. United Nations Education Scientific and Cultural Organization. Paris.

Uri, N. (2001). A note on soil erosion and its environmental consequences in the United States. *Water, Air, and Soil Pollution, 129,* 181-197.

USBC (2002). *Statistical Abstract of the United States 2002.* Washington, DC: U.S. Bureau of the Census, U.S. Government Printing Office.

USDA. (1999). Agricultural Phosphorus and Eutrophication. U.S. Department of Agriculture. Washington, DC: Agricultural Research Service.

USDA. (2002). *Agricultural Statistics.* Washington, DC: U.S. Department of Agriculture.

USGS. (2003). Birds as Indicators of Riparian Vegetation Condition in the Western U.S. U.S. Geological Society. Northern Prairie Wildlife Research Center. 3 pages.

Vandermeer, J., Lawerence, D., Symstad, A., and Hobbie, S. (2002). Effect of biodiversity on ecosystem functioning in managed ecosytems. In: *Biodiversity and Ecosystem Functioning: Synthesis and Perspectives.* Pp 221- 236. Oxford, UK: Oxford University.

Vietmyer, N. (1995). Applying biodiversity. *Journal of the Federation of American Scientists, 48,* (4), 1-8.

Wagner, T. (2003). Seasonality of canopy beetles in Uganda. In Y. Basset, V. Novotny, S.E. Miller, and R.L. Kitching (Eds), *Arthropods of Tropical Forests: Spatio-temporal Dynamics and Resource Use in the Canopy* (146-158). Cambridge: Cambridge University Press.

Walter, H. (1985). *Vegetation of the Earth and Ecological Systems of the Geo-biosphere.* Berlin: Springer-Verlag.

Walter, K.S. and Gillett, H.J. (1998). *1997 IUCN Red List of Threatened Plants.* Complied by the World Conservation Monitoring Centre. Cambridge, UK: IUCN/The World Conservation Union.

Warters, M.E. (1969). A Faunistic and Ecological Study of the Insects Found on the Corn Plant in Minnesota. Ph.D. Thesis, University of Minnesota, St. Paul, Minnesota.

Wayne, L.G., Brenner, D.J., Colwell, R.R., Grimont, P.A.D., Kandler, O., Krichevsky, M.I., Moore, L.H., Moore, W.E.C., Murry, R.G.E., Stackebrant, E., Starr, M.P. and Truper, H.G. (1987). Report of the ad hoc committee on reconciliation of approaches to bacterial systematics. *International Journal of Systematic Bacteriology, 37*(4), 463-464.

Western, D. (1994). Ecosystem conservation and rural development: the case in Amboseli. In D. Western, R.M. Wright, and S.C. Strum. (Eds.), *Natural Connection: Perspectives in Community-Based Conservation* (pp. 15-53). Washington, DC: Island Press.

Western, D. and Pearl, M.C. (1989). *Conservation for the Twenty-first Century.* New York: Oxford University Press.

Westman, W.E. (1990). Managing for biodiversity. *BioScience 40*, 26-33.Whalon, D. (2002). Integrated Crop Management in Apples. Clarksville Horticultural Research Station, Michigan State University. East Lansing, Michigan. 5 pages.

Whitcomb, W.H. and Godfrey, K.E. (1991). The use of predators in insect control. In D. Pimentel (Ed.), *Handbook of Pest Management in Agriculture* (Vol. II, pp. 215-241). Boca Raton, Florida: CRC Press.

WHO. (2000). Malnutrition Worldwide. World Health Organization. *http://www.who.int/nut/malnutrition_worldwide.htm (July 27, 2000).*

Wiebe, K. (1997). Resources, Sustainability, and Food Security. Food Security Assessment, International Agriculture and Trade Report No. GFA-9, pp. 36-42. Updated: April 24, 2002.

Wilson, E.O. (1987). The little things that run the world. *Conservation Biology 1,* 344-346.

Wilson, E.O. (Ed.) (1988). *Biodiversity.* Washington, DC: National Academy of Sciences.

Wilson, E.O. (2003). Conserving biodiversity: a global view. *BirdScope, 17* (4), 3-4.

World Bank. (2002). *World Bank approves new forest policy and strategy.* Increasing livelihoods for poor people while better protecting forests. The Board of Executive Directors. The World Bank. Washington, DC. 5 pages.

World Wildlife Fund. (1999). America's Global Warming Solutions. Study prepared for World Wildlife Fund and Energy Foundation. *Tellus Institute: Resource and Environmental Strategies*. August 1999. Pages 1-43.

WRI. (2002). Mechanisms for the Loss of Biodiversity. World Resources Institute. *http://www.wri.org/diodiv/b05-gbs.html*. (2/1/04)

Wright, D.H. (1983). Species-energy theory: an extension of species-area theory. *Oikos, 41*, 496-506.

Xerces Society. (2001). Why are Invertebrates Important? The Xerces Society. *http:// exerces.org/why_conserve_inverts.htm*. (10/23/03).

Yohn, C.W. and Rayburn, E.B. (2004). *Forage Management*. West Virginia University, Extension Service. 4 pages.

In: Focus on Biodiversity Research
Editor: Jan Schwartz, pp. 27-41

ISBN: 1-60021-372-3

Chapter 2

BIODIVERSITY AND CLIMATE CHANGE: A MODELING PERSPECTIVE

Joerg Tews*
Geomatics and Landscape Ecology Research Laboratory (GLEL),
Department of Biology, Carleton University, Ottawa, Ontario, Canada
Current address: Biology Department, Acadia University, Wolfville, Nova Scotia, Canada

ABSTRACT

Global climate change has been widely recognized as a major threat to biodiversity. A significant number of empirical studies provide evidence of species decline or extinction due to climatic change. By applying different modeling frameworks, the scientific community has ambitiously tried to provide politicians, stakeholders, and the public with predictions of the response of biodiversity to climate change. For example, a recent analysis based on a climate envelope model hypothesized that 15-37% of species in various regions of the world may go extinct within the next 50 years. However, despite the general appeal of such quantitative predictions, empirical distribution models have been criticized as being highly uncertain since they do not reflect local species interactions, genetic responses, or spatial processes. In contrast, bottom-up approaches such as mechanistic simulation models often consider species' life-history processes in a spatially explicit, stochastic context. However, mechanistic models rarely consider multi-species interactions and predictions may be invalid for other populations within the species' range. On the other hand, analytical food web models often do consider the dynamics of such multi-species interactions. However, the majority of such models is non-spatial and based on the assumption of a non-linear environment. In this chapter, I discuss the constraints and advantages of these three major modeling frameworks (empirical distribution models, bottom-up mechanistic simulation models, and analytical food web models) and synthesize how they have been applied to study the effects of

* Phone: 902-697-2377, E-mail: joerg.tews@gmail.com

climate change on biodiversity. I conclude that the combined consideration of a stochastic, spatial, and non-linear environment applied within the context of multiple species interactions may be useful in assessing the potential consequences of climate change for biodiversity. However, the current lack of empirical data, the complexity of such potential models, and the propagation of uncertainty remains a great challenge to be overcome in the future.

Keywords: analytical model, bottom-up model, climate envelope model, complexity, extinction, food web model, global change, individual-based model, mechanistic model, spatially explicit simulation model, species diversity

1. Introduction

The Earth's biological diversity and climate are ever-changing. Over the past millions of years, changes in global biodiversity have been a natural phenomenon. Species and mass extinction events (where entire taxonomic groups are wiped out in relatively short periods of time) have occurred as a result of environmental catastrophes, climatic changes, and the evolutionary advantages of one species over another. However, with the rise of human civilization, the natural environment has experienced directed changes, instead of random alterations or long-term, system-inherent changes. Anthropogenic habitat destruction, conversion, and fragmentation, as well the introduction of alien (and sometimes invasive) species are viewed as the primary threats to biodiversity on our planet (Pimm et al. 2001).

However, in the past decades, another factor has come into play: anthropogenic climate change. According to the IPCC (2001), the average global surface temperature increased by 0.6 C during the 20th century. Even if we were able to stop all CO_2 emissions right now, it is expected that temperatures could increase by an additional 0.6 C (IPCC 2001). If the trend in current greenhouse gas emissions continues, the average temperature is expected to increase by 1.4 to 5.8 ^{o}C before the year 2100 (IPCC 2001). In addition, precipitation amounts are predicted to increase in variability in many parts of the world. Large-scale effects of climate change may include rising sea levels and flooding. This will have major impacts on the current distribution and dynamics of ecosystems, leading to the expansion of some ecosystems into new areas and the retraction of others. Effects of climate change on biodiversity will vary from region to region, with significant changes anticipated in the far north and south of the planet and in mountainous areas, where species will be trapped in increasingly unfavorable conditions.

Climate change is projected to impact different levels of biological organization: individuals, populations, communities, and ecosystems will all experience the direct and indirect effects of climate change (Figure 1). Climate is an important determinant for many species' life-cycles. Climatic conditions at a given location and point in time may have direct implications for the rate of reproduction or survival. Assuming that a species does not go extinct, the biotic response to changes in climate may be either persistence *in situ* (if the change is within the tolerance limit of a species) or a shift in the species' range (if climate change at a given location is outside the species' tolerance limit). Both responses may include evolutionary adaptation processes on a time scale comparable to changes in climate (Davis et al. 2005). However, there is uncertainty as to whether species can adapt quickly enough to respond to anthropogenic climatic change.

Many scientific studies have revealed temperature-related shifts in various traits of a variety of species and populations (Root et al. 2003). Species densities may change and distributional ranges may shift (mostly towards higher latitudes or altitudes), the genetic structure of a species may shift, or changes may occur in the timing of events (such as bird migration). Generally, climate change effects can be grouped into changes in phenology (e.g., egg laying dates in birds), demography (i.e., changes in fecundity or survival), and population (e.g., changes in population size over time). There is ample empirical evidence of possible climate change effects resulting from changes in phenology (e.g., a delay in egg laying may reduce chances of offspring survival). Far more difficult is the detection of climate-induced changes in demographic rates and longer term population changes because these may be confounded with effects of habitat or density dependent effects.

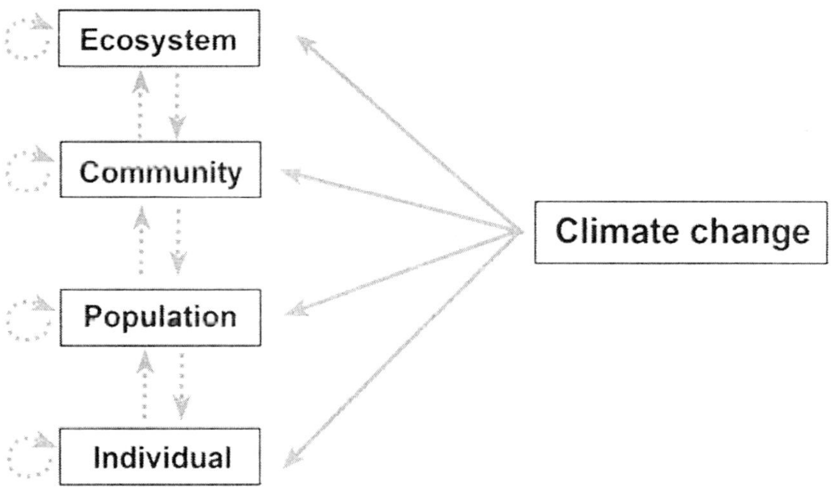

Figure 1. Climate change impacts can be divided into direct (bold) and indirect (dashed) effects. Direct effects can affect all levels of biological organization: the individual, population, community, and ecosystem levels. For example, a rise in sea level may have a direct impact on tidal communities, or higher temperature can alter the metabolic rates of individuals. Indirect effects can occur within one level or can be mediated either top-down or bottom-up. For example, if one species in a multi-species community or food web is affected, this may have indirect impacts on other species within that community. Such intra-community affects then can further translate into indirect effects on other communities within an ecosystem. On the population level, indirect effects can be felt at that level (e.g., source-sink population dynamics in a metapopulation context) or at lower or higher levels (e.g., if one major population of one species is extirpated by a stochastic event this may have indirect consequences for other species on the community level).

The effects of climate change on population, community, and overall ecosystem integrity may pose a threat for global biodiversity (Thomas et al. 2004). In this chapter, I will first provide some examples of recent empirical evidence of the effects of climate change on biodiversity. I will then discuss the application of ecological models in assessing and predicting current and possible future impacts of climate change on biodiversity. In particular, I will focus on the general principles underlying current empirical, mechanistic, and analytical modeling approaches. I will furthermore discuss the advantages and limitations of the different modeling strategies. Finally, I will suggest a potential direction for future modeling of the effects of climate change on biodiversity.

2. The Effects of Climate Change on Biodiversity

As previously noted, the detection of climate-induced changes in demographic rates is a challenging task. One of the few studies investigating the impact of climate change on demography examined survival rates of Emperor Penguins (*Aptenodytes forsteri*) (Barbraud and Weimerskirch 2001). This study showed that an increase in sea surface temperatures resulted in lower penguin survival rates, as warmer temperatures reduced the availability of the Antarctic krill used by fish and squid, the main penguin prey. A similar phenomenon occurred in the southern elephant seal (*Mirounga leonine*) at the Macquarie Island isthmus. In this area, years dominated by the El-Nino southern oscillation are characterized by cooler sea temperatures and more extensive sea ice. First-year survival of elephant seals was found to be highest during these years, indicating that increases in sea-surface temperatures may negatively affect this species (Mahon and Burton 2005). Similarly, for the Arctic it is believed that climate warming will lead to a decrease in survival and fecundity for some polar bear (*Ursus maritimus*) populations, as reduction in the extent of sea ice, earlier annual thaws, and delayed freezing limit the bears' access to seals, their main prey (see Derocher et al. 2004). This has been particularly well documented for the western side of Hudson Bay (e.g., Stirling et al. 1999).

Effects of climate change on the species level are not necessarily negative. A recent discussion at a conference revealed that, as expected, the vast majority of studies have found negative effects. However, it was suggested that this may be due to a bias in favor of studies suggesting negative effects of climate change, although there may also be positive effects. For example, Chamaille-Jammes et al. (2006) reported that regional warming in southern France resulted in a positive fitness response in mountain populations of the common lizard *Lacerta vivipar*: individual body size increased dramatically for four populations studied over 18 years, resulting in an increase in reproductive output. Effects of anthropogenic-induced climate warming may also be felt through trophic levels. Durant et al. (2003) found that breeding success of Atlantic puffins (*Fratercula arctica*) was positively affected by higher sea temperatures, which impacted the size and abundance of the puffins' main prey, the Norwegian spring-spawning herring (*Clupea harengus*).

Generally, both positive and negative climate change effects may arise from species interactions (Hulme 2005), and individual climate change effects may be different under varying environmental conditions. Climate change may also have different effects on different species within the same taxonomic groups. In the previous paragraph, I mentioned the positive impact of increased temperatures for puffins. However, even within the taxonomic group of alcids (e.g., puffins and auklets), responses may differ among species. For example, Kitaysky and Golubova (2000) showed that planktivorous auklets showed increased fecundity with lower sea-surface temperatures because they feed on macrozooplankton. The effects of climate change on biodiversity are diverse. It is therefore impossible to generalize across species or populations.

From a bottom-up point of view, climate change affects species primarily on an individual level. As indicated earlier, if the amplitude of environmental change ranges within the tolerance level of a species, individuals may be able to buffer this variation in various ways. Such 'buffer mechanisms' are most often behavioral (Grimm et al. 2005b). For example, based on an individual-based simulation model for the Lesser Spotted Woodpecker (*Picoides minor*), Rossmanith et al. (2006) showed that polyandry (here the mating of one female with more than one male due to a surplus of males) and general flexibility in the mating system is a buffer mechanism that can significantly reduce the impact of environmental change in this species. Other examples include behavioral adaptation such as the ability of predatory birds to increase territory sizes if resources become scarce (Wichmann et al. 2005), increases in the pack sizes of wolves to increase hunting efficiency when there is more than the usual amount of snow (Post et al. 1999), and the general ability of predators to shift prey species. Net effects of climate change may therefore vary depending on the ability of the species to buffer environmental noise. For longer-lived species, behavioral adaptation may be an effective strategy. However, if environmental conditions change more rapidly than the mitigation effect of behavioral adaptation, species may suffer from climatic changes.

In the previous paragraphs, I outlined and presented some examples of how species may respond to changes in climatic conditions. As we have seen, these responses can be positive or negative, but they can also be neutral, or they may change from positive to negative or vice versa depending on the intensity of climate change (for example, even if climate change initially has a positive effect on certain species, this may shift to an overall negative effect if conditions exceed the optimum).

Because the pace of current climate change is believed to be more rapid than past non-anthropogenic changes, researchers generally believe that it may increase the risk of extinction for many species worldwide. However, detecting real extinction events as a result of climate change is a challenging task because of the confounding negative effects of habitat loss, human land use, and development. Moreover, the effects of climate change often occur over relatively long periods of time, as opposed to relatively short-term causes of extinction such as overexploitation through intense harvest (as in poor fisheries management). As a result, there is little empirical evidence of extinction events driven by climate change.

However, there are a few good examples of the negative effects of climate change. In the mountains of Costa Rica, Pounds et al. (2005) found that a sharp decline in amphibian diversity over the past two decades was directly linked to epidemics driven by climate change. Temperatures at highland sites were shifting towards optimum conditions for the growth of the chytrid fungus *Batrachochytrium dendrobatidis*, thus encouraging outbreaks of this deadly disease. In California, McLaughlin et al. (2002) showed that the extinction of two populations of the checkerspot butterfly, *Euphydryas editha bayensis*, were caused by a combination of habitat loss and regional climate change in the form of increasing variability in precipitation. The authors suggested that extremes in annual precipitation most likely reduced the temporal overlap of larvae and plants, which increased larval mortality and the magnitude of population fluctuations. Increasing frequency and severity of weather was suggested to be the main factor that drove both populations to extinction within three decades.

Due to the potential threat of climate change on global biodiversity, politicians, stakeholders, and the public are eager for projections of future trends. Undoubtedly, we need empirical evidence on which to base these projections. However, empirical research is often very time consuming and may only document changes that have already occurred. This leads us to the use of models to project the future effects of climate change.

In biodiversity research, different types of models have been used with different types of underlying assumptions. These models can be classified as bottom-up and top-down approaches, or as empirical, mechanistic, or analytical models. In most cases, top-down approaches are represented by empirical, large-scale species distribution or analytical models. In contrast, bottom-up modeling approaches are often represented by mechanistic, single-species, spatially explicit simulation models that consider life-history processes and species interactions. In the following section, I discuss the constraints and advantages of three main modeling strategies. In addition, I discuss the methodological gap that seems to exist between these modeling approaches and suggest ways to bridge this gap.

3. Modeling Approaches

In ecology, many different types of models exist, and they have been classified according to many different criteria. According to Levins' (1966) classification, I distinguish between empirical, mechanistic, and deterministic models. Empirical models are derived from statistical relationships between parameters and variables and are often based on regression and correlation analyses. Deterministic or analytical models are based on mathematical equations. Mechanistic models use probabilistic uncertainty and are applied under the assumption of non-linear system dynamics. In this synthesis, I do not intend to provide a complete overview of available modeling techniques, but instead to focus on three major modeling approaches that are applied in assessing the potential impact of environmental change on biodiversity: (1) empirical distribution models, (2) mechanistic, probabilistic simulation models, and (3) analytical models of community dynamics.

3.1. Empirical Distribution Models

Predictive species distribution models make use of a wide array of statistical techniques such as regression analysis, neural networks, and Bayesian analyses. A common method to predict the response of species to climate change is through so called climate envelope models (also called ecological niche models) (Pearson and Dawson 2003, Ibanez et al. in press). In this approach, an 'envelope' is drawn around the current distribution of a species, and climatic variables are identified that correspond to this area. Predicted changes in these variables are then used to delineate the species' distribution in the future.

Recently, Thomas et al. (2004) used this approach in combination with the species-area relationship (an empirical power-law relationship describing how the number of species relates to area) to project species extinction risks due to climate change. For this method, the envelope under climate change is predicted, and, based on the projected contraction or extension of this area, the species-area relationship is used to assess extinction risk. Based on mid-range climate-warming scenarios for 2050, the latter study predicted that 15–37% of the species in various regions of the world would be 'committed to extinction'. In another study, using a range of niche-based modeling techniques, Thuiller et al. (2005) assessed the potential impact of future climate change on plant diversity in Europe. They found that more than half of the 1350 plants studied could be vulnerable or threatened by 2080. Erasmus et al. (2002) studied the potential responses of 179 South African animal species to climate change. Using a multivariate climate envelope approach, the authors predicted that approximately 17% of the species investigated would expand their ranges, 78% would display range contractions, 3% would show no response, and 2% would become locally extinct.

What general patterns can be seen from these and other climate envelope modeling studies? As indicated earlier, the effects of climate change on biodiversity will vary from region to region, with significant changes anticipated at extreme altitudes and latitudes. For example, Konvicka et al. (2003) found that butterflies had shifted their habitat range uphill as a result of recent climate change. Other examples from the northern hemisphere indicate range shifts into more northerly regions with a warming climate. However, recent studies suggest that range contraction may also occur in other marginal areas: Wilson et al. (2005) show that lower elevational limits for 16 butterfly species in central Spain rose by an average of 212 m in three decades, accompanying a 1.3 C rise in mean annual temperature. Generally, the distributional range of a species can be distinguished between rear edge (i.e., populations residing at the current low-latitude margins of the species distribution), continuous range, and leading edge. Previous studies have largely focused on the importance of leading edges. However, there is increasing evidence that rear edge populations are often disproportionately important for the survival and evolutionary potential of species as they harbor a higher genetic diversity (Hampe and Petit 2005).

There are several disadvantages and criticisms that have been raised with respect to the application of climate envelope models. First and foremost, static distribution models assume that an equilibrium state exists between environmental conditions and the observed species distribution patterns. However, in most cases, ecological systems are subject to non-equilibrium dynamics; such models are unable to cope with stochastically and dynamically changing environments (Guisan and Zimmermann 2000). Secondly, empirical distribution models do not consider biotic interactions, which should be an important model component because shifts in distributional ranges on the species level result in changes in community composition and, thus, changes in biotic interactions. Thirdly, dispersal is often neglected or only considered in very broad categories in climate envelope models. Yet, species-specific dispersal rates and the structure of the landscape over which individuals disperse may determine how fast a species may be able to respond to changes in environmental conditions by colonizing new areas (Pearson and Dawson 2003). Finally, the notion of 'extinction risk' in the context of climate envelope models is generally flawed (see Thomas et al. 2004), because extinction is a process that is related to changes in population abundance or density leading to local or regional species extirpation, and not to expansion or contraction of species ranges. A contraction of a species' range does not necessarily mean that the risk of extinction increases, as populations may still thrive within that reduced range if local abundances are high. Therefore, if we aim to make sound predictions about the effects of climate change on individual species, it is crucial to consider the spatial patterns of abundance within the distribution and not just the overall range of the species (Shoo et al. 2005).

3.2. Mechanistic, Bottom-up Simulation Models

A second modeling approach that I will discuss in this chapter is mechanistic, bottom-up stochastic simulation models, which are frequently used in biodiversity research and have become a powerful tool to study ecological systems. These models generally differ from the empirical distribution models discussed above in that they focus on the effects of environmental change on the population dynamics of single (or multiple) species instead of inferring changes in spatial distribution based on statistical relationships. Such mechanistic models are primarily rooted in a bottom-up approach and may be individual- or population-based (both can be cell-based and spatially explicit).

Individual-based (or agent-based) models are built from the perspective of individual organisms and their behavior (Grimm and Railsback 2005). The bottom-up strategy of such models is to compile relevant information about entities at a lower level of the system (i.e. agents or individuals), formulate theories about their behavior, incorporate these theoretical behaviors into a computer simulation model, and observe the emergence of system-level properties related to particular questions (Grimm et al. 2005a). As climate change generally has direct effects on the population or individual level and is often related to stochastic events (e.g., precipitation), stochastic models such as individual-based models have been toted as a promising tool for bottom-up studies of environmental change. Most individual-based simulation models are spatially explicit because the rules for the behavior of an agent are often governed by space.

Mechanistic, bottom-up simulation models are popular because they are based on simple rules, but can generate complex, real-world patterns. Two premises underlie the frequent use of such models. Firstly, simple sets of model rules that, when followed by a computer model, may generate extremely complex patterns. Secondly, the natural environment (including biodiversity) also shows many extremely complex patterns. Thus, it is concluded that simple rules underlie many extremely complex phenomena, and with the help of powerful computers, scientist are able to detect the simple rules that may underlie complex phenomena observed in nature (Richardson 2002). In this regard, the strategy of 'pattern-oriented modeling' has been become a frequently used modeling framework (Wiegand et al. 2003, Grimm et al. 2005a). The central question that is addressed through this approach is to analyze which observed patterns characterize the system and its dynamics, and what variables and processes (or rules) must be incorporated into the model so that these patterns emerge.

The study of Clark et al. (2001) coupled an individual-based model of brook trout (*Salvelinus fontinalis*) and rainbow trout (*Oncorhynchus mykiss*) with a geographic information system to assess the effects of climate-related changes in water temperature, stream flow, and mortality episodes on southern Appalachian stream populations. They found changes in flow to be at least as important as changes in temperature in predicting trout responses to climate change. Even though this study provides a meaningful application, the model is not a good predictor of species composition changes, as pointed out by the authors. Climate-related changes in the species composition of food webs (either directly or indirectly through changes in the abundance of one species affecting another species) can have major implications for the persistence of individual species, particularly in aquatic systems.

In the study of Tews et al. (2004), a grid-based simulation approach was used to model climate change implications for the savanna shrub *Grewia flava* (Figure 2). This population- and grid-based model has a grid cell as the smallest entity. Cell-based models (or cellular automata) are often used for vegetation modeling, as plants are stationary and only "move" across a landscape while dispersing their seeds. In this study, projected climate change scenarios were incorporated to estimate the potential response of shrub abundance and density to changes in precipitation. It was found that a possible increase in precipitation may facilitate an increase in shrub cover and lead to declines in regional biodiversity and in the carrying capacity of rangelands.

Mechanistic, bottom-up simulation models have several advantages and may be useful tools in studying the impact of climate change on species diversity. Such simulation models show their scientific value by facilitating the re-conceptualization of theoretical assumptions, the generation of hypotheses, and the proof of concepts. However, they are also subject to significant limitations. Some argue that there are many potential paths that may lead to the same pattern generated with such models. In other words, for a given scenario, there may be several qualitatively different, non-overlapping explanations for any particular phenomena resulting from nonlinear interactions (Richardson 2002). A particular challenge for such spatially explicit modeling frameworks, as for most modeling approaches, is the trade-off between complexity and simplicity. On the one hand, complexity may increase the degree of reality. On the other hand, it also increases the level of uncertainty. The trade-off between the model's pay-off (i.e., how much we can learn from it) and the complexity is usually highest in a zone of intermediate complexity (Grimm and Railsback 2005)

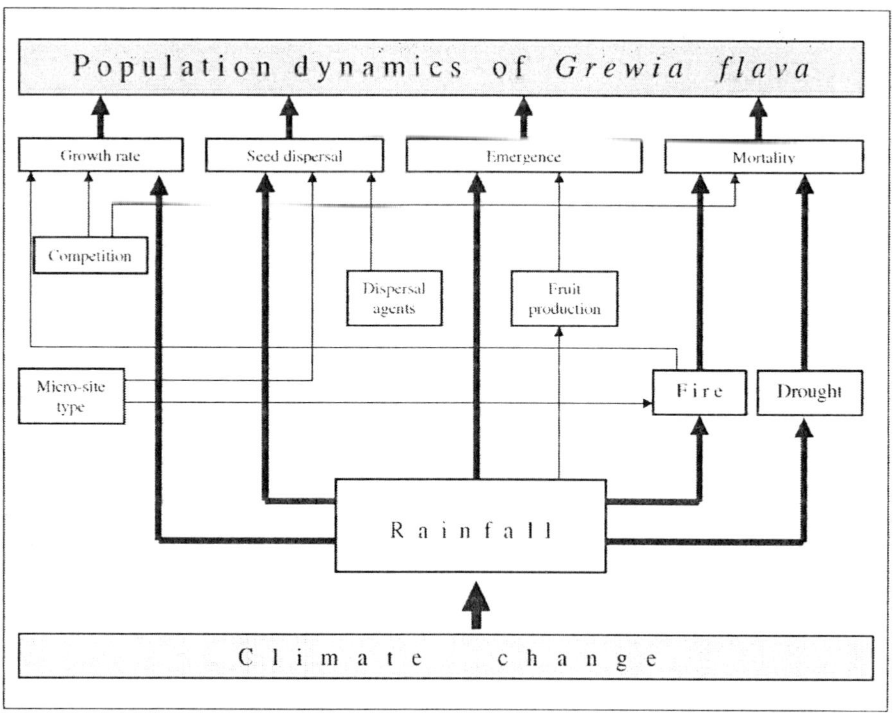

Figure 2. Example of a flow chart for a mechanistic, spatially explicit, population-based simulation model of the savanna shrub *Grewia flava*. Bold arrows indicate processes where model parameters and variables are affected by climate change via the annual amount of rainfall (see Tews et al. 2004). Such diagrams help to decide on the structure of a model and the processes to be included.

As a result of the inherent complexity of spatially explicit models and the computational power needed, the large majority of such models are applied to single species only, generally neglecting the dynamics of biotic interactions. For example, many mechanistic simulation models incorporate the effects of inter-specific competition or mortality (e.g., by a predator) as model parameters. However, if, for example, a predator underlies species-specific population dynamics, the risk of mortality for a simulated prey species may significantly change over time and therefore model predictions may be inconsistent.

3.3. Food Web Models

Generally, it is axiomatic in ecology that the dynamics of one set of species cannot be understood without reference to the dynamics of other species and processes (Raffaelli et al. 2005). Based on historic records, there is strong evidence that climate change will lead to the reorganization of food webs. For example, a recent study from the Canadian High Arctic found that sedimentary records from ponds spanning the last several thousand years showed major shifts in food web communities within only the last 200 years (Quinlan et al. 2005). Generally, the potential to adapt to climate change (and therefore the probability of survival) increases from the top to the bottom of a food web as average life-cycle length decreases. The adaptability of species to environmental change also co-varies within trophic levels and, therefore, the order in which species are lost ultimately governs the ecosystem-level consequences of biodiversity loss (Solan et al. 2004).

Direct climate change effects on one species may translate into indirect effects for interrelated species. Such effects may eventually lead to extinction of one species and possibly further secondary extinction events. For example, a decrease in fecundity leading to a lower abundance of a prey species may affect predator species as well as other prey species (when the predator switches to other prey species). Such food web relations can be grouped into bottom-up and top-down effects. In the Isle Royale National Park in Lake Superior, Post et al. (1999) found that wolves responded to an increase in the amount of snow by hunting in larger packs. This, in turn, increased the killing efficiency by wolves and brought a subsequent decline in the moose population, which eventually resulted in less browsing pressure on balsam fir saplings. There are a relatively large number of empirical studies that provide evidence of how changes in climate may affect smaller food chains, predator-prey dynamics, or single-trophic food webs (e.g., Kishi et al. 2005, Wilmers et al. 2006). For larger networks, empirical data are currently lacking, and therefore simulation models are used to approach this issue in a rather theoretical framework.

Generally, there are three main types of food web models that are frequently applied: static models (which assign links between species according to simple rules), dynamic models (which incorporate population dynamics among interacting species, e.g. by means of Lotka-Volterra type models), and assembly and evolutionary models (which build food webs starting from a few species by adding additional species through a process of invasion or speciation). One question that has sparked a large number of studies in the past several decades is that of whether more complex and diverse communities are more stable. This question is relevant to the study of climate change effects: if a particular species inhabiting two differently structured food web communities goes extinct as a direct result of climate change, which food web will be more threatened by further loss of species and stability?

Research conducted in the early 70's counter-intuitively predicted that the more species a food web contains, the higher will be its instability (see May 1972). This view has changed as examples have emerged of more complex food webs (with a higher degree of species diversity and connectance) showing higher stability than less complex food webs.

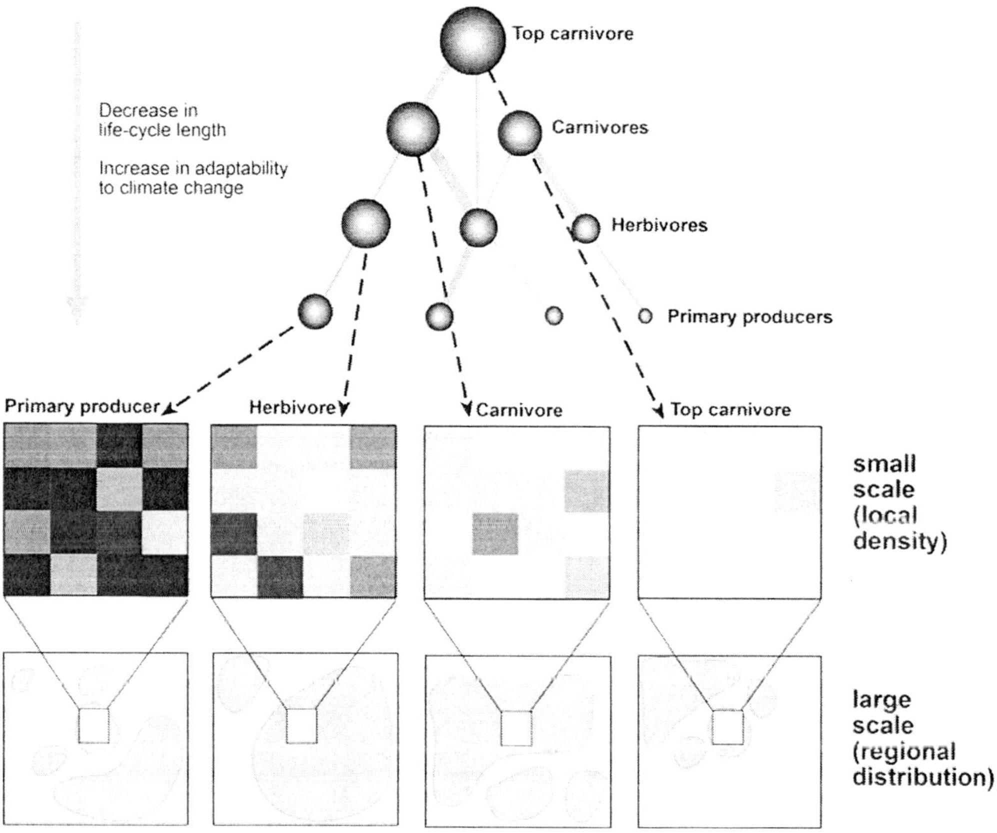

Figure 3. The adaptability of species to climate change co-varies both between and within trophic levels (indicated by the circles). As average life-cycle length decreases from the top to the bottom of a food web, the potential to adapt to climate change increases as well, although it may vary within a trophic level. As indicated on the smaller scale, larger-bodies species such as carnivores occur in lower densities, and top predators may even be absent in some areas. Thus, link or interaction strength among species in a community varies across space (indicated by the differently sized grey lines). In essence, a local food web may be similar in terms of species assemblage to those in adjacent areas, but the inherent dynamics may be different because of different interaction strengths. The same pattern occurs on a larger scale: climatic change is anticipated to change species' ranges, leading to changes in the species composition of communities.

The structural characteristics of a food web are of crucial importance in predicting how vulnerable a system will be to environmental change and potential changes in biodiversity. The consideration of space has long been neglected in food web modeling. Undoubtedly, considering the spatial structure of a food web inflates the degree of complexity and uncertainty. However, as several recent studies have shown, space is a crucial component in modeling and describing food web structure. Food webs are influenced by processes in adjacent systems, and they are spatially heterogeneous and dynamic in space and time (Brose et al. 2005). The fact that larger bodied species occur in lower densities/abundances than smaller bodied species illustrates this (see Figure 3). Species also change their position in space and time, or their distributions may be heterogeneous due to habitat fragmentation or natural barriers. Therefore, the strength of interaction among species (i.e., link strength) varies across space and time. These considerations are of particular importance for future studies using food web models to assess the consequences of climate change on biodiversity.

4. Synthesis and Outlook

Assessing and predicting the consequences of climate change on biodiversity remains a challenging task. Different modeling frameworks have been applied to approach this important field of research. There is no unifying modeling recipe to help us address the issue. Each research question and biological system requires specific methodologies, and each modeling technique has its advantages and disadvantages. In this chapter, I have provided a short overview of the major approaches currently applied. Large-scale, empirical research is difficult due to the different temporal and spatial scales of response for species with different home ranges and reproductive cycles. Therefore, different modeling frameworks are used to address the issue and learn about general processes and patterns.

Empirical distribution models, including climate envelope models, are useful for large scale, continental predictions of shifts in species distribution. Unfortunately, such models neglect species-specific, non-linear responses to environmental changes and are unable to cope with stochastically and dynamically changing environments. Moreover, such models do not consider the spatial structure of landscapes (which determines how species shift their distributions), and they consider dispersal in only very broad categories. As shown, mechanistic, bottom-up simulation models may be useful in overcoming these difficulties. However, the large majority of such models are only applied to single species, and they often incorporate biotic interactions in a static way. The failure to account for multiple trophic levels in biodiversity studies may lead to under- or overestimations of the effects of climate change. To solve this dilemma, several population models of interacting species would have to be run simultaneously. Food web models consider such multiple species interactions. However, these models usually lack important spatial 'real world' components, as well as the stochasticity and non-linearity of nature. The combined consideration of a stochastic, spatial, and non-linear environment applied within the context of multiple species interactions may be a useful way to assess climate change effects. However, the current lack of empirical data, the complexity of such models, and the propagation of uncertainty present great challenges to this approach. If a single-species approach is necessary, it may be useful to model keystone species, i.e., species that have a disproportionately high influence on other species relative to their body size.

Assessing the potential risk of species loss due to human-induced climatic change should also incorporate current trends in habitat loss. Many researchers believe that over-harvesting, environmental contamination, and habitat loss and fragmentation present even greater threats to biodiversity and the resilience of natural systems than climate change. However, it is clear that anthropogenic climatic change is yet in its infancy and may well become the most important threat to the earth's biological diversity in the coming decades. One might argue that species have always adapted to climatic changes and that species extinctions are frequent events of evolutionary history. However, I believe that the combined effects of a relatively rapid shift in climatic conditions and the current threats of habitat destruction create a deadly cocktail for the natural dynamics of biological diversity on earth.

ACKNOWLEDGEMENTS

The author gratefully acknowledges financial support by the German Academy of Naturalists Leopoldina by funds of the German Ministry of Education and Research (BMBF-LPD 9901/8-110). Many thanks also to Rebecca Tittler for helpful comments and suggestions.

REFERENCES

Barbraud C, Weimerskirch H (2001) Emperor penguins and climate change. *Nature* 411: 183–186.

Brose U, Pavao-Zuckerman M, Eklof A, Bengtsson J, Berg M, Cousins SH, Mulder C, Verhoef HA, Wolters V (2005) Spatial aspects of food webs. Pages 463-469 in: *Dynamic Food Webs: Multispecies Assemblages, Ecosystem Development and Environmental Change.* de Ruiter PC, Wolters V, Moore JC (eds.) Academic Press.

Chamaille-Jammes S, Massot M, Aragon P, Clobert J (2006) Global warming and positive fitness response in mountain populations of common lizards *Lacerta vivipara*. *Global Change Biology* 12: 392-402

Clark ME, Rose KA, Levine DA, Hargrove, WW (2001) Predicting climate change effects on Appalachian trout: combining GIS and individual-based modeling. *Ecological Applications* 11: 161–178.

Derocher AE, Lunn NJ, Stirling I (2004) Polar bears in a warming climate. *Integrative and Comparative Biology* 44: 163–176

Durant JM, Anker-Nilssen T, Stenseth NC (2003) Trophic interactions under climate fluctuations: the Atlantic puffin as an example. *Proceeding of the Royal Society London B* 270: 1461–1466.

Erasmus BFN, Van Jaarsveld AS, Chown SL, Kshatriya M, Wessels KJ (2002) Vulnerability of South African animal taxa to climate change. *Global Change Biology* 8: 679-693.

Grimm V, Railsback SF (2005) *Individual-based modeling and ecology.* Princeton University Press, Princeton, New Jersey

Grimm V, Revilla E, Berger U, Jeltsch F, Mooij WM, Railsback SF, Thulke HH, Weiner J, Wiegand T, DeAngelis DL (2005a) Pattern-oriented modeling of agent-based complex systems: lessons from ecology. *Science* 310: 987-991.

Grimm V, Revilla E, Groeneveld G, Kramer-Schadt S, Schwager M, Tews J, Wichmann MC, Jeltsch F (2005b) Importance of buffer mechanisms for population viability analysis. *Conservation Biology* 19: 578-580.

Guisan A, Zimmermann NE (2000) Predictive habitat distribution models in ecology. *Ecological Modelling* 135: 147–186.

Hampe A, Petit RJ (2005) Conserving biodiversity under climate change: the rear edge matters. *Ecology Letters* 8: 461–467.

Hulme P (2005) Adapting to climate change: is there scope for ecological management in the face of a global threat? *Journal of Applied Ecology* 42: 784-794.

Ibanez I, Clark JS, Dietze MC, Feeley K, Hersh M, LaDeau S, McBride S, Welch NE, Wolosin MS (in press) Predicting biodiversity change: Outside the climate envelope, beyond the species-area curve. *Ecology*.

IPCC (2001) Intergovernmental Panel on Climate Change's Synthesis Report [www.ipcc.ch]

Kishi D, Murakami M, Nakano S, Maekawa K (2005) Water temperature determines strength of top-down control in a stream food web. *Freshwater Biology* 50: 1315–1322.

Kitaysky A, Golubova EG (2000) Climate change causes contrasting trends in reproductive performance of planktivorous and piscivorous alcids. *Journal of Animal Ecology* 69: 248-262.

Konvicka M, Maradova M, Benes J, Fric Z, Kepka P (2003) Uphill shifts in distribution of butterflies in the Czech Republic: effects of changing climate detected on a regional scale. *Global Ecology and Biogeography* 12: 403-10.

Levins R (1966) The strategy of model building in population biology. *American Scientist* 54:421-431.

McMahon CR, Burton HR (2005) Climate change and seal survival: evidence for environmentally mediated changes in elephant seal, *Mirounga leonina*, pup survival *Proceedings of the Royal Society London* B 272: 923–928.

May RM (1972) Will a large complex system be stable? *Nature* 238: 413–414.

McLaughlin J F, Hellmann JJ, Boggs CL, Ehrlich PR (2002) Climate change hastens population extinction. *Proceedings of the National Academy of Sciences* 99: 6070-6074.

Pearson RG, Dawson TP (2003) Predicting the impacts of climate change on the distribution of species: are bioclimate envelope models useful? *Global Ecology and Biogeography* 12: 361–371

Pimm SH, et al. (2001) Can we defy nature's end? *Science* 293: 2207–2208.

Post ES, Peterson RO, Stenseth NC, McLaren BE (1999) Ecosystem consequences of wolf behavioural response to climate. *Nature* 401: 905–907.

Pounds, JA, et al. (2006) Widespread amphibian extinctions from epidemic disease driven by global warming. *Nature* 439: 161-167.

Quinlan R, Douglas MSV, Smol JP (2005) Food web changes in arctic ecosystems related to climate warming. *Global Change Biology* 11: 1381–1386.

Raffaelli D (2005) How extinction patterns affect ecosystems. *Science* 306: 1141-1142.

Richardson, KA (2002). Methodological implications of complex systems approaches to sociality': some further remarks," *Journal of Artificial Societies and Social Simulation* 5: 2.

Root TL, Price JT, Hall KR, Schneider SH, Rosenzweig C, Pounds A (2003) Fingerprints of global warming on wild animals and plants. *Nature* 421: 57-60.

Rossmanith E, Grimm V, Blaum N, Jeltsch F (2006) Behavioural flexibility in the mating system buffers population extinction: lessons from the Lesser Spotted Woodpecker (*Picoides minor*) *J. Animal Ecology,* in press.

Shoo LP, Williams SE, Hero J-M (2005) Potential decoupling of trends in distribution area and population size of species with climate change. *Global Change Biology* 11: 1469–1476.

Solan M, Cardinale BJ, Downing AL, Engelhardt KAM, Ruesink JL (2004) Extinction and ecosystem function in the marine benthos. *Science* 306: 1177-1180.

Stirling I, Lunn NJ, Iacozza J (1999) Long-term trends in the population ecology of polar bears in western Hudson Bay in relation to climate change. *Arctic* 52: 294-306.

Tews J, Jeltsch F (2004) Modelling the impact of climate change on woody plant population dynamics in South African savanna. *BMC Ecology* 4:17 [online]

Thomas CD, Cameron A, Green RE, Bakkenes M, Beaumont LJ, Collingham YC, Erasmus BF, De Siqueira MF, Grainger A, Hannah L, Hughes L, Huntley B, Van Jaarsveld AS, Midgley GF, Miles L, Ortega-Huerta MA, Peterson AT, Phillips OL, Williams SE (2004) Extinction risk from climate change. *Nature* 427: 145-148.

Thuiller W, Lavorel S, Araujo MB, Sykes MT, Prentice IC (2005) Climate change threats to plant diversity in Europe. *PNAS* 102: 8245-8250.

Wichmann MC, Groenveld J, Jeltsch F, Grimm, V (2005) Mitigation of climate change impacts on raptors by behavioural adaptation: ecological buffering mechanisms. *Global and Planetary Change* 45: 273-281.

Wiegand T, Jeltsch F, Hanski I, Grimm V (2003) Using pattern-oriented modeling for revealing hidden information: a key for reconciling ecological theory and application. *Oikos* 100: 209-222.

Wilmers CC, Post E, Peterson RO, Vucetich JA (2006) Predator disease out-break modulates top-down, bottom-up and climatic effects on herbivore population dynamics. *Ecology Letters* 9: 383–389.

Wilson RJ, Gutierrez D, Gutierrez J, Martinez D, Agudo R, Monserrat VJ (2005) Changes to the elevational limits and extent of species ranges associated with climate change. *Ecology Letters* 8: 1138–1146.

In: Focus on Biodiversity Research
Editor: Jan Schwartz, pp. 43-62

ISBN: 1-60021-372-3

Chapter 3

DRUG DISCOVERY, BIODIVERSITY CONSERVATION, AND ECONOMIC DEVELOPMENT: A MICRO-LOAN PROGRAM AT CUC PHUONG COMMUNE (VIETNAM)

***Bui Minh Vu*[1] *and Le Thi Xuan*[2]**

[1] Institute for International Business Management and Training, Vietnam.

[2] Vietnamese Academy of Science and Technology, Vietnam and
Dinh Duy Hai et al., People's Committee, Vietnam

ABSTRACT

The 1992 United Nations Convention on Biological Diversity urges nations to conserve, to value, and to use biodiversity in a sustainable manner. Commercial developers of products based on biodiversity are urged to share equitably the benefits that may arise as a result of the effort with the steward of the biodiversity. In line with these objectives, a reciprocity measure as part of a plant-based drug discovery and development endeavor using the plant diversity of Cuc Phuong National Park, Vietnam, in the form of a micro-loan program was implemented for poor families of the Muong ethnic minority in the Cuc Phuong Commune located just outside of the park. Families identified by an Economic Development Village Board were provided a no-interest loan to be returned to the Commune People's Committee within 3 years, which the Committee then used to provide a rotating loan to other identified poor families in the community. Under this program, for the period of 2000 to 2006, 45 poor Muong families participating in the program have improved their household economy as a result of increased productivity, primarily, through the purchase of farming animals (cows, water buffaloes) to work their lands. This improvement of household economy has a beneficial effect in helping reduce the human pressure on the forest, and the biodiversity therein, of the Cuc Phuong National Park, aiding in the effort to preserve, value and use the Park's biodiversity for the future.

INTRODUCTION

Article 1 ("Objectives") of The United Nations Convention on Biological Diversity (UN CBD) states: "The objectives of this Convention, to be pursued in accordance with its relevant provisions, are the conservation of biological diversity, the sustainable use of its components and the fair and equitable sharing of the benefits arising out of the utilization of genetic resources, including by appropriate access to genetic resources and by appropriate transfer of relevant technologies, taking into account all rights over those resources and to technologies, and by appropriate funding" (Convention text, 1992, at http://www.biodiv.org/convention/articles.asp). In a plant-based natural product drug discovery and development endeavor, a process known as bioprospecting (Reid *et al.*, 1993; Soejarto *et al.*, 2005), the terms "fair and equitable sharing of the benefits arising out of the utilization of genetic resources" signifies the implementation of reciprocity measures toward the owners or the steward of the biodiversity – the forest and the plant resources within. Often reciprocity is only thought of in terms of very large sums of money in the form of royalties paid by a pharmaceutical company that develops and commercializes the drug. However, immediate, short-term benefits that accrue to the steward of the biodiversity and the holders of traditional knowledge associated with that biodiversity as part of the research process are as important, if not more important, than the long-term, high-yield benefits, which, often times, never materialize. This paper serves as a paradigm of small scale, but genuine short-term benefits, which a plant-based drug discovery effort can deliver, while the research process is being undertaken.

Thus, as part of a reciprocity measure in the International Cooperative Biodiversity Groups (ICBG) Program "Studies on Biodiversity of Vietnam and Laos" based administratively at the University of Illinois at Chicago (UIC) (also known as UIC ICBG or Vietnam-Laos ICBG; Soejarto *et al.*, 1999, 2006), a family scale money-loaning initiative was implemented. In this initiative, referred to as a "micro-loan" program, money was provided as a loan to selected poor ethnic Muong families in the Cuc Phuong Commune, located at the south-eastern end of Cuc Phuong National Park, Ninh Binh Province, Vietnam. This micro-loan program adopted the policy of the Vietnamese government, originally implemented at provincial level in 1992 in Ho Chi Minh City under the name of Hunger Eradication and Poverty Reduction (HEPR) Program, which was focused on increasing poor households' opportunity to get access to preferential credit, accommodation, food, basic social services and infrastructure at communal level. The Government recognized the movement and assigned the Vietnamese Ministry of Labor, Invalids and Social Affairs (MOLISA) to guide and supervise HEPR activities and to set the national poverty line. By 1995, the HEPR Program had spread over the nation, and similar HEPR activities were implemented in other provinces such as Song Be, Dong Nai, Hai Phong, and Ha Tay. The 8th Congress of the Vietnamese Communist Party (June 28, 1996) set the official HEPR goals, followed by the signing of Decision No. 531/TTg on August 8th 1996 by the Prime Minister on a unified management of national programs (Dac and Huu, 2004, p. 6). These programs included measures to ensure poor households get access to land; to provide preferential credit to 90-95% of poor households; to provide free healthcare services, free vocational training, and free education; to provide subsidy of products that serve mountainous and isolated areas; and to develop infrastructure in the poorest communes.

The contributions and success of micro-loan and micro-finance programs to alleviate conditions of local poverty and to provide poor people access to credit have been extensively documented elsewhere (Morduch, 1999), and there is a growing body of literature especially on the success of such programs in South East Asia (especially, Bangladesh) (Robinson, 2001; Weiss and Kurmanalieva, 2003; Burra *et al.*, 2005)

In 1998, HEPR officially became a National Target Program by strengthening and merging a large number of national programs into several significant foci. Poverty alleviation became both a social policy and an important development goal. It is also an indispensable part of the national strategy for socio-economic development in the period of 2001-2010 (Dac and Huu, 2004, p. 6).

Implementation of the Vietnamese National Target Program on HEPR and Job Creation

The National Target Program on HEPR (also called Program 133) was officially launched in July 1998 and was continued into the period of 2001-2005. This program was merged with the program on job creation (Program 120) to form the national target program on HEPR and job creation (Program 143). The goals of the national program on HEPR were to eradicate chronic hunger and to reduce the poor household rate to 10% in the year 2000.

The objectives of Program 133 and 143 for the period of 2001-2005 were as follows (Dac and Huu, 2004, p. 10):

a. To decrease the poor household rate to below 10% and to apply a new poverty line, which stipulates an annual decrease of poor household rate at 1.5-2.0% and the eradication of chronic hunger.
b. To develop basic infrastructure such as small-scale irrigation systems, roads, markets, schools, clinics, and to improve access to electricity and clean water for poor communes.
c. To create 1.4-1.5 million jobs each year.
d. To lower unemployment rate to 6% in urban areas.

The criteria for selecting poor communes for Program 133 and 143 were set as follows (Dac and Huu, 2004, p. 27):

a. More than 25% of households in the commune are poor.
b. Infrastructure quality: It satisfies less than 3 of 6 requirements below:
   - Clean water: Less than 30% of households in the commune can get access to clean water.
   - Electricity: Less than 50% of households in the commune can get access to electricity.
   - Transportation: There is no road to commune center or it cannot be accessed at a certain time of the year.

- Education facility: Number of classroom cannot satisfy more than 70% of the total number of pupils, or classrooms are made of simple material for temporary use.
- Health care: Lack of commune public healthcare center or poorly equipped healthcare center.
- Public market: Lack of permanent market place or unsettled market.

THE UIC ICBG MICRO-LOAN PROGRAM TARGETED TO MUONG ETHNIC COMMUNITY AT CUC PHUONG COMMUNE

One of the main objectives of the UIC-based Vietnam-Laos ICBG ("Studies on Biodiversity of Vietnam and Laos") is to discover new, biologically active molecules from plants for development into drugs for therapies against cancer, AIDS, malaria, and tuberculosis (Soejarto *et al.*, 1999, 2002, 2006). In Vietnam, the primary biodiversity target for the drug discovery effort is the vegetation found in the 20,000-hectare forests of Cuc Phuong National Park, a forest-covered limestone formation located approximately 100 km (straight line) southwest of Hanoi, the capital city of Vietnam (Figure 1).

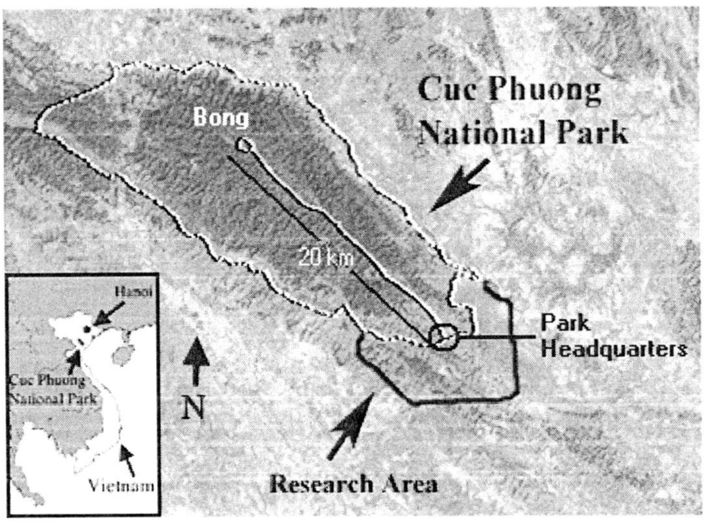

Figure 1. Outline of Cuc Phuong National Park based on satellite image, with an insert showing the geographic location of the Park in Vietnam. Satellite image was provided through the courtesy of White Point Systems, Inc., Friday Harbor, WA.

For the first 5-year period (Phase I) of the project (1998-2003), both biodiversity-based or "random" collections of plants throughout the Park, and ethnobotany-based collections were implemented. Plants from the Cuc Phuong area, collected using both approaches, together with plants collected from Laos, supplied the biodiversity pool for the UIC ICBG drug discovery effort. Field interviews to acquire ethnobotanical samples with healers and general community members of the ethnic Muong communities located around the park were performed, on the medicinal uses of plants of Cuc Phuong National Park, but concentrating the effort at the Cuc Phuong Commune located adjacent to the Park's entrance (Figure 2). Some villagers of the Cuc Phuong Commune represent new communities, resulting from the Vietnamese government's effort (1990-1996) to relocate ethnic Muong who were originally settled in various hamlets or villages inside the preserve boundaries since before the establishment of the park as a national park (1962).

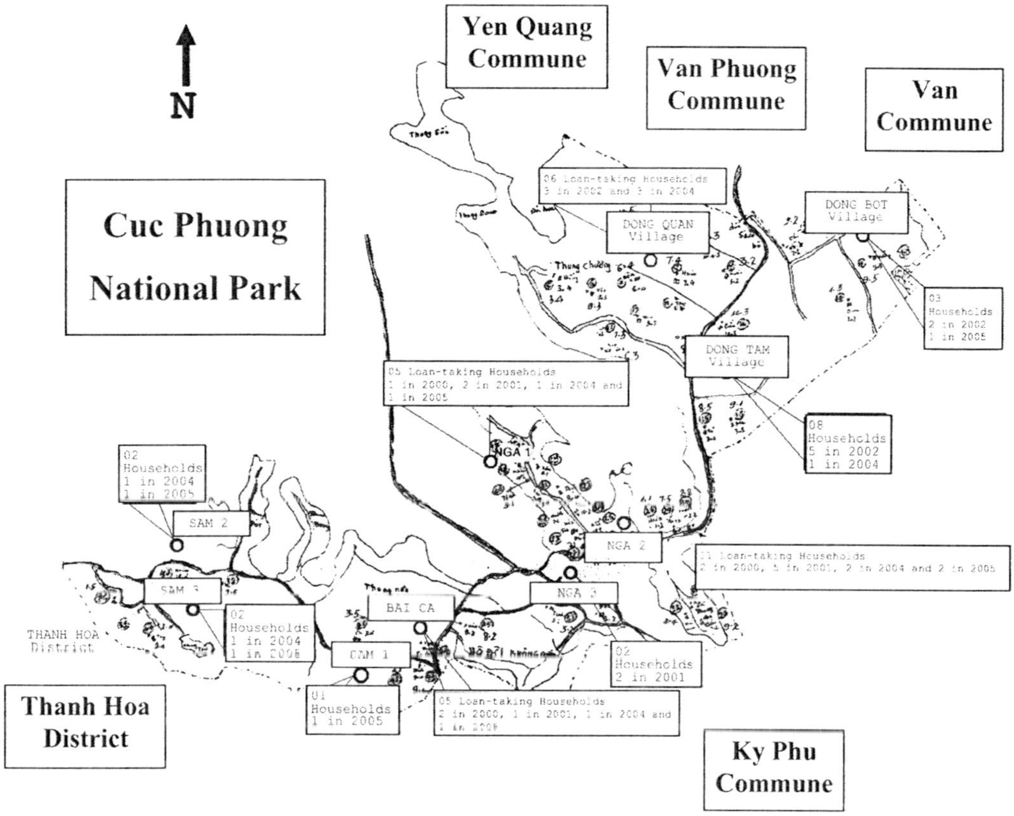

Figure 2. Locations of hamlets/villages participating in the Micro-loan Program at Cuc Phuong Commune, showing numbers of families who initiated participation in different years. See Appendix 1 for the full list of the families.

In Vietnam, permission for plant collection, and for the export of plant samples and/or their extracts to the USA, was granted by the Ministry of Agriculture and Rural Development (MARD) of Vietnam, and by Cuc Phuong National Park. Permission for performing interviews from healers and members of the commune was granted, initially, by the commune, through a process of consultations and meetings between the UIC ICBG Principal Investigator, the Commune Head, and the Project Leader of UIC ICBG Associate Program 4 (subsequently referred to in this paper as AP-4). Associate Program 4, a subprogram of the UIC ICBG, is based at the Institute of Biotechnology at the Vietnamese Academy of Science and Technology (VAST) charged with ethnobotanically based bioprospecting and with promoting economic development. Further meetings were held between UIC ICBG project personnel, commune members, and other commune officers (People's Committee of the Cuc Phuong Commune), who eventually referred the request to the Vietnam National People's

Committee in Hanoi. In the actual interviews, the purpose of the study and the interviews was explained to each interviewee and individual informed consent was obtained in each case. No payment was made for the interviews. Field interviews were performed under a protocol approved by the Institutional Review Board of the University of Illinois at Chicago (Protocol number H-97-1056).

As compensation for the permission and for the information on medicinal plants provided by the Cuc Phuong Commune, various measures of reciprocity were implemented, which included enhancement of the village kindergarten, various forms of support to the village school, establishment of a medicinal plant garden for the community, the micro-loan program, and community education on various aspects of their day-to-day living (for example, agricultural practices in sloping lands, fruit trees for the garden, Cuc Phuong biodiversity conservation). This paper focuses on the micro-loan program effort of this ICBG, which was implemented by AP-4 personnel.

Objectives of the UIC ICBG Micro-loan Program

In 2000, the UIC ICBG Principal Investigator (D. D. Soejarto) and the AP-4 Project Leader (Le Thi Xuan) set down a strategy for the implementation of the Micro-loan Program with the financial support of UIC ICBG funds. A Fund for hunger eradication and poverty reduction (HEPR) was established to provide no-interest loans to selected poor Muong families in Cuc Phuong Commune, with the specific objectives of helping these families improve their household economy. The project aims at alleviating hunger and poverty for these families, so that they do not need to cut down trees or to hunt animals in the Cuc Phuong National Park in order to earn their living. Loan-receiving households have to make a commitment to pay back the borrowed money to the Cuc Phuong Commune. The returned funds are used to create a rotating loan pool that will be given to other poor households according to the regulations set by a support committee, the Economic Development Village Board (EDVB) appointed by the UIC ICBG Principal Investigator (Board President - D.D. Hai, Head of Cuc Phuong Commune; Members - Dao Van Khuong, Director, Cuc Phuong National Park; Le Thi Xuan, Project Leader, ICBG AP-4; Bui Minh Vu, Professor, Forest Research Institute, Hanoi and Co-Project Leader, AP-4; D. D. Soejarto, UIC ICBG Principal Investigator, Adviser).

Implementation of the Program

Principles and Criteria for Providing Loans

EDVB set the following eight requirements and criteria for the eligibility of poor households (referred to as the beneficiaries of the program) to receive loans:

- Heads of households must submit a petition to the Commune People's Committee that clearly specifies the purpose of borrowing money.

- The list of loan-receiving families must be approved by EDVB and by local authorities (People's Committee).
- Each family receives a loan without interest ranging from two million to three million VND (Vietnam dongs; approximately US $133-$200) and pays back the whole amount within three years, according to the following schedule: 50% of the borrowed money within 24 months; the remaining amount within 36 months. Money is to be returned to the Cuc Phuong Commune. The returned money will be given to other poor families in a rotating manner. This method is called "creating rotating loans".
- When receiving loans, families (borrowers) will have to sign a Borrower Register, which is monitored by a Sub-Board in charge of finance management.
- Loan-taking families must sign a Borrower Contract, using a printed form issued by the Commune People's Committee.
- EDVB monitors the loan of each family.
- The head of the family must affix his signature in the list of beneficiaries (loan receivers) issued by the People's Committee of Cuc Phuong Commune.
- If loan-receiving families do not abide by these principles, they will not have the right to borrow money or the money will be taken back in cash or in kind, if the money has been received.

Criteria for Receiving Loans

Poor Muong families will be considered eligible for receiving loans if they fulfill the following criteria:

- Their period of food shortage must be more than two months a year.
- They have sick, old or disabled members or they are receiving support through government welfare policy.
- They have the ability to pay the borrowed money back and have determination to improve their living conditions.

Steps for the Delivery of the Loaned Money

Step 1: Preparation

- Informing local households about goals and plan of the ICBG Project, namely, the specific aims of Associate Program 4 based at the Institute of Biotechnology.
- People's Committee of Cuc Phuong Commune, in cooperation with EDVB, prepares a list of households that are eligible for getting loans.
- Preparation of required documents and papers, such as application for borrowing, loaning contract, etc.
- Setting aside an adequate budget.

Step 2: Conducting Survey and Interviews and Taking Photos of Families to be Given a Loan

- The number of families applying for loans has been normally 20% higher than the number to be given a loan. For example, 10 families were scheduled to receive a loan in the 5th period (in 2005); a list of 12 families was prepared. Following on-site surveys and interviews conducted with the 12 families, the actual 10 families each to be given a loan were selected.
- Conducting the survey: Survey sheets with printed questions were provided to each family; questionnaires attempt to determine the complexity/difficulty level of each family through assessment of the state of its living conditions, education for children, the length of food shortage period in a year, and to determine whether the loan-receiving families understand the purpose of the loan, know how to use the borrowed money, and are able to pay back the money on maturity.

Step 3: Selecting Households to Receive Loans

- EDVB selects poorest household to provide loans.
- Eligible households are invited to the People's Committee of Cuc Phuong Commune to complete procedure for receiving the loans.

Step 4: Inspection, Assessment of how Loan-taking Families Use the Loans following Receipt of Money

Visits, initially, every three months, for re-interviews and inspection were performed. More detailed re-interviews were conducted as follows:

- 1st visit: Preliminary inspection three months after release of loans. The intention of the inspection is to ascertain whether the loan-receiving families are using the money for the targeted purpose, to rectify wrong use of the loans (if any), and to help the families use the money correctly when they are confused on how the money is to be used.
- 2nd visit: Six months after loan-taking families received loans -- to determine and assess the effectiveness of the loans.
- 3rd visit: one year after the receipt of a loan -- to assess the effectiveness for the last time and to remind the families to pay back the borrowed money to EDVB.
- Subsequent visits every six months – to review and record general progress.

Step 5: Preparation to Select Loan-takers for the Next Loan Period

Statistics on Number of Families of Loan-takers, 2000-2006

Up to March, 2006, the EDVB of the Cuc Phuong Commune has selected and provided loans to 45 poor Muong families in 10 villages (Figure 2) to help develop houschold animal breeding and crop cultivation practices. In March, 2006, an additional 10 families were provided loans, but data for these 10 families were not available for inclusion in the present analysis.

As can be seen in Table 1, the number of families that received loans was smallest (5) in 2000, when the program was initiated. Subsequently, the number of loan-takers was increased to 10, per year, through 2005. Every village of Cuc Phuong Commune has at least one poor family that received loans. Among ten villages, Nga 2 has the largest number of loan-taking families. The number of villages that have loan-receiving families increased from three in 2000 to seven in 2004, and further to eight in 2005.

Table 1. Families of Muong community in Cuc Phuong Commune that received loans from the UIC ICBG AP-4 Project in the period of 2000-2005

Village	Total number of families	Number of families that received loans				
		2000	2001	2002	2004	2005
Nga 1	5	1	2	-	1	1
Nga 2	11	2	5	-	2	2
Nga 3	2	-	2	-	-	-
Dong Tam	8	-	-	5	1	2
Dong Quan	6	-	-	3	3	-
Dong Bot	3	-	-	2	-	1
Sam 1	1	-	-	-	-	1
Sam 2	2	-	-	-	1	1
Sam 3	2	-	-	-	1	1
Bai Ca	5	2	1	-	1	1
Total in all villages	45	5	10	10	10	10

Money Provided as Loan in Five Loan Periods

The total amount of money that has been delivered to selected poor Muong households in Cuc Phuong Commune through five loan periods is presented in Table 2.

Table 2. Total amount of money delivered as loans, 2000-2005

Village	Amount	Money provided to households in each period (in million VND)[a]				
		2000	2001	2002	2004	2005
Nga 1	12	2	4	-	3	3
Nga 2	26	4	10	-	6	6
Nga 3	4	-	4	-	-	-
Dong Tam	19	-	-	10	3	6
Dong Quan	15	-	-	6	9	-
Dong Bot	7	-	-	4	-	3
Sam 1	3	-	-	-	-	3
Sam 2	6	-	-	-	3	3
Sam 3	6	-	-	-	3	3
Bai Ca	12	4	2	-	3	3
Total	110	10	20	20	30	30

[a] One million VND = US $66.

As can be seen in Table 2, the total amount of loans that have been provided to 45 poor Muong households in the five loan periods, 2000-2005, is 110 million VND (US $7,330). The number of villages that have benefited from the program is as follows:

Period 1 (2000): three villages borrowed 10 million VND (10% of the total loaned fund).
Period 2 (2001): four villages borrowed 20 million VND (18% of the total loaned fund).
Period 3 (2002): three villages borrowed 20 million VND (18% of the total loaned fund).
Period 4 (2004): seven villages borrowed 30 million VND (27% of the total loaned fund).
Period 5 (2005): eight villages borrowed 30 million VND (27% of the total loaned fund).

The villages that received the benefits may be summarized as follows: (a) The amount of money provided as a loan to each village has averaged at about 11 million VND (10% of the total loaned fund); (b) The largest amount provided as a loan to one village was 26 million VND (Nga 2 village) (24% of the total fund loaned); (c) The smallest amount provided as a loan to one village was three million VND (Sam 1 village) (3% of the total fund loaned).

RESULTS

For the duration of the Micro-loan Program, the loan-receiving families have strictly adhered to the procedures set down by EDVB of the Cuc Phuong Commune on getting loans. After receiving the loans, each family actively used the money for the purposes stated in their petitions. The forty-five families have used the loan to buy farm animals (cow and buffalo) to work the land and to breed, as well as to establish crop plant cultivation. (a) *Animal Breeding*: An amount of 107,600,000 VND (98.87% of the total loaned fund) has been used to buy 26 buffaloes, 29 cows, two deer, four goats, and to raise bees for honey. (b) *Crop Cultivation*: 2,400,000 VND has been used to grow pineapples, sugar cane, and the herb *Solanum hainanense* Hance (Solanaceae), a species of plant used in the treatment of liver-related diseases, and to purchase new rice varieties, fertilizer, and pesticide. The full data on the results of this micro-loan program are presented as Appendices 1-4 (SUPPORTING DATA AND DOCUMENTS).

Why are cows and buffaloes as the first thing that borrowers purchased? The investment is safe and dependable. For cows, a baby cow is produced every year, strengthening animal power for working the land. As an additional bonus, cows produce fertilizer (in the form of the dung); they are easy to manage; and they have high sale value. Buffaloes have more-or-less the same value. Goats are more difficult to control and can destroy vegetation, while chickens provide a low return.

Periodic re-interviews of loan-takers by ICBG AP-4 investigators were performed and workshops were held with the attendance of members of EDVB and the loan-taking families to evaluate the progress of the program (data compilation and analysis). The data indicate that the turnover (the dollar yield or revenue) from the loan-taking families participating in the program during the six-year period reached 220,100,000 VND (US $14,673), broken down, as follows:

Animal Breeding

- Cows: 94,643,000 VND (43% of the total turnover). This money does not include the annual turnover from cow's dung, which is estimated to be

  2 m^3/cow/year x 29 cows x 100,000 VND/m^3 of dung = 5,800,000 VND (US $387).
- Buffaloes: 74,834,000 VND (34% of the total turnover). This money does not include the annual turnover from buffalo's dung which is estimated to be:

  3 m3/buffalo/year x 26 buffaloes x 100,000 VND/m3 = 7,800,000 VND.

  (The turnover from dung is not included in the amount of total turnover computation, because some families sell the dung, while others use it as fertilizer for their own cultivation needs. Interestingly, the period of the micro-loan project was the first time this animal breeding by-product appeared on the markets at Cuc Phuong Commune and Nho Quan District.)
- Deer: 13,206,000 VND (6% of the total turnover).
- Goats: 2,201,000 VND (1% of the total turnover).
- Chicken and pigs: 2,201,000 VND (1% of the total turnover).

Crop Cultivation

Compared with breeding, the turnover from cultivation is small. The turnover from growing pineapple, sugar cane and *Solanum hainanense* is 33,015,000 VND (US $2,201), making up only 15% of the total turnover. As is shown in Figure 3, breeding practices proved to be most effective for the use of the loaned money, with the turnover from cows, buffaloes, deer, goats, pigs and chickens making up 85% of the total, while the revenue from cultivation made up only 15% of the total. Also shown in Figure 3, cows and buffaloes provided the largest return, making up 77% of the total (43% from cows, 34% from buffaloes). Thus, cows rank highest in generating revenue (43%), followed by buffaloes (34%), deer (6%), and goats, chicken and pigs (as a group) (3%).

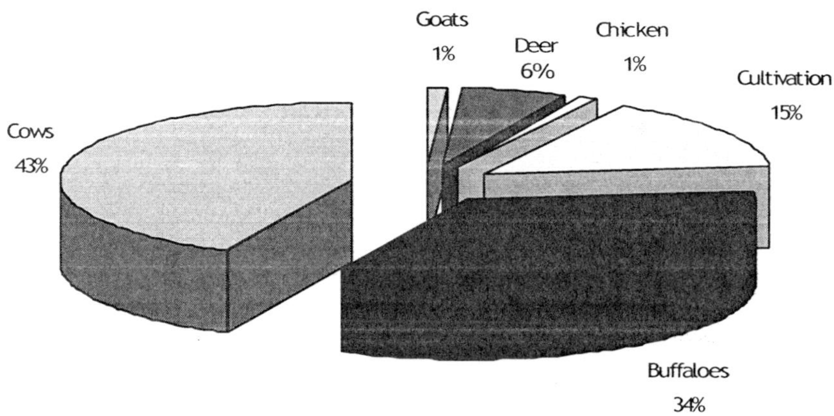

Figure 3. Turnover from household animal breeding and crop cultivation activities of 45 families in five years (2000-2005).

Figure 4. Mr. Dinh Van Hoa, head of household that generated the highest amount of revenue/profit (23,000,000 VND or US $1,533) as a result of participation in the UIC ICBG micro-loan program, starting on September 13, 2000, when the program was initiated.

The data clearly indicates that, households in the Cuc Phuong commune should invest in breeding domestic animals, if they wish to improve their economic standing. In terms of "profits", animal breeding yielded a profit of 110,000,000 VND (US $7,330). Averaging over the entire period of the project, each family has so far made a profit of 2,446,000 VND (US$163). For those families who received loans at the start of the program, September 13, 2000, the highest amount of profit gained was 23,000,000 VND (US $1,533) (Mr. Dinh Van Hoa, Figure 4; Appendix 2), while the lowest was 5,000,000 VND (US $313).

Thanks to these good results, the loan-taking families have risen above the poverty line, set up as the minimal condition for taking the loan, and does not anymore suffer from food shortages. Some have spent the profit to repair or rebuild their houses (Appendix 3), to purchase motorbikes, TV sets, or radios. Some families furthered their children's education by sending them to high school and university. Of the four loan-taking families with the lowest assets by the end of the observation period, three initiated their loans in Period 4, and have thus had less time to improve their household economic status. These families also experienced difficulties in their animal breeding.

ASSESSMENT

At a meeting held on April 14, 2005, with the participation of representatives from People's Committee of Cuc Phuong Commune, Farmer Association of Cuc Phuong, HEPR Board of Cuc Phuong, ICBG Associate Program 4, and Cuc Phuong National Park (CPNP) personnel, the implementation of the project was assessed as follows.

Economic results

- Although only a small number of families received loans, the implementation of the project has been successful as a result of good management and close supervision, and guidance from the EDVB of the Cuc Phuong Commune.
- The beneficiaries of the program were very poor and in need of improving their living conditions; other forms of loan (such as, government HEPR Programs 120, 133, 135, 143, 661) require complicated procedures that these families could not afford to implement.
- The loan-taking families strictly abided by the EDVB's simple rules and regulations on receiving the loans, learned loan management from training courses organized by the ICBG project, and were able to improve their household economies.

Environmental impact

- Human pressure on the forest biodiversity of Cuc Phuong National Park has been reduced. Continuing observations and individual interviews by the park's personnel belonging to five Forest Protection Units of the park located in the research site, with members of the Cuc Phuong Commune, indicate that the loan-takers did not return to the forest to cut down trees and hunt animals, because they were kept busy to tending their animal breeding and running crop plant cultivation activities in the implementation of the micro-loan program.
- The availability of cheap natural fertilizer (animal dung) for plant cultivation helps keep the environment tidy and free of contamination from chemical fertilizer.

Social impact

- Creation of jobs, eradication of hunger and alleviation of poverty.
- Reduction of the gap between the rich and the poor.
- Stabilization of social conditions, strengthening of neighborhood relations.

Impact on biodiversity

- From the perspective of Cuc Phuong National Park
  - Job creation raised awareness on the importance of protecting the forest biodiversity.
  - Conservation awareness provided incentives to the local population to assist Cuc Phuong National Park in protecting the forest.
- From the perspective of the Muong communities
  - Strengthening solidarity
  - Strengthening awareness of the need to assist the Cuc Phuong National Park in the protection of its forests and the biodiversity within.

CONCLUSION

Based on the assessment of the micro-loan program as stated above, and on interviews with the 45 beneficiary households, as well as with 20 local officials, and on discussions in a seminar held with the Forest Protection Service of Cuc Phuong National Park (CPNP), a consensus surfaced, namely, that the ICBG-sponsored micro-loan program has been timely and effective. Local authorities, such as People's Committee of Nho Quan District, HEPR Board of Nho Quan District, People's Committee of Ninh Binh Province, and HEPR of Ninh Binh Province all expressed agreement that the UIC ICBG micro-loan program has implemented a *new and effective form of loan* to help poor families through developing household animal breeding and crop plant cultivation programs. Compared with other forms of loan programs, the ICBG micro-loan is simpler in terms of procedure, and it requires no mortgage and no interest. The rotating loan process is practical and feasible, giving every family a chance to get a loan. Thus, we feel that this program should continue, in support of the implementation of the Vietnamese government's HEPR program.

From a different perspective, an important achievement of the micro-loan program is the fact that 100% of the beneficiaries of the program have been gainfully employed in the implementation of the program, hence, have had no time, or need, to return into the forest to cut trees or to hunt for animals. An interview on April 14, 2006 by the lead author of this paper (Bui Minh Vu) with Mr. Pham Cong Hoan, Head of the Cuc Phuong National Park Forest Protection Office (based on reports received from five Forest Protection Units of the park), confirmed that none of these families had returned to the park illegally. Thus, in concert with the effort of ICBG Associate Program 4 in providing community education, and in concert with the effort of Associate Program 1 in educating the communities on the awareness of protecting the biodiversity of Cuc Phuong National Park through its CAP (Conservation Awareness Program) (Soejarto *et al.*, 2006), the micro-loan effort has positively contributed to the protection of the park's biodiversity. In the context of our ICBG project, the program serves as a model to implement economic development as a measure of reciprocity in a plant-based drug discovery and development program.

APPENDIX 1. LIST OF 45 POOR MUONG FAMILIES APPROVED FOR MICRO-LOAN PROGRAM OF ICBG-AP4 PROJECT IN THE PERIOD OF 2000-2005 (LISTED BY HAMLET/VILLAGE)

No.	Name of loan-taking household	Village	Group no.	Date	Amount of loan (million VND)
	DVL	Nga 1	Group 1	Sep. 13, 2000	2
2	BTX	Nga 1	Group 2	Apr. 02, 2001	2
3	DVK	Nga 1	Group 2	Apr. 02, 2001	2
4	NTT	Nga 1	Group 4	Mar. 13, 2004	3
5	DVT	Nga 1	Group 5	June 13, 2005	3
6	DVH	Nga 2	Group 1	Sep. 13, 2000	2
7	DVD	Nga 2	Group 1	Sep. 13, 2000	2
8	DVB	Nga 2	Group 2	Apr. 02, 2001	2
9	DVT	Nga 2	Group 2	Apr. 02, 2001	2
10	DVH	Nga 2	Group 2	Apr. 02, 2001	2
11	DVN	Nga 2	Group 2	Apr. 02, 2001	2

APPENDIX 1. (CONTINUED)

No.	Name of loan-taking household	Village	Group no.	Date	Amount of loan (million VND)
12	DTG	Nga 2	Group 2	Apr. 02, 2001	2
13	DVB	Nga 2	Group 4	Mar. 13, 2004	3
14	DTB	Nga 2	Group 4	Mar. 13, 2004	3
15	DTD	Nga 2	Group 5	June 13, 2005	3
16	DVI	Nga 2	Group 5	June 13, 2005	3
17	DTT	Nga 3	Group 2	Apr. 02, 2001	2
18	DVS	Nga 3	Group 2	Apr. 02, 2001	2
19	DTM	Dong Tam	Group 3	May 11, 2002	2
20	DVK	Dong Tam	Group 3	May 11, 2002	2
21	DTV	Dong Tam	Group 3	May 11, 2002	2
22	DVU	Dong Tam	Group 3	May 11, 2002	2
23	DVN	Dong Tam	Group 3	May 11, 2002	2
24	DVB	Dong Tam	Group 4	Mar.13, 2004	3
25	BVB	Dong Tam	Group 5	June 13, 2005	3
26	DTR	Dong Tam	Group 5	June 13, 2005	3
27	DVC	Dong Quan	Group 3	May 11, 2002	2
28	DVB	Dong Quan	Group 3	May 11, 2002	2
29	DVH	Dong Quan	Group 3	May 11, 2002	2
30	DVT	Dong Quan	Group 4	Mar.13, 2004	3
31	DVH	Dong Quan	Group 4	Mar.13, 2004	3
32	DVH	Dong Quan	Group 4	Mar.13, 2004	3
33	DMC	Dong Bot	Group 3	May 11, 2002	2
34	DTT	Dong Bot	Group 3	May 11, 2002	2
35	DVN	Dong Bot	Group 5	June 13, 2005	3
36	DVD	Bai Ca	Group 1	Sep. 13, 2000	2
37	DVH	Bai Ca	Group 1	Sep. 13, 2000	2
38	DTX	Bai Ca	Group 2	Apr. 02, 2001	2
39	DVT	Bai Ca	Group 4	Mar. 13, 2004	3
40	BTT	Bai Ca	Group 5	June 13, 2005	3
41	DVC	Sam 1	Group 5	June 13, 2005	3
42	DVH	Sam 2	Group 4	Mar. 13, 2004	3
43	DVT	Sam 2	Group 5	June 13, 2005	3
44	DVT	Sam 3	Group 4	Mar. 13, 2004	3
45	DVV	Sam 3	Group 5	June 13, 2005	3

Appendix 2. List of Poor Muong Families Who were Highly Successful in the Implementation the Micro-loan Program through Improvement of Their Household Economy

Name of loan-taking household[a]	Village	Period number	Start of micro-loan program participation	Asset in 2005 million VND)
DVH	Nga 2	Period 1	Sep. 13, 2000	23
DVD	Nga 2	Period 1	Apr. 2, 2001	20
DVH	Nga 2	Period 2	Apr. 2, 2001	15
DVT	Nga 2	Period 2	Mar. 13, 2004	12
DVN	Nga 2	Period 2	June 13, 2005	12
DVB	Dong Quan	Period 3	Sep. 13, 2000	20
DVK	Nga 1	Period 2	Sep. 13, 2000	20

[a] Economic standing at last interview in 2005:

- The Household DVH

  By selling one buffalo for 4,000,000 VND and some real-estate for 10,000,000 VND, he earned 14,000,000 VND in total which he invested in the purchase and breeding of deer. At time of the interview, his asset was 23,000,000 VND at current value (he owned three deer and one buffalo).
- The Household DVD

  By selling two cows for 6,800,000 VND, he continued to invest this money into cattle breeding. The current value he acquired at time of interview was around 20,000,000 VND (two cows and three buffaloes).
- The Household DVH

  By selling two cows for 6,000,000 VND, he used some of the money for rebuilding his house, and the remainder for continuing cattle breeding. At time of interview, his asset was around 15,000,000 VND at current value (two cows and one buffalo).
- The Household DVT

  At time of interview, he owned three buffaloes with current value of 12,000,000 VND.
- The Household DVN

  By selling one cow for 3,800,000 VND, he used some of the money to rebuild his house, the remainder to continue cattle breeding. At time of interview he owned four cows at current value of 12,000,000 VND.
- The Household DVB

  By selling one buffalo for 5,000,000 VND, he invested this amount of money into deer breeding. At time of interview he owned three deer at current value of 20,000,000 VND.
- he Household DVK

  He sold one buffalo to allow him to build a house. At time of interview he owned three buffaloes and one deer at current value of 20,000,000 VND.

APPENDIX 3.
LIST OF POOR MUONG FAMILIES IN THE MICRO-LOAN PROGRAM WHO BUILT A NEW AND BETTER HOUSE AT LAST INTERVIEW IN 2005

No.	Name of loan-taking households	Village	Period number	Date of participation	Amount of loan (million VND)
	DVL	Nga 1	Period 1	Sep. 13, 2000	2
	BTX	Nga 1	Period 2	Apr. 02, 2001	2
	DVK	Nga 1	Period 2	Apr. 02, 2001	2
	NTT	Nga 1	Period 4	Mar.13, 2004	3
	DVH	Nga 2	Period 1	Sep. 13, 2000	2
	DVD	Nga 2	Period 1	Sep. 13, 2000	2
	DVB	Nga 2	Period 2	Apr. 02, 2001	2
	DVT	Nga 2	Period 2	Apr. 02, 2001	2
	DVH	Nga 2	Period 2	Apr. 02, 2001	2
	DVN	Nga 2	Period 2	Apr. 02, 2001	2
	DTG	Nga 2	Period 2	Apr. 02, 2001	2
	DVB	Nga 2	Period 4	Mar. 13, 2004	3
	DTB	Nga 2	Period 4	Mar. 13, 2004	3
	DTM	Dong Tam	Period 3	May 11, 2002	2
	DVK	Dong Tam	Period 3	May 11, 2002	2
	DTV	Dong Tam	Period 3	May 11, 2002	2
	DVU	Dong Tam	Period 3	May 11, 2002	2
	DVN	Dong Tam	Period 3	May 11, 2002	2
	DVB	Dong Tam	Period 4	Mar. 13, 2004	3
	DVC	Dong Quan	Period 3	May 11, 2002	2
	DVH	Dong Quan	Period 3	May 11, 2002	2
	DVT	Dong Quan	Period 4	Mar. 13, 2004	3
	DVH	Dong Quan	Period 4	Mar. 13, 2004	3
	DVH	Dong Quan	Period 4	Mar. 13, 2004	3
	DTT	Dong Bot	Period 3	May 11, 2002	2
	DVD	Bai Ca	Period 1	Sep. 13, 2000	2
	DVH	Bai Ca	Period 1	Sep. 13, 2000	2
	DTX	Bai Ca	Period 2	Apr. 02, 2001	2
	DVT	Bai Ca	Period 4	Mar. 13, 2004	3
	DVH	Sam 2	Period 4	Mar. 13, 2004	3

Appendix 4. List of Unpublished TECHNICAL Reports in Support of the Paper (Original Documents in Vietnamese in Brackets)

Hoan, Pham Cong (2005) Report on activities of Forest Protection Department of Cuc Phuong National Park from 2000 to 2005. Cuc Phuong, April 2005, 5 pp. [Phạm Công Hoan (2005) Báo cáo về kết quả hoạt động của Hạt Kiểm lâm Vườn Quốc Gia Cúc Phương từ năm 2000 đến năm 2005. Cúc Phương , tháng 4 năm 2005, 5 trang.]

People's Committee of Cuc Phuong Commune (1999) Traditional conventions of Cuc Phuong Commune. Cuc Phuong, 11 Pages. [Ủy Ban Nhân dân Xã Cúc Phương (1999) Hương ước của Xã Cúc Phương. Cúc Phương, 11 trang.]

People's Committee of Cuc Phuong Commune(2000) Direction of socio-economic development of Cuc Phuong commune in the period of 2000-2010, Cuc Phuong. People's Committee document. Cuc Phuong, 2000. 5 Pages. [Ủy Ban Nhân dân Xã Cúc Phương (2000) Phương hướng phát triển Kinh tế - Xã hội của Xã Cúc Phương trong giai đoạn từ năm 2000 đến năm 2010. Tài liệu của Ủy Ban Nhân dân Xã Cúc Phương . Cúc Phương năm, 5 trang.]

People's Committee of Cuc Phuong Commune (2005) Annual reports on socio-economic situation of Cuc Phuong commune. People's Committee document, April 2005, 23 pp.

[Ủy Ban Nhân dân Xã Cúc Phương (2005) Báo cáo hàng năm về tình hình thực hiện kế hoạch phát triển kinh tế - xã hội của Xã Cúc Phương. Ủy Ban Nhân dân Xã Cúc Phương, Tháng 4 năm 2005, 23 trang.]

Sinh, Dinh Cong (2005) Annual report on activities of Forestry Board of Cuc Phuong Commune. Cuc Phuong, April 2005, 3 pp. [Đinh Công Sính (2005) Báo cáo hàng năm về hoạt động của Ban Lâm nghiệp Xã Cúc Phương. Cúc Phương, tháng 4 năm 2005, 3 trang.]

Vu, Bui Minh (2005) Documents on the investigation of the implementation of micro-loan program for 45 poor Muong families in CucPhuong Commune in 6 years (2000-2005). Cuc Phuong, 4/2005, 20 pp. [Bùi Minh Vũ (2005) Tài liệu điều tra về tình hình thực hiện kế hoạch vay vốn của 45 hộ gia đình nghèo người Mường Xã Cúc Phương trong 5 đợt vay vừa qua(2000-2005). Cúc Phương, tháng 4 năm 2005, 20 trang.]

Vu, Bui Minh (2005) AP4 Project Annual Report on training results. Hanoi, July 2005,15 pp. [Bùi Minh Vũ (2005) Báo cáo hàng năm về tình hình thực hiện công tác Đào tạo của Dự án AP4. Hà nội, tháng 7 năm 2005, 15 trang.]

Acknowledgments

The authors express sincere thanks to the Muong communities, to the People's Committee of the Cuc Phuong Commune, to the District authorities of Nho Quan, and to the Director and staff of Cuc Phuong National Park for assistance and support in the implementation of this micro-loan program. Thanks are also expressed to the ICBG Associate Program 4 staff based at the Institute of Biotechnology, Vietnamese Academy of Science and Technology, for their hard work and dedication in the implementation of the program. The program is funded by a US government grant, NIH 1-UO1-TW-01015-01 (1998-2003) and 2-UO1-TW001015-06 (2003-2008), International Cooperative Biodiversity Groups Program, at the Fogarty International Center, NIH (National Institutes of Health) to the University of Illinois at Chicago, through funds from NIH, NSF (National Science Foundation) and USDA-AFS (Foreign Agriculture Service of the United States Department of Agriculture).

REFERENCES

Burra, N., Deshmukh-Ranadive, J. and Murthy, R. K. (2005). *Micro-credit, Poverty and Empowerment: Linking the Triad.* Thousand Oaks, CA: Sage Publications.

Dac, D. H. and Huu, N. H. (2004). A Draft Report on Assessment of the National Target Program on HEPR and the Program 135. VIE02/001 Project. Hanoi, June 2004, 99 pages (in Vietnamese).

Morduch, J. (1999). "The Microfinance Promise." *Journal of Economic Literature* 37(4), 1569-1614.

Reid, W. A., Laird, S. A., Gamez, R., Sittenfeld, A., Janzen, D. H., Gollin, M. A., Juma, C. (1993). A new lease on life. In W. A. Reid, S. A. Laird, C. A. Meyer, R. Gamez, A. Sittenfeld, D. H. Janzen, A.H., M. A. Gollin and C. Juma (Eds.), Biodiversity Prospecting: *Using Genetic Resources for Sustainable Development*, pp. 1-52. Washington, DC.: World Resources Institute.

Robinson, M. S. (2001). *The Microfinance Revolution: Sustainable Finance for the Poor,* vols. 1-3. Washington, D.C.: World Bank.

Soejarto, D. D., Gyllenhaal, C., Regalado, J. C., Pezzuto, J. M., Fong, H. H. S., Tan, G. T., Hiep, N. T., Xuan, L. T., Binh, D. Q., Hung, N. V., Bich, T. Q., Thin, N. N., Loc, P. K., Vu, B. M., Southavong, B. H., Sydara, K., Bouamanivong, S., O'Neill, M. J., Lewis, J., Xie, X.-M. and Dietzman, G. (1999). Studies on biodiversity of Vietnam and Laos: The UIC-based ICBG program. *Pharmaceutical Biology* 37 (Suppl.), 100-113.

Soejarto, D. D., Gyllenhaal, C., Regalado, J. C., Pezzuto, J. M., Fong, H. H. S., Tan, G. T., Hiep, N. T., Xuan, L. T., Hung, N. V., Bich, T. Q., Loc, P. K., Vu, B. M., Southavong, B. H., Sydara, K., Bouamanivong, S., O'Neill, M. J. and Dietzman, G. (2002). An international collaborative program to discover new drugs from tropical biodiversity of Vietnam and Laos. *Natural Product Science* 8(1), 1-15.

Soejarto, D. D., Fong, H. H. S., Tan, G. T., Zhang, H. J., Ma, C. Y., Franzblau, S. G., Gyllenhaal, C., Riley, M. C., Kadushin, M. R., Pezzuto, J. M., Xuan, L. T., Hiep, N. T., Hung, N. V., Vu, B. M., Loc, P. K., Dac, L. X., Binh, L. T., Chien, N. Q., Hai, N. V., Bich, T. Q., Cuong, N. M., Southavong, B., Sydara, K., Bouamanivong, S., Ly, H. M., Thuy, T. V., Rose, W. R. and Dietzman, G. R. (2005). Ethnobotany/Ethnopharmacology and mass bioprospecting: Issues on intellectual property and benefit-sharing. *Journal of Ethnopharmacology* 100(1), 15-22.

Soejarto, D. D., Zhang, H. J., Fong, H. H. S., Tan, G. T., Ma, C. Y., Gyllenhaal, C. , Riley, M. C., Kadushin, M. R., Franzblau, S. G., Bich, T. Q., Cuong, N. M., Hiep, N. T., Loc, P. K., Xuan, L. T., Hai, N. V., Hung, N. V., Chien, N. Q., Binh, L. T., Vu, B. M., Ly, H. M., Southavong, B., Sydara, K., Bouamanivong, S., Pezzuto, J. M., Rose, W. C., Dietzman,

G. R., Miller, B. M. and Thuy, T. V. (2006). "Studies on biodiversity of Vietnam and Laos" 1998-2005: Examining the impact. *Journal of Natural Products* 69, 473-481.

Weiss, H. M. and Kurmanalieva, E. (2003). Microfinance and Poverty Reduction in Asia: What is the Evidence? ADB Institute Research Paper series, No. 53. Tokyo: Asian Development Bank Institute.

In: Focus on Biodiversity Research
Editor: Jan Schwartz, pp. 63-100
ISBN: 1-60021-372-3

Chapter 4

HISPANIOLAN SPIDER BIODIVERSITY AND THE IMPORTANCE OF COMBINING NEONTOLOGICAL AND PALAEONTOLOGICAL DATA IN ANALYSES OF HISTORICAL BIOGEOGRAPHY

David Penney*
Earth, Atmospheric and Environmental Sciences,
The University of Manchester, Manchester, UK

ABSTRACT

Conflicting opinions regarding the relative importance of dispersal versus vicariance in understanding Caribbean biogeography continue to stimulate lively debate, despite recent advances in regional tectonics, geology, palaeogeography and palaeontology. The Greater Antilles are young geographical features, probably mid-Miocene and Hispaniola has a rich terrestrial invertebrate fossil record in the form of Dominican Republic amber inclusions. Spiders are common in Dominican amber and all fossil species belong in extant families. The island is unique in terms of its known spider fauna, in that a similar number of families have described species recorded on Hispaniola from both fossils in amber and from the extant fauna. It is also the region of the world where the amber fauna is most similar to the Recent fauna, and Miocene Hispaniolan spider biodiversity was presumably similar to that at present.

Fossils form empirical data with both phylogenetic and temporal implications and are of paramount importance for understanding historical biogeography. They can play a decisive role in falsifying hypotheses proposed solely on the distributions of extant taxa. It is also possible to generate hypotheses for both fossil and extant faunas that are ultimately falsibiable through the discovery of new Hispaniolan taxa. The amber spider fauna has clear affiliations to the present South American fauna, so how did they colonize the island from the mainland: dispersal or vicariance? The high Miocene biodiversity

* E-mail: david.penney@manchester.ac.uk

(including taxa with poor dispersal capabilities) on a geologically young island would seem to add more weight in favour of the vicariance model for explaining on-island spider lineages, although dispersal cannot be excluded for some of the highly dispersive taxa. However, a case-by-case approach is required in order to determine the relative contributions of the different models. The future research direction for understanding the biogeographical patterns will lie in the application of cladistic and phylogenetic biogeographical approaches. It is interesting to note that preliminary obersvations based on pedipalp morphology suggest some fossil taxa are more derived than extant forms. The contribution that the Dominican (and Mexican) amber fossil fauna can make to our understanding of the origins of Hispaniolan biodiversity, and Caribbean biogeography in general, should not be underestimated.

Keywords: Amber, Araneae, Caribbean, Dispersal, Fossil, Greater Antilles, Palaeogeography, Palaeontology, Vicariance.

Introduction

The present rate of global climate change, habitat destruction through anthropogenic and other factors, and the subsequent rate of species extinctions means it is likely that many taxa will disappear before they have been scientifically described. Briggs [1991] likened the current episode of species demise to a present day mass extinction. He suggested that the rate of species loss is greater and more significant than the end-Cretaceous extinction that wiped out the dinosaurs. In order to assess the effects of this current global change it is necessary to inventory as much of the global biota as possible, before components of it disappear forever. However, in addition to recognizing existing biodiversity and biogeographic patterns, it is important that we understand their origins. Thus, we also need to examine past distributions using palaeontological evidence from the fossil record (and consider these in relation to past global climate change over deep time) in order to help us interpret the potential consequences of the current and ongoing episode.

Island faunas pose interesting biogeographical questions because of difficulties associated with immigration and emigration, when compared to continental landmasses. The high biodiversity of the Caribbean, coupled with its complex geological history, has attracted interest in the historical biogeography of the region [Woods and Sergile, 2001], although most studies have focussed on vertebrates [Hedges, 2001]. The Caribbean is considered a biodiversity hotspot and the West Indies is one of the largest and most diverse archipelagos in the world; home to thousands of endemic species. Recent advances in West Indian geology and plate tectonics [Iturralde-Vinent and MacPhee, 1999; Iturralde-Vinent and Lidiak, 2006], increasing availability of fossils (particularly in amber from the Dominican Republic) [Poinar, 1992], and the application of DNA sequencing in the study of evolutionary history [e.g. Ogden and Thorpe, 2002], provides the basis for a new approach to understanding the complex, evolutionary history of the region's biota.

A Catalysis Meeting (held in the Dominican Republic, July 2006) was recently sponsored by the US National Evolutionary Synthesis Center (NESCent) and initiated and organized by Drs R.E. Glor (University of California), R.E. Ricklefs (University of Missouri-St. Louis) and J.B. Losos (Washington University). The primary purpose of the meeting was to facilitate future collaborative, interdisciplinary research by scientists in diverse fields, with the aim of developing a comprehensive new perspective on the evolution of West Indian biodiversity (past and present). Participants included geologists, palaeontologists (including the present author), biogeographers, systematists, and population/evolutionary biologists from six different countries. The following question for the delegates was one of particular interest: *How have historical connections among islands, and to the mainland, contributed to patterns of diversity?*

The origins of the West Indian biota have long been debated and the last century has seen many differences of opinion regarding the various hypotheses provided to explain the distributions observed today, e.g. former landbridge connections, vicariance, and overwater dispersal. One major aim of the NESCent meeting was to produce synergistic collaborations with geological, palaeontological and phylogenetic data (including specialists on numerous, diverse taxonomic groups) combining to result in a fuller understanding of the West Indies' history than would be possible relying on a single type of information. In this paper, I examine the potential role that fossil spiders preserved in Miocene Dominican Republic amber can add to the debate.

Fossiliferous Amber from the Dominican Republic

Amber is a fossil Lagerstätte famous for its exceptional preservation of entombed animal, usually arthropod, inclusions, which are often preserved with life-like fidelity. This is particularly true of Dominican amber, which probably exhibits the best quality of preservation of all know ambers [Grimaldi and Engel, 2005], despite there being more than 160 deposits known worldwide [Martínez-Delclòs *et al.*, 2004]. Although traces of amber are present in various other countries of the Caribbean, e.g., Haiti, Puerto Rico and Jamaica, it occurs in exploitable quantities only in the Dominican Republic [Iturralde-Vinent, 2001]. Dominican amber has been known to Europeans since the end of the fifteenth century [Baroni-Urbani and Saunders, 1980], but was only brought to the attention of scientists in the twentieth [Sanderson and Farr, 1960]. Despite this late onset of palaeontological study, Hispaniola has a rich terrestrial invertebrate fossil record in the form of species described from Dominican Republic amber inclusions.

The amber is found in two main regions of the country: the Cordillera Septentrional in the north, and the Cordillera Oriental in the north-east (Figure 1). The amberiferous region in the north consists of the upper 300m of an Oligocene to Middle Miocene suite of clastic rocks called the La Toca Formation and consists mainly of sandstone containing thin lamellae of lignite, and occasional conglomerate. In the north-east the amber is found embedded in lignite and sandy clay in the 100m thick Yanigua Formation. These two amber deposits are thought to have been deposited in the same sedimentary basin 15–20Ma (but probably closer to 16Ma), prior to their displacement along major faults [Iturralde-Vinent and MacPhee, 1996; see Iturralde-Vinent, 2001 for a comprehensive review of the geology of these deposits]. For a discussion of alternative ages for this amber deposit see Poinar and Poinar [1999].

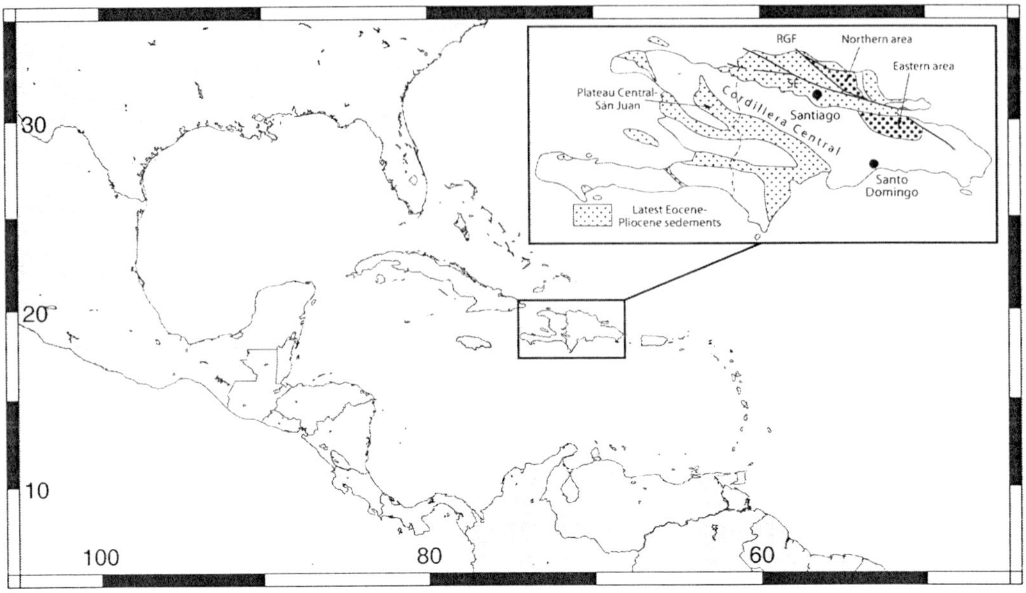

Figure 1. Current geography of Caribbean islands and adjacent continental landmasses, with inset (after Itturalde-Vinent and MacPhee, 1996) showing the location of the main amber mining districts (northern and eastern areas) in Hispaniola.

Poinar [1991] described the extinct tree *Hymenaea protera* (Leguminosae, Caesalpinioideae) as the botanical source of Dominican Republic amber; see also Hueber and Langenheim [1986]. This generic placement was confirmed by Langenheim [1995], through evidence based on resin chemistry and amber plant inclusions. The amber was formed in a tropical climate similar to that in the region today [e.g. Poinar and Poinar, 1999].

Resin as a Trap and the Bias of Amber Preservation

The preservative and antibiotic qualities of resin, which kill or retard fungal and bacterial growth have been known since antiquity [Poinar and Hess, 1982]. Preservation in amber basically represents an extreme case of mummification with natural embalming by the entrance of plant sap into the organisms' body cavity [Poinar and Hess, 1982]. Despite the exceptional preservation, a major problem faced by palaeontologists is that the fossil record is incomplete and many of the biases with regard to which organisms did, and which did not, get trapped remain unidentified or unquantified. Few authors have investigated the ecology of amber inclusions and quantitative studies of amber palaeobiology are in their infancy. At present, what we do know as a result of quantitative investigations of amber faunas, is that different amber-forming resins appear to have functioned as a trap in the same way [Penney and Langan, 2006], and that they are biased towards sampling active, trunk-dwelling faunas, which are more prone to wandering onto an exposed, sticky resin surface, getting stuck, and which are then entombed by a subsequent resin flow [Penney, 2002—in disagreement with the conclusions of Henwood [1993], which is assessed critically by Penney]. This is easily observed in many specimens as a layering effect in the amber, with inclusions trapped at the interface of the layers. However, organisms were also trapped by becoming engulfed in a single, more rapid flow of less viscous resin [Penney, 2005a].

By far the commonest inclusions encountered in amber are insects, but this is hardly surprising given their global dominance today. Spiders are also common in amber faunas and have a long and diverse geological history extending to the deep Paleozoic, although many Cenozoic spiders belong in extant families [Penney *et al.*, 2003]. Despite this overabundance of insects, our knowledge of fossil spiders in Dominican amber tends to exceed that of insects at ordinal level. For example, combined fossil and Recent Hispaniolan faunal checklists exist for a number of arthropod orders, e.g. Trichoptera [Flint and Pérez-Gelabert, 1999], Neuroptera [Pérez-Gelabert and Flint, 2000], Blattodea [Gutiérrez and Pérez-Gelabert, 2000], Orthoptera [Pérez-Gelabert, 2001] and Diplopoda [Pérez-Asso and Pérez-Gelabert, 2001], but when compared to spiders [Penney and Pérez-Gelabert, 2002; updated here, see Appendix] these other groups are poorly represented as fossils (Table 1; see also Pérez-Gelabert [1999]). This is not to suggest that these smaller orders cannot play a decisive role in studies of historical biogeography, but the justification for an analysis based on the comparison of extant and fossil spider distributions is clearly evident.

Table 1. Comparison of extant and fossil (*excluding subfossils in copal) species richness values from orders for which data are available (see text for references)

Order	Total spp.	Extant spp.	Fossil spp.	Extant:fossil ratio
Trichoptera	99	77	22	1:0.29
Neuroptera	55	52	3	1:0.06
Blattodea	92	85	7	1:0.08
Orthoptera	111	103	8	1:0.08
Diplopoda	162	146	16	1:0.11
Araneae	467*	313	154*	1:0.49

Biodiversity of Hispaniolan Spiders

Hispaniola is unique in terms of its known spider fauna, in that almost the same number of families have described species recorded on Hispaniola from fossils in Miocene Dominican Republic amber (39) and from the extant fauna (42). Furthermore, additional families are also known for the fossil fauna, but have not had formal species described (see later). It is also the region of the world where the amber fauna is most similar to the Recent fauna [Penney, 2005b]. A checklist of known fossil and Recent Hispaniolan spiders was given by Penney and Pérez-Gelabert [2002; with additions by Penney, 2004a]. Since then, a number of authors have provided new records for Hispaniola, including the first identifications of certain families for the island, and as a result of advances in taxonomy and systematics a number of the taxa listed by Penney and Pérez-Gelabert [2002] have undergone name changes or have even been transferred to different families. A complete and updated checklist is provided in the Appendix.

The Extant Fauna

Prior to the twentieth century, only a small number of large, common spiders had been described from Hispaniola. The first important contribution to the knowledge of Hispaniolan spiders was a short paper by Banks [1903], based on specimens collected by R. J. Crew. He recorded 63 species, primarily from the vicinity of Port-au-Prince, Haiti. The principal collectors of Hispaniolan spiders were P. R. Uhler, 1873; W. M. Mann, 1902, 1912–1913; Bates and Darlington, 1934; P. J. Darlington, 1934, 1938; A. Audant and D. Hurst. Uhler's main interest was entomology, although he also collected a number of spiders during his visit to the western part of Haiti, some of which were described over the years by the Lithuanian arachnologist Keyserling. Some of the jumping spiders (Salticidae) collected from Haiti by Mann in 1902 were described by the Peckhams. Bates and Darlington collected a few spiders along with their entomology collections during their visit to Haiti in 1934, and Darlington returned to Hispaniola for more intensive arachnological collecting later that year and also in 1938, during which time he visited eastern and central parts of the island, including some of the highest altitudes in the Caribbean.

Elizabeth Bryant described most of the known Recent spiders from Hispaniola [Bryant, 1943: Salticidae; 1945: Argiopidae = Araneidae, Tetragnathidae, Theridiosomatidae; 1948: numerous families], most of which are held in the Museum of Comparative Zoology, Harvard University. In her final publication on Hispaniolan Araneae, Bryant [1948] considered the total described fauna to consist of 224 species, but recognized that this was probably a gross underestimate of the true total. Since Bryant's synopsis, several authors have reported or described new species for the island [e.g., Muma, 1953; Archer, 1958; Gertsch, 1958; Exline and Levi, 1962; Berman and Levi, 1971; Levi, 1955, 1957, 1959, 1962, 1963ab, 1964, 1977, 1986, 1991, 1993, 1995, 1999, 2001; Chickering, 1967; Gertsch and Ennik, 1983; Dondale and Redner, 1984; Opell, 1979; Maddison, 1986; Smith, 1986; Wunderlich, 1986; Platnick and Shadab, 1974a, 1980, 1989; Richman, 1989]. In more recent years Brescovit [1993, 1999] has described two species of Anyphaenidae, Coyle [1995] two new species of Dipluridae, Edwards and Wolff [1995] reported three new species of jumping spiders, Alayón-García [1992, 1995, 2002, 2004] has described several new spider species, Alayón-García *et al.* [2001] identified the Old World *Cyrtophora citricola* (Araneidae) as a new record for the island, Piel [2001] listed a new Araneidae, Huber [2000] identified one new species in his revision of Neotropical Pholcidae, and Sánchez-Ruíz [2005] described a new caponiid. Rheims and Brescovit [2004] described the first extant Hersiliidae from Hispaniola and Platnick and Penney [2004] identified the first Prodidomidae from from the island, which also represented the first New World record of the genus *Zimiris*. Penney [2004a] provided eight additional new species records for the Island based on a small, opportunistically collected sample of 23 identifiable species. Such a large proportion of new records in such a small sample demonstrates that the extant spider fauna of Hispaniola is poorly known and worthy of further investigation.

The Fossil Fauna

It is only a quarter of a century since Ono [1981] described the first spider preserved in Miocene amber from the Dominican Republic. Subsequently, Dominican amber spiders were described by Schawaller [1981a, 1982, 1984], Wunderlich [1981, 1982, 1986, 1987, 1988, 2004], Reiskind [1989], Wolff [1990] and Penney [2000a, 2001, 2005cd]. Penney [2000b] revised the Dominican amber Anyphaenidae and Penney [in press] revised the Neotropical fossil Hersiliidae. The spiders described as *Mysmena dominicana* Wunderlich, 1998 (Mysmenidae) and *Ceratinopsis deformans* (Wunderlich, 1998) (sub *Grammonota*, Linyphiidae) by Wunderlich [1998] and the specimen reported as Archaeidae [Wunderlich, 1999] from Dominican Republic amber are all subfossils preserved in Madagascan copal [Wunderlich, 2004: 39]. The undescribed specimen identified as Amaurobiidae by Schawaller [1981b] is actually preserved in Baltic amber [Wunderlich, 1988] as is the linyphiid *Custodela lamellata* (Wunderlich, 1988) (sub *Lepthyphantes*) [Wunderlich, 2004: 1342]. Poinar [1992] reported the families Oecobiidae: Urocteinae and Archaeidae as present in Dominican Republic amber, but they are unknown as fossils from this source.

Additional spider taxa reported, but without formally described species, and which are otherwise unrecorded in Dominican amber include: Symphytognathidae *sensu lato* [Schawaller, 1981b], Agelenidae, Philodromidae [Schawaller, 1981b; Wunderlich, 1988], Liocranidae, Pisauridae, Palpimanidae: *Otiothops* sp. [Wunderlich, 1988], Hahniidae [Penney, 1999, based on a juvenile and possibly a misidentification] and Lycosidae [Penney, 2001]. Genera recorded but not identified to species, from families with additional described species in Dominican amber include: Dipluridae: *Ischnothele*?; Oonopidae: *Heteroonops*?, *Opopaea*?; Tetragnathidae: *Leucauge*; Theridiidae: *Craspedisia*; Salticidae: *Descanso* [Wunderlich, 1988], ?*Sarinda* [Wunderlich, 2004] and Salticidae: *Nebridia* [Cutler, 1984]. The spider described as *Anelosimus clypeatus* Wunderlich, 1988 was removed from that genus by Penney [2001] and currently remains unassigned. A complete systematic catalogue, including synonymies, transfers, etc. of fossil spiders preserved in Dominican Republic amber described up until the end of 2005 was provided by Penney [2006a].

COMPARISON OF THE EXTANT AND FOSSIL SPIDER FAUNAS

Quantitative comparisons of inventories of hyperdiverse taxa, such as spiders, from different regions (including through geological time) can be problematic because of undersampling bias. Even intensive inventories of spiders from poorly studied regions are typically composed of many singletons (species represented by only one specimen). Extrapolative richness estimates from such data typically indicate that many species have been missed and comparing overlap between two such inventories can multiply these difficulties, because of uncertainty over whether taxa absent from one site were present and not sampled, or were truly absent. Palaeontological data are scarce by their nature and new species in amber are often described from singletons and only occasionally from a series of specimens, which is usually few in number. Thus, species numerical abundance data are limited in terms of quantitative investigations. Traditionally, analysis of terrestrial arthropod palaeocommunities has been at family level, which is a good predictor of underlying species diversity and is applied extensively as the 'higher taxon approach' by neontologists investigating distribution, ecological correlates, and diversity patterns of extant tropical insects [Labandeira, 2005]. Similar issues relate to qualitative investigations, but not to the same degree and it is reasonably easy to generate falsifiable hypotheses for the fossil and the extant faunas based on family (and also genus) presence and absence data.

There are currently 49 spider families with formal species described from Hispaniola when both Miocene and Recent faunas are considered together (see Appendix). When the families listed without formally described fossil species (see earlier) are included, this value rises to 50, with 32 and 36 shared families respectively. This equates to similarity values for the Miocene and Recent Hispaniolan spider families of 69.6% for families with described species and 72.0% when all fossils are considered. The similarity values for genera are considerably lower at 13.0% for those with described species and 15.1% when those without described species are included; 26 genera are currently considered strictly fossil. However, such genera have, in the past, been subsequently discovered in the extant fauna (see later) or synonymized with recent genera (e.g. Raven [2000: 573] synonymized *Microsteria* Wunderlich, 1988 = *Masteria* L. Koch, 1873 [Dipluridae]), so this value is considered an upper limit for the taxa described to date. The species similarity is 0% because all described Dominican amber fossil and sub-fossil spiders are considered extinct [Penney, 2006a].

For families recorded from 15 or more species interesting similarities and dissimilarities exist between the Miocene and Recent faunas (Appendix). The frequency of genera and species is similar between the Miocene amber and Recent faunas for Pholcidae and Theridiidae, they are distinctly dissimilar for Tetragnathidae, Araneidae and Salticidae and are also slightly dissimilar for Anyphaenidae. These similarities may be an indication that those families were as diverse in the Miocene as they are at present and the dissimilarities may reflect the opposite. However, it must be remembered that the Recent fauna is poorly known and some of the dissimilarities may be purely the result of taphonomic biases associated with the amber fauna, i.e., there may be some mechanism related either directly to the resin formation or the ecology of these spider families that reduced their chances of getting caught and preserved as amber inclusions [e.g., Penney, 2002; Penney and Langan, 2006]. In this case, it may be argued that these families may also have been as diverse on Hispaniola in the Miocene as they are today, but that they are under-represented as amber inclusions.

Araneidae and Tetragnathidae are usually relatively large, conspicuous, aerial web-spinning spiders. Therefore, they are often common in collections of Recent spiders because they are easily seen by collectors. In addition, their large size may facilitate their escape from the sticky resin trap, preventing them occurring frequently as amber inclusions. The jumping spiders (Salticidae) form the largest extant spider family, with more than 5,035 species in 553 genera [Platnick, 2006] and have a global distribution (excluding Antarctica). However, salticids appear to be geologically young (see known fossil range in Figure 2). The active predatory behavior of salticids predisposes them to entrapment in resin [Penney, 2002] and they are currently unknown from Cretaceous ambers, so it may be that Salticidae is a recently evolved family [e.g., García-Villafuerte and Penney, 2003] and this may explain why there are considerably more species known from the Recent fauna. However, this 'over-representation' of salticids for the extant fauna may also be due to their diurnal, relatively rapid, start–stop, hunting behaviour, which makes them conspicuous to collectors. Alternatively, the Salticidae are a notoriously difficult family from a taxonomic perspective and many taxonomists are reluctant to study palaeontological material. Except for the few well-resolved groups, fossil salticid taxonomy is beyond the scope of the current author, so it may well be that fossil amber salticids exist, but that they await formal description.

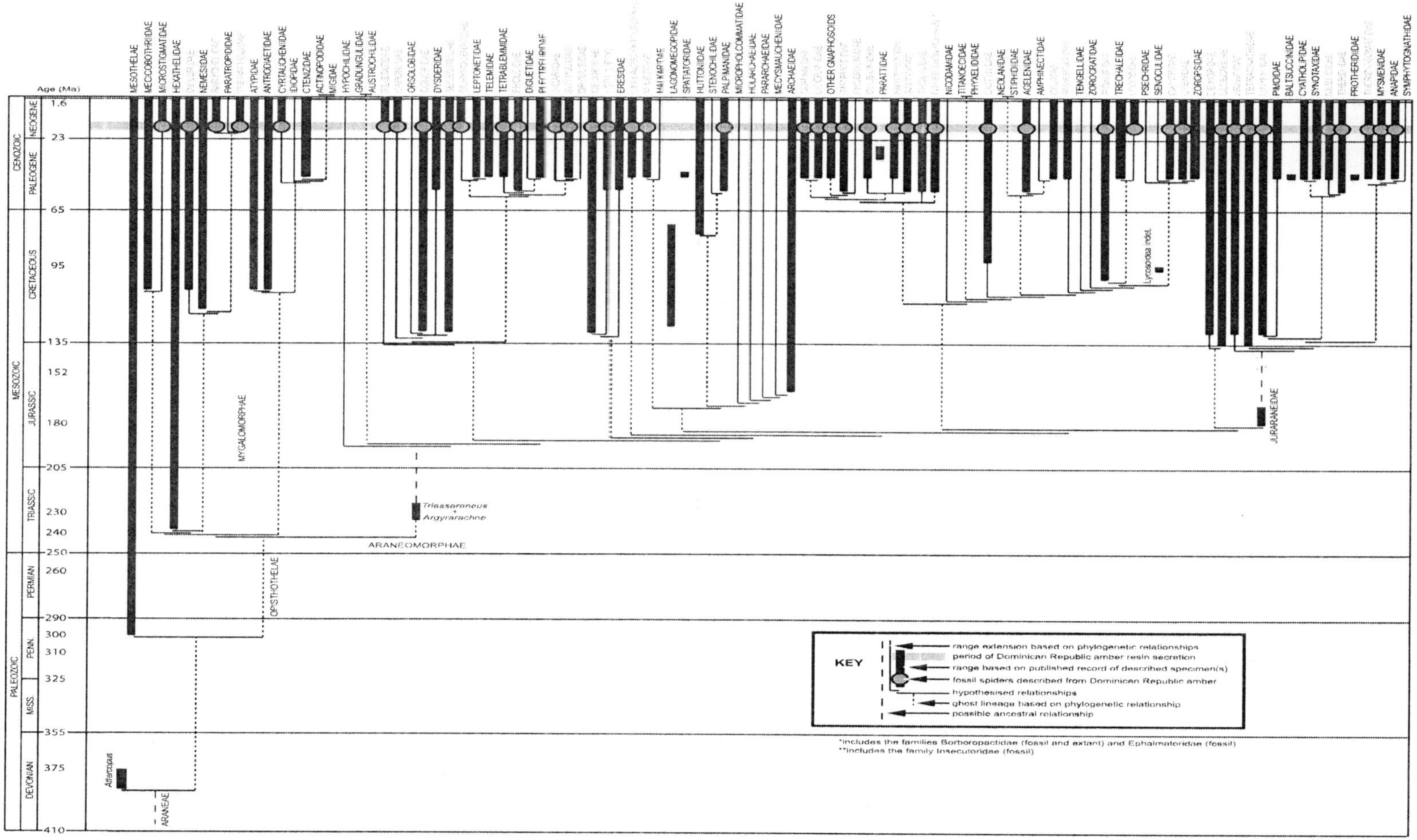

Figure 2. Evolutionary tree of spiders (after Penney *et al.*, 2003), highlighting the fossil (red circles) and Recent (green text) Hispaniolan faunas.

The only obvious case that does not fit the above scenarios is the family Dictynidae. These appear to have been considerably more diverse on Hispaniola in the Miocene than they are at present, with 15 known fossil species and only one Recent example (although the author has collected a second extant species but currently only from a single female, making an identification difficult at this stage). All the fossil dictynids appear to be valid species because Wunderlich [1988] based his diagnoses on distinct differences in male pedipalp morphology, except for *P. singularis*, which is known from only one female. All known Dictynidae amber fossils belong to strictly fossil genera (Appendix), so it may be the case that they were more diverse in the past, at least on Hispaniola. However, the Recent Hispaniolan spider fauna is not known well enough to support this conclusion at present, and future collections of extant material may reveal more species of Dictynidae and maybe also examples of these genera currently known only from fossils.

When Bryant [1948] published her synopsis of the Recent Hispaniolan spider fauna she listed 224 species for the island. Since then few authors have focused specifically on this fauna, which now numbers 313 species (Appendix). As a comparison to Hispaniola Bryant [1948] noted that 280 extant spider species occured in Cuba. In contrast to Hispaniola, arachnological studies have progessed with regard to Cuban spiders and Alayón-García [2000] listed 567 species (more than twice the number cited by Bryant). Thus, the currently known spider fauna for Hispaniola is probably a gross under-representation of the true total. Although Hispaniola is smaller than Cuba (76,071km^2 c.f. 114,525km^2) it has a considerably greater altitudinal variation with a maximum of 3,174m c.f. 1,999m for Cuba [Flint and Pérez-Gelabert, 1999] providing a broad range of habitat variation.

There is general agreement that there was a very substantial rise in global biodiversity through the Cenozoic and especially the Neogene (i.e. the last 23 million years) [Crame and Rosen, 2002]. This was a time of crucial plate tectonic movement that led to climate change and essentially gave the tropics their modern form [Crame and Rosen, 2002]. Given the young age of Dominican amber (15–20Ma) and the similarity of the extant and amber spider faunas (Figure 2, Appendix), the *a priori* assumption is that the Hispaniolan spider fauna had a similar diversity in the Miocene to that observed/expected today. However, our knowledge of Cenozoic Hispaniolan biodiversity will always be far from complete because not all Miocene species will have been preserved in amber, and of those that were, only a relatively small proportion will be found as fossils. Within the amber forest, resin secretion was restricted to a single tree species and would therefore be selective in the organisms it trapped, maybe temporally (e.g. seasonal) as well as spatially. Thus, amber inclusions will represent only a subsample of the total fauna present at the time. Nonetheless, given the similarity of both faunas (Figure 2; Appendix) they do afford an excellent basis for comparison in studies of historical biogeography.

It should be noted that some differences between the Hispaniolan fossil and Recent faunas do occur in other groups. For example, certain extant genera of Hymenoptera, Isoptera, Lepidoptera and Coleoptera were present on Hispaniola in the Miocene but are absent from the island today [Wilson, 1985; Nagel, 1997; Poinar and Poinar, 1999; Hall *et al.*, 2004]. Many insects are more susceptible to extinction than spiders because they form specialized associations with other organisms, such as plants (as herbivores or pollinators), or as parasites, whereas most spiders are generalist predators and if one prey item becomes extinct they can easily switch to another source [Penney *et al.*, 2003; Penney, 2004b]. In addition, extinction resistance in spiders may be facilitated by their ability to survive extended periods of starvation, or other adverse conditions, by entering a state of metabolic torpor.

Poinar [1999] proposed that the demise of certain groups known from Dominican amber but absent from Hispaniola (and other islands in the Greater Antilles) today resulted from a cool period associated with increased ariditiy during the Plio-Pleistocene. However, evidence from sedimentary and geomorphological data, alluvial terraces and albedo reflectivity indices suggest that the Dominican Republic was not drastically affected by the Pleistocene glaciations [Schubert, 1988]. Curtis *et al.* [2001] proposed that both the continental and insular Caribbean regions were cool and arid during the Late Pleistocene, based on oxygen isotope analysis of carbonate shells from lake sediment cores. However, their only Antillean sample was from the south-western tip of Haiti and their results [Curtis *et al.*, 2001: fig. 2] show considerably less variation in this sample than those from the mainland, and indeed, not a great deal of aridity in comparison to that at present, although they support their ideas with previously published palynological data. Their conclusion that analyses of West Indian biogeography must be cognizant of Pleistocene and Holocene climatic changes is certainly a valid one.

The Importance of Amber Fossils in Studies of Historical Biogeography

Fossils are of paramount importance in studies of historical biogeography. They can play a decisive part in the falsification of hypotheses proposed solely on the distributions of extant taxa [e.g., Eskov, 1990]. As an arachnological example, the current Gondwanan distribution of the extant spider family Archaeidae supports the theory of mobilistic biogeography, i.e., that the fragmentation of Gondwanaland and the subsequent continental drift can explain the current distribution of this family, with extant fauna restricted to South Africa and Madagascar (an apochoric distribution; e.g., Kraus [1978]). However, because fossils of this family are highly diverse in Baltic amber [Wunderlich, 2004], and also occur in Tertiary French amber [Penney, 2006b], Cretaceous Burmese amber [Penney, 2003] and in sediments from the Jurassic of Kazakhstan [Eskov, 1987], the palaeontological data demonstrate that their current distribution is plesiochoric, and so a different explanation is required. Another arachnological example concerns the family Huttonniidae, with extant species restricted to New Zealand, but with fossils described in Cretaceous amber from Canada [Penney and Selden, 2006].

The theory of ousted relicts [e.g., Eskov and Golovatch, 1986] proposes that austral disjunctions result from a formerly pancontinental distribution followed by the extinction of 'intermediate links' from the northern continents. There is a considerable amount of palaeontological data, in the form of northern hemisphere fossil representatives of Recent austral taxa of numerous different animal groups in support of this theory [e.g., Eskov, 1987: table 2; Eskov, 1992: table 1], and this tends to be the rule rather than the exception. However, the above authors did not consider the phylogenetic placement of the fossil taxa, and from the viewpoint of competing theories in historical biogeography, these amber fossils are only important if they are not plesiomorphic or the sister group to the extant taxa [Grimaldi, 1992].

Additional problems that may occur when generating biogeographical hypotheses without considering fossils concern non-indigenous taxa. For example, one might be tempted to explain Hispaniolan spider families known in the Recent fauna from only synanthropic or widespread continental species as recent introductions or colonizers (e.g., the families Oecobiidae and Filistatidae: see Appendix). The only extant oecobiid on Hispaniola is the synanthropic *Oecobius concinnus* Simon, 1893, which is distributed from USA to Brazil, the West Indies, the Galapagos Islands and introduced elsewhere [Santos and Gonzaga, 2003] and the only extant filistatid on the island is *Kukulcania hibernalis* (Hentz, 1842), which has a widespread distribution in the New World, so is also presumably an introduced species. However, the possibility that these families are recent introductions can be exluded when the fossil record is taken into consideration because both are described from fossil species in amber [Wunderlich, 1988; Penney, 2005d], so were already present on Hispaniola in the Miocene.

The main models for explaining historical biogeography (dispersal and vicariance) differ with regard to the age of barriers and disjunctions, thus both appear open to falsification through statements about the age of one or the other [Platnick and Nelson, 1978]. The minimum age of a taxon is that of the oldest fossil attributed to it or its sister taxon. A vicariance correlation can be falsified if one of the disjunct taxa is older than the barrier and a dispersal correlation can be falsified if one of the taxa is the same age as, or older, than the barrier. However, both models include protection from falsification based on palaeontological evidence through rejection of a particular barrier in favour of an older barrier, which can be postulated if not demonstrated [Platnick and Nelson, 1978].

It is imperative that palaeontological data are considered when generating hypotheses to explain the biogeographic patterns observed today because fossils form empirical data with both phylogenetic and temporal implications. As pointed out by Platnick and Nelson [1978], the kinds of biogeographic analyses found relegated to the back pages of systematic revisions cannot be justified, yet they continue to be published without taking into consideration palaeontological or geological data.

Greater Antillean Origins, Palaeogeography and Consequences for Biogeography

Prior to plate tectonics providing a mechanism for vicariance in the latter part of the twentieth century, the 'land bridge' was the primary alternative mechanism to dispersal in biogeography [Hedges, 2001]. Indeed, Petrunkevitch [1928] considered the Greater Antillean spider fauna to represent an eastern outgrowth of the Central American fauna by way of a presumed earlier land connection and subsequent continent–island vicariance. However, such a land connection appears never to have existed [Ross and Scotese, 1988; Iturralde-Vinent and MacPhee, 1999]. The most recent hypothesis proposes that during the Eocene–Oligocene transition, the developing northern Greater Antilles and northwestern South America were briefly (33–35Ma) connected by a landspan (a subaerial connection between a continent and one or more off-shelf islands) centered on the emergent Aves Ridge, but the massive uplift that apparently permitted these connections was finished by 32Ma [Iturralde-Vinent and MacPhee, 1999]. However, this idea is not accepted by all workers, e.g., Hedges [2001] criticised the GAARlandia hypothesis of Iturralde-Vinent and MacPhee [1999] on a point by point basis, and favoured a dispersal model as the primary means for explaining the origins of West Indian biodiversity. The criticims of Hedges [2001] were refuted convincingly by MacPhee and Iturralde-Vinent [2005], and so the debate continues.

Nonetheless, the Greater Antilles in their current guise appear to be relatively young geographical features, probably no older than the middle Miocene [Iturralde-Vinent and MacPhee, 1999], rather than having evolved from the Proto-Greater Antillean and subsequently the Greater Antillean landmass formed on the west of the Proto-Caribbean region during the late Lower Cretaceous as proposed by Ross and Scotese [1988]. Therefore, all on-island lineages forming the Recent fauna must be younger than Middle Eocene [Iturralde-Vinent and MacPhee, 1999]. It is likely, and even probable, that Caribbean islands existed during the Mesozoic and that they doubtless had a biota as well, but so far as can be determined, none of the terrestrial environments that may have supported such life continued into the Cenozoic [MacPhee and Iturralde-Vinent, 2000]. The Proto-Antilles were located only one to three crater diameters away [Hedges, 2001] from the Chicxulub impact site in Yucutan at the K/T extinction event (65Ma). Whatever land masses were emergent in this region at the time would have had all flora and fauna anihilated as a result of associated mega-tsunamis and hypercanes (giant hurricanes) and the knock-on consequences in terms of climate and environmental change (see Robertson *et al.* [2004] for additional local effects). During the period of amber-forming resin secretion (15–20Ma; Iturralde-Vinent and MacPhee [1996]) Hispaniola was a distinct island. There may have been a connection to Puerto Rico via a narrow neck of land, but this is not certain [Iturralde-Vinent and MacPhee, 1999]. Since then, tectonic movements of the Caribbean Plate have carried Hispaniola slowly eastwards to its current position.

Biodiversity Predictions and Origins of Hispaniolan Spiders

It is not necessarily possible to predict the presence of undiscovered fossil spiders from families currently unknown from the amber fauna, based on what we observe in the Recent fauna. However, we can hypothesise their existence on Hispaniola during the time of resin secretion during the Miocene and that they evaded capture because of their habitat preference or ecological niche [Penney, 1999]. On the other hand, families present in the amber fauna but unrecorded from the Recent fauna can be predicted to occur on the island. For example, based on presence and absence data of endemic and non-endemic spider species in the fossil and extant Hispaniolan faunas, Penney [1999] made the following predictions: the families Cyrtaucheniidae, Microstigmatidae, Nemesiidae, Ochyroceratidae, Tetrablemmidae, Palpimanidae, Hersiliidae, Symphytognathidae *s.l.*, Anapidae, Mysmenidae and Hahniidae would form components of the Recent fauna; Filistatidae and Desidae colonized Hispaniola following the Miocene amber formation; Drymusidae, Amaurobiidae and Deinopidae were present on Hispaniola during the Tertiary, but avoided capture in resin or have yet to be found in the amber; and Scytodidae, Oecobiidae, Uloboridae, Dictynidae and Clubionidae have colonized Hispaniola since the Miocene amber formation but these families, which were present on Hispaniola during the period of amber formation, also contain undiscovered endemic species.

All these hypotheses are falsifiable or supportable through future discoveries of new Hisaniolan taxa. Indeed, extant Hersiliidae have been described from the island by Rheims and Brescovit [2004] and a new extant species of the genus *Arachnolithulus* Wunderlich, 1988 (Ochyroceratidae) originally known only from fossil species has also recently been found on Hispaniola [Hormiga, pers. comm. 2005, in prep.]. The idea that Filistatidae first colonized the island post-Miocene has been falsified by the discovery of fossil species in amber by Penney [2005d]. The remaining hypotheses still stand, except possibly for the presence of Hahniidae which was based on a dubious identification [Penney and Pérez-Gelabert, 2002]. The same basic principles may be applied to presence and absence data for genera to generate higher resolution hypotheses, but caution should be exercised because higher taxonomic levels will be more susceptible to extinction than lower ones, unless of course they are monotypic.

It is clearly evident that many different types of organismal associations have a long geological history and many were already well established by the Mioecene, such as, plant–pollinator, plant–herbivore, predator–prey, host–parasite relationships, phoresy, commensalism, etc. Thus, it is also possible to predict the presence of other taxa from expected associations with observed taxa. This is possibly a more realistic proposition for the extant fauna, but nonetheless, fossil evidence supports this idea for the Miocene fauna also. For example, Wunderlich [1988] described the extant spider genus *Ischnothele* Ausserer, 1875 (Dipluridae) and Penney [2000a] described a new species of Mysmenidae from the extant genus *Mysmenopsis* Simon, 1897, both from Dominican amber. Of the 17 extant species of *Mysmenopsis* known at the time, Platnick and Shadab [1978] reported five of them as kleptoparasites in the webs of diplurid spiders. The two Recent species (*M. monticola* and *M. furtiva*) to which the fossil species is most closely related were described by Coyle and Meigs [1989] from Jamaica; they reported a probable kleptoparasitic mode of life because they were collected from the webs of *Ischnothele* spp., and Coyle *et al.* [1991] provided direct evidence for the kleptoparasitic and commensualistic nature of *M. furtiva* in the web of its diplurid host. Based on the idea of behavioural fixity/uniformitariansim [Boucot, 1989], which infers that extinct organisms behaved in a similar manner to their Recent relatives at genus and often family level, there is no reason to believe that the fossil species behaved any differently; because it is morphologically similar to the Jamaican species (with only minor differences in genitalic structure and tibial spine numbers). Thus, the presence of Dipluridae in the above example could have been used to predict the presence of Mysmenidae on Hispaniola during the Miocene, but even more so vice versa.

Patterns of endemism in other groups have been invoked in discussions of West Indian biogeography [e.g. Genaro and Tejuca, 2001; Hedges, 2001], but it is unclear at present what this adds to the debate on the grand scale. The problem lies in the range of variation between groups, for example, regional endemism in vertebrates ranges from 35% (Aves) to 99% (Amphibia) with an average of 74% [Hedges, 2001: table 1] and in Cuba insect ordinal endemism may range from 0% to >90% with an average range of 40–60% [Genaro and Tejuca, 2001]. The latter authors note a lack of information for Hispaniolan extant insect biodiversity. In their paper on Hispaniolan spiders, Penney and Pérez-Gelabert [2002] listed 116 of the 296 extant species (39%) as endemic. However, given the poorly known nature of the extant fauna and the fact that many of the so-called endemics were described by Bryant (see earlier), who has subsequently had many of her species synonymized by other authors, the accuracy of this value as a true representation is highly dubious. Until the existing taxa are revised and the Recent fauna is better known, little can be derived from general studies of spider endemism with regard to Caribbean (especially Hispaniolan) biogeography.

Once such work is undertaken, interesting patterns can be expected and particularly so when the extant taxa are compared with closely related fossil species. For example, *Trachelas* (Corinnidae) spiders from North and Central America were revised by Platnick and Shadab [1974ab] and placed into four species groups separated on the grounds of genitalia, somatic morphology and geography. The *bicolor* group is the only one found in the West Indies. Penney [2001] described the new fossil species *Trachelas poinari* from Dominican amber, which also belongs in this group and is most closely related to *T. borinquensis* Gertsch, 1942 (Recent, endemic to Puerto Rico), and the only other *Trachelas* species to possess a retrolateral patellar apophysis. This may represent an example of island–island vicariance following the separation of Puerto Rico from the northern Hispaniola block, presumably sometime within the last 14Ma [e.g. Iturralde-Vinent and MacPhee, 1999].

As discussed above, the competing arguments regarding dispersal versus vicariance in understanding Caribbean biogeography continue to stimulate lively debate and conflicting opinions. In both cases a barrier is involved. In the vicariance model, dispersal takes place in the absence of a barrier (the appearance of the barrier fragments the range of an ancestral species), whereas in the dispersalist model, dispersal occurs across an existing barrier [Platnick and Nelson, 1978]. The dispersal capabilities of organisms vary and this is clearly going to have important consequences for those with poor disperal abilities when trying to cross a barrier (not to mention the additional problems of finding the right habitat and establishing a founder population once it/they arrive). This is particularly true of spiders, in which some species can disperse great distances at high altitudes through ballooning (for a review of spider families that balloon see Bell *et al.* [2005]), whereas some fossorial, burrow-dwelling species may disperse only very short distances from the maternal burrow. For example, this concept is evident in the extremely isolated Hawaiian Islands, where the few spider families that have colonized the archipelago, have been from highly dispersive taxa and appear to have done so repeatedly [Gillespie *et al.*, 1998]. Consequently, one might expect a relatively low familial diversity for islands on which dispersal is the primary mode of immigration. This is not the case for Hispaniola now, or back in the Miocene [Penney and Pérez-Gelabert, 2002; Penney, 2005b; Appendix], and the significance of this observation increases when one considers the relatively young age of the island. This would seem to add more weight in favour of the vicariance model for explaining on-island spider lineages, although dispersal cannot be ruled out for some of the highly dispersive taxa.

Evidence for links with the South American fauna can be found in the Recent distributions of numerous extant spider genera known as fossils in Dominican amber (see Appendix). For example, *Masteria* L. Koch, 1873 (Dipluridae), *Psalistops* Simon, 1889 (Barychelidae), *Misionella* Ramírez and Grismado, 1997 (Filistatidae), *Coryssocnemis* Simon, 1893 and *Modisimus* Simon, 1893 (Pholcidae), *Nops* MacLeay, 1839 (Caponiidae), *Monoblemma* Gertsch, 1941 (Tetrablemmidae), *Stenoonops* Simon, 1891 (Oonopidae), *Mysmenopsis* Simon, 1897 (Mysmenidae), *Azilia* Keyserling, 1881 and *Cyrtognatha* Keyserling, 1881 (Tetragnathidae), *Anyphaenoides* Berland, 1913 and *Lupettiana* Brescovit, 1996 (Anyphaenidae), *Corinna* C. L. Koch, 1841 (Corinnidae), *Pseudosparianthis* Simon, 1887 (Sparassidae), *Corythalia* C. L. Koch, 1850 and *Lyssomanes* Hentz, 1845 (Salticidae) and *Molinaranea* Mello-Leitão, 1940 [Penney, in prep.] (Araneidae) have extant species restricted to, or primarily distributed within, South America, the Greater and Lesser Antilles and Central America (Panama). The remaining genera are more widespread and some include cosmopolitan taxa, but all are well represented in South America or the Neotropics. None of the extant spider genera recorded from Miocene Dominican amber present any reasoning to detract from a South American origin for this fauna.

The future research direction for understanding these biogeographical patterns will lie in the application of cladistic and phylogenetic approaches to biogeography. Cladistic biogeography emphasizes the search for congruent biogeographic patterns using cladograms, disregarding both dispersal and viariance explanations *a priori*. It assumes correspondence between taxonomic relationships and area relationships is biogeographically informative (see Reiskind [1989] for an example concerning Caribbean spiders including a species in Dominican amber) and that general area cladograms can be derived from individual area cladograms of different taxa [Morrone and Crisci, 1995]. Phylogenetic biogeography suggests that if different monophyletic groups show the same biogeographic pattern, then they probably share the same biogeographic history and infers that primitive members of a taxon are found closer to its centre of origin than more apomorphic ones, which are found at the periphery [Morrone and Crisci, 1995]. Indeed, it is interesting to note that preliminary obersvations based on pedipalp morphology of Miocene Dominican amber fossils, e.g., *Misionella* (Filistatidae) and *Molinaranea* (Araneidae; Penney [in prep.]), when compared to extant taxa restricted to South America, suggest that the amber species are more derived and thus further removed from the centre of origin, which in this case would be South America. Clearly, there is great potential for additional research in this area.

Mexican Amber as an Additional Comparative Palaeontological Resource

Another potentially informative and comparative palaeontological resource for understanding the historical biogeography of Caribbean spiders (and other groups) is amber from Chiapas, Mexico. However, the Early Miocene–Late Oligocene age of this amber [e.g., Poinar, 1992] is not as well constrained as previously thought, and it may actually be the same age as Dominican amber [Solórzano Kraemer and Rust, in prep.; Solórzano Kraemer, pers. comm. 2006], which would obviously have interesting implications for direct comparisons with the fossil Hispaniolan fauna. The degree of preservation is not quite as good as that in Dominican amber, but spiders are relatively common as inclusions. Unfortunately, only a few spiders have been described from this deposit [Petrunkevitch, 1963, 1971; García-Villafuerte and Vera, 2002; Garcia-Villafuerte and Penney, 2003; Penney, in press; García-Villafuerte, in prep.; Dunlop *et al.*, in prep.] but the taxonomy of those described by Petrunkevitch is poor and his taxa are in need of revision [Penney, in press]. Penney [in press] revised the Neotropical fossil Hersiliidae, which included fossil taxa in Mexican amber. A number of genera (including extant and fossil) were synonymized and a number of extant species had to be transferred to a previously strictly fossil genus. This highlights the importance of palaeontologists considering Recent taxa when describing fossils, and likewise the need for neontologists to consider fossil taxa. Without such, cross-disciplinary awareness and consideration, which unfortunately is often lacking by palaeontologists and neontologists alike, comparisons of fossil and extant assemblages will be pointless and uniformative at best, but misleading at worst.

The potential importance of Mexican amber for understanding the origins of Hispaniolan spiders derives from the fact that the Yucatan Peninsula did not have land connections to South America (post K/T and prior to the Mexican amber formation). Thus, any similarities of Mexican amber spiders to the South American fauna must be a result of dispersal, despite the recent Panamanian isthmus formation. Thus, a comparative analysis of Mexican and Dominican amber spiders with proximate extant faunas will aid our understanding of the relative contributions of vicariance and dispersal for explaining the origins of current Caribbean biogeography.

Conclusion

Fossils form empirical data with both phylogenetic and temporal implications and are of paramount importance for understanding historical biogeography. They can play a decisive role in falsifying hypotheses proposed solely on the distributions of extant taxa. It is also possible to generate hypotheses for both fossil and extant faunas that are ultimately falsibiable through the discovery of new Hispaniolan taxa. The amber spider fauna has clear affinities to the present South American fauna, so how did they colonize the island from the mainland: dispersal or vicariance? The high Miocene biodiversity (including taxa with poor dispersal capabilities) on a geologically young island would seem to add more weight in favour of the vicariance model for explaining on-island spider lineages, although dispersal cannot be excluded for some of the highly dispersive taxa. However, a case-by-case approach is required in order to determine the relative contributions of the different models. The future research direction for understanding the biogeographical patterns will lie in the application of cladistic and phylogenetic biogeographical approaches and it is interesting to note that preliminary obersvations based on pedipalp morphology suggest some fossil taxa are more derived than extant forms.

Both the fossil and Recent Hispaniolan spider faunas are worthy of further investigation, particularly because of their similarity, which means they can be used as directly comparable, complementary data sets to address various questions relating to Caribbean biogeography. Clearly, the greater the completeness of the faunal inventories and the larger the datasets in such analyses the more reliable the conclusions. The contribution that the Dominican Republic (and Mexican) amber fossil fauna can make to our understanding of the origins of Caribbean biogeography should not be underestimated.

Appendix: Checklist of Hispaniolan Spider Species

Families are listed in systematic order following Platnick [2006]. Only full species are included in the list. Species from Navassa, a tiny island (5.2km^2) 64km west of Hispaniola are not included; a list of the 58 known spider species was provided by Alayón-García [2001]. In numerous old arachnological works the dates printed on the journals do not actually represent the correct publication year. All publication dates for the Recent fauna and generic placements follow Platnick [2006]. Names in square brackets are nomina dubia; * = fossil genus; † = fossil species; [†] = subfossil species. This list is based primarily on that of Penney and Pérez-Gelabert [2002] and Penney [2006a] with emendations and updates as required. References for the extant taxa are not cited but these can be located easily using the bibliography in Platnick [2006]. In total, the Hispaniolan spider fauna currently comprises 470 named species (excluding Theridiidae: gen. indet. *clypeatus* [Wunderlich, 1988)). The fossil fauna consists of 154 named species in 83 genera and 39 families. None of the families, 31% of the genera and all the species are extinct. Three subfossil species have been described from copal. The extant fauna consists of 313 named species, 175 genera and 42 families.

Order Araneae
Suborder Opisthothelae
Infraorder Mygalomorphae

Family *Dipluridae*

Ischnothele jeremie Coyle, 1995
Ischnothele garcia Coyle, 1995
†*Masteria sexoculata* (Wunderlich, 1988)

Family *Cyrtaucheniidae*

†*Bolostromus destructus* Wunderlich, 1988

Family *Microstigmatidae*

*†*Parvomygale distincta* Wunderlich, 2004

Family *Barychelidae*

Psalistops fulvus Bryant, 1948
†*Psalistops hispaniolensis* Wunderlich, 1988
Psalistops maculosus Bryant, 1948
Trichopelma nitidum Simon, 1888

Family *Theraphosidae*

Citharacanthus spinicrus (Latreille, 1819)
Cyrtopholis agilis Pocock, 1903
Cyrtopholis cursor (Ausserer, 1875)
Holothele sericea (Simon, 1903)
*†*Ischnocolinopsis acutus* Wunderlich, 1988
Phormictopus cancerides (Latreille, 1806)
Infraorder Araneomorphae

Family *Filistatidae*

Kukulcania hibernalis (Hentz, 1842)
†*Misionella didicostae* Penney, 2005

Family *Sicariidae*

†*Loxosceles aculicaput* Wunderlich, 2004
Loxosceles caribbaea Gertsch, 1958
Loxosceles cubana Gerstch, 1958
†*Loxosceles defecta* Wunderlich, 1988
†*Loxosceles deformis* Wunderlich, 1988
Loxosceles taino Gerstch and Ennik, 1983

Family *Scytodidae*

Scytodes fusca Walckenaer, 1837
Scytodes longipes Lucas, 1844
†*Scytodes piliformis* Wunderlich, 1988
†*Scytodes planithorax* Wunderlich, 1988
†*Scytodes stridulans* Wunderlich, 1988

Family *Drymusidae*

Drymusa simoni Bryant, 1948

Family *Ochyroceratidae*

*†*Arachnolithulus longipes* Wunderlich, 2004
*†*Arachnolithulus pygmaeus* Wunderlich, 1988

Family *Pholcidae*

Artema atlanta Walckenaer, 1837
†*Coryssocnemis velteni* Wunderlich, 2004
Leptopholcus hispaniola Huber, 2000
Micropholcus fauroti (Simon, 1887)
Modisimus femoratus Bryant, 1948
†*Modisimus calcar* Wunderlich, 1988
†*Modisimus calcaroides* Wunderlich, 1988
†*Modisimus crassifemoralis* Wunderlich, 1988
Modisimus fuscus Bryant, 1948
Modisimus glaucus Simon, 1893
†*Modisimus oculatus* Wunderlich, 1988
†*Modisimus tuberosus* Wunderlich, 1988
Modisimus vittatus Bryant, 1948
†*Pholcophora brevipes* Wunderlich, 1988
†*Pholcophora gracilis* Wunderlich, 1988
†*Pholcophora longicornis* Wunderlich, 1988
Physocyclus globosus (Taczanowski, 1874)
Tainonia serripes (Simon, 1893)
*†*Serratochorus pygmaeus* Wunderlich, 1988

Family *Caponiidae*

Caponina darlingtoni Bryant, 1948
Nops blandus (Bryant, 1942)

Nops ernestoi Sánchez-Ruíz, 2005
Nops gertschi Chickering, 1967
†*Nops lobatus* Wunderlich, 1988

Family *Tetrablemmidae*

†*Monoblemma? spinosum* Wunderlich, 1988

Family *Segestriidae*

Ariadna multispinosa Bryant, 1948
†*Ariadna paucispinosa* Wunderlich, 1988

Family *Oonopidae*

*†*Fossilopaea sulci* Wunderlich, 1988
Ischnothyreus peltifer (Simon, 1891)
†*Oonops seldeni* Penney, 2000
Oonops validus Bryant, 1948
†*Orchestina dominicana* Wunderlich, 1981
†*Orchestina tibialis* Wunderlich, 1988
†*Stenoonops incertus* (Wunderlich, 1988)
Triaeris stenaspis Simon, 1891

Family *Mimetidae*

†*Mimetus? bituberculatus* Wunderlich, 1988
Mimetus hispaniolae Bryant, 1948
Mimetus syllepsicus Hentz, 1832

Family *Oecobiidae*

Oecobius concinnus Simon, 1893
†*Oecobius piliformis* Wunderlich, 1988

Family *Hersiliidae*

†*Prototama maior* (Wunderlich, 1988)
†*Prototama media* (Wunderlich, 1988)
†*Prototama minor* (Wunderlich, 1987)
Yabisi guaba Rheims and Brescovit, 2004

Family *Deinopidae*

Deinopis lamia MacLeay, 1839

Family *Uloboridae*

†*Miagrammopes dominicanus* Wunderlich, 2004
Miagrammopes latens Bryant, 1936
Philoponella semiplumosa (Simon, 1893)
Uloborus glomosus (Walckenaer, 1842)
Uloborus trilineatus Keyserling, 1883
Zosis geniculata (Oliver, 1789)

Family *Nesticidae*

Eidmannella pallida (Emerton, 1875)
*†*Hispanonesticus latopalpus* Wunderlich, 1986
Nesticus maculatus Bryant, 1948

Family *Theridiidae*

†*Achaearanea extincta* Wunderlich, 1988
Anelosimus studiosus (Hentz, 1850)
†*Argyrodes crassipatellaris* Wunderlich, 1988
Argyrodes elevatus Taczanowski, 1873
Argyrodes nephilae Taczanowski, 1873
†*Argyrodes parvipatellaris* Wunderlich, 1988
Ariamnes haitensis (Exline and Levi, 1962)
†*Chrosiothes biconigerus* Wunderlich, 1988
†*Chrosiothes curvispinosus* Wunderlich, 1988
†*Chrosiothes emulgatus* Wunderlich, 1988
†*Chrosiothes longispinosus* Wunderlich, 1988
†*Chrosiothes monoceros* Wunderlich, 1988
†*Chrosiothes tumulus* Wunderlich, 1988
†*Chrosiothes unicornis* Wunderlich, 1988
Chrysso albomaculata O.P.-Cambridge, 1882
†*Chrysso? conspicua* Wunderlich, 1988
†*Chrysso? dubia* Wunderlich, 1988
Chrysso pulcherrima (Mello-Leitão, 1917)
Coleosoma floridanum Banks, 1900
*†*Cornutidion elongatum* Wunderlich, 1988
Craspedisia spatulata Bryant, 1948
Dipoena dominicana Wunderlich, 1986
†*Dipoenata altioculata* Wunderlich, 1988
†*Dipoenata cala* Wunderlich, 1988
†*Dipoenata clypeata* Wunderlich, 1988
†*Dipoenata globulus* Wunderlich, 1988
Dipoenata morosa (Bryant, 1948)
[†]Dipoenata praedominicana (Wunderlich, 1986)
†Dipoenata stipes Wunderlich, 1988
†Dipoenata yolandae Wunderlich, 1988
[†]Episinus antecognatus Wunderlich, 1986
†Episinus brevipalpus Wunderlich, 1988
†Episinus cornutus Wunderlich, 1988
Episinus dominicus Levi, 1955
Episinus gratiosus Bryant, 1940
†Episinus praecognatus Wunderlich, 1982
†Episinus tuberosus Wunderlich, 1988
Faiditus americanus (Taczanowski, 1874)
Faiditus caudatus (Taczanowski, 1874)
Faiditus darlingtoni (Exline and Levi, 1962)

†Lasaeola puta Wunderlich, 1988
†Lasaeola pristina (Wunderlich, 1988)
†Lasaeola vicina (Wunderlich, 1982)
†Lasaeola vicinoides Wunderlich, 1988
Latrodectus geometricus C.L. Koch, 1841
Latrodectus mactans (Fabricius, 1775)
Neospintharus furcatus (O.P.-Cambridge, 1894)
Nesticodes rufipes (Lucas, 1846)
Spintharus flavidus Hentz, 1850
†Spintharus longisoma Wunderlich, 1988
Steatoda erigoniformis (O.P.-Cambridge, 1872)
Steatoda grossa (C. L. Koch, 1838)
†Stemmops incertus Wunderlich, 1988
†Stemmops prominens Wunderlich, 1988
†Styposis pholcoides Wunderlich, 1988
Theridion antillanum Simon, 1894
Theridion australe Banks, 1899
†Theridion contrarium Wunderlich, 1988
Theridion dilucidum Simon, 1897
†Theridion erectoides Wunderlich, 1988
†Theridion erectum Wunderlich, 1988
[*Theridion fuesslini* Simon, 1894]
Theridion hassleri Levi, 1963
†Theridion inversum Wunderlich, 1988
Theridion melanostictum O.P.-Cambridge, 1876
Theridion positivum Chamberlin, 1924
†Theridion variosoma Wunderlich, 1988
†Theridion wunderlichi Penney, 2001
Theridula gonygaster (Simon, 1873)
Thymoites banksi (Bryant, 1948)
Thymoites guanicae (Petrunkevitch, 1930)
Thymoites pallidus (Emerton, 1913)
Tidarren sisyphoides (Walckenaer, 1842)
Wamba congener O.P.-Cambridge, 1896

Family *Theridiosomatidae*

[*Allototua guttata* Bryant, 1945]
Ogulnius fulvus Bryant, 1945
Ogulnius latus Bryant, 1948
*†*Palaeoepeirotypus iuvenis* Wunderlich, 1988
*†*Palaeoepeirotypus iuvenoides* Wunderlich, 1988
†Theridiosoma incompletum Wunderlich, 1988
Wendilgarda clara Keyserling, 1886

Family *Anapidae*

*†*Palaeoanapis nana* Wunderlich, 1988

Family *Mysmenidae*

*†*Dominicanopsis grimaldii* Wunderlich, 2004
†*Mysmenopsis lissycoleyae* Penney, 2000

Family *Linyphiidae*

Ceraticelus paludigenus Crosby and Bishop, 1925
Eperigone serrata Ivie and Barrows, 1935
†*Floricomus fossilis* Penney, 2005
Florinda coccinea (Hentz, 1850)
Frontinella bella Bryant, 1948
Frontinella communis (Hentz, 1850)
Grammonota calcarata Bryant, 1948
Lomaita darlingtoni Bryant, 1948
†*Meioneta bigibber* (Wunderlich, 1988)
†*Meioneta fastigata* (Wunderlich, 1988)
†*Meioneta separata* (Wunderlich, 1988)
†*Selenyphantes flagellifera* (Wunderlich, 1986)
Tutaibo anglicanus (Hentz, 1850)

Family *Tetragnathidae*

Agriognatha argyra Bryant, 1945
Agriognatha espanola Bryant, 1945
Agriognatha rucilla Bryant, 1945
Antillognatha lucida Bryant, 1945
†*Azilia hispaniolensis* Wunderlich, 1988
Azilia montana Bryant, 1940?
Chrysometa bigibbosa (Keyserling, 1864)
Chrysometa conspersa (Bryant, 1945)
Chrysometa cornuta (Bryant, 1945)
Chrysometa maculata (Bryant, 1945)
Chrysometa obscura (Bryant, 1945)
Chrysometa sabana Levi, 1986
†*Cyrtognatha weitschati* Wunderlich, 1988
Glenognatha mira Bryant, 1945
Hispanognatha guttata Bryant, 1945
†*Homalometa fossilis* Wunderlich, 1988
Leucauge argyra (Walckenaer, 1842)
Leucauge regnyi (Simon, 1897)
Leucauge venusta (Walckenaer, 1842)
Leucage venustella Strand, 1916
†*Nephila breviembolus* Wunderlich, 1986
Nephila clavipes (Linnaeus, 1767)
†*Nephila dommeli* Wunderlich, 1982
†*Nephila furca* Wunderlich, 1986
†*Nephila longembolus* Wunderlich, 1986
†*Nephila tenuis* Wunderlich, 1986

Tetragnatha elongata Walckenaer, 1842
Tetragnatha nitens (Audouin, 1826)
Tetragnatha orizaba (Banks, 1898)
Tetragnatha pallescens F.O.P.-Cambridge, 1903
†*Tetragnatha pristina* Schawaller, 1982
Tetragnatha tenuissima O.P.-Cambridge, 1889

Family *Araneidae*

Acacesia hamata (Hentz, 1847)
Acanthepeira stellata (Walckenaer, 1805)
Aculepeira busu Levi, 1991
Aculepeira visite Levi, 1991
Alcimosphenus licinus Simon, 1895
Allocyclosa bifurca (McCook, 1887)
*†*Araneometa excelsa Wunderlich, 1988*
*†*Araneometa herrlingi Wunderlich, 1988*
*†*Araneometa spirembolus Wunderlich, 1988*
Araneus bryantae Brignoli, 1983
Araneus elizabethae Levi, 1991
Araneus hispaniola (Bryant, 1945)
Araneus hotteiensis (Bryant, 1945)
†*Araneus? nanus* Wunderlich, 1988
[*Araneus perplexus* (Walckenaer, 1842)]
Argiope argentata (Fabricius, 1775)
Argiope trifasciata (Forskål, 1775)
Cyclosa berlandi Levi, 1999
Cyclosa bifurcata (Walckenaer, 1842)
Cyclosa caroli (Hentz, 1850)
Cyclosa haiti Levi, 1999
Cyclosa turbinata (Walckenaer, 1842)
Cyclosa walckenaeri (O.P.-Cambridge, 1889)
Cyrtophora citricola (Forskål, 1775)
†*Enacrosoma verrucosa* (Wunderlich, 1988)
Eriophora ravilla (C.L. Koch, 1844)
Eustala bisetosa Bryant, 1945
Eustala delasmata Bryant, 1945
Eustala devia (Gerstch and Mulaik, 1936)
Eustala fuscovittata (Keyserling, 1864)
Eustala perdita Bryant, 1945
Eustala vegeta (Keyserling, 1865)
*†*Fossilaraneus incertus* Wunderlich, 1988
Gasteracantha cancriformis (Linnaeus, 1758)
Gea heptagon (Hentz, 1850)
Kapogea sellata (Simon, 1895)
Larinia minor (Bryant, 1945)
Mangora fascialata Franganillo, 1936

Mecynogea martiana (Archer, 1958)
Metazygia cienaga Levi, 1995
Metazygia crewi (Banks, 1903)
Metazygia dubia (Keyserling, 1864)
Metazygia gregalis (O.P.-Cambridge, 1889)
Metepeira compsa (Chamberlin, 1916)
Metepeira datona Chamberlin and Ivie, 1942
Metepeira jamaicensis Archer, 1958
Metepeira triangularis (Franganillo, 1930)
Metepeira vigilax (Keyserling, 1893)
Micrathena forcipata (Thorell, 1859)
Micrathena militaris (Fabricius, 1775)
Micrathena similis Bryant, 1945
Neoscona arabesca (Walckenaer, 1842)
Neoscona marcanoi Levi, 1993
Neoscona moreli (Vinson, 1863)
Neoscona nautica (L. Koch, 1875)
Neoscona oaxacensis (Keyserling, 1864)
Ocrepeira darlingtoni (Bryant, 1945)
Ocrepeira serrallesi (Bryant, 1947)
Parawixia tredecimnotata F.O.P.-Cambridge, 1904
*†*Pycnosinga fossilis* Wunderlich, 1988
Verrucosa arenata (Walckenaer, 1842)
Wagneriana tauricornis (O.P.-Cambridge, 1889)
Wagneriana undecimtuberculata (Keyserling, 1865)
Wagneriana vegas Levi, 1991
Witica crassicaudus (Keyserling, 1865)

Family *Lycosidae*

Agalenocosa bryantae (Roewer, 1951)
Arctosa inconspicua (Bryant, 1948)
Hogna reducta (Bryant, 1942)
Hogna tantilla (Bryant, 1948)
Pardosa hamifera F.O.P.-Cambridge, 1902
Pardosa portoricensis Banks, 1902
Pirata sedentarius Montgomery, 1904
Trochosa reichardtiana Strand, 1916

Family *Pisauridae*

Thaumasia annecta Bryant, 1948
Thaumasia marginella (C.L. Koch, 1847)
Tinus connexus (Bryant, 1940)

Family *Oxyopidae*

Hamataliwa communicans (Chamberlin, 1925)
Hamataliwa haytiana (Chamberlin, 1925)

Hamataliwa nigritarsa Bryant, 1948
Hamataliwa rana (Simon, 1897)
Oxyopes crewi Bryant, 1948
†*Oxyopes defectus* Wunderlich, 1988
Peucetia viridans (Hentz, 1832)

Family *Zoridae*

Odo abudi Alayón-García, 2002

Family *Ctenidae*

Ctenus avidus Bryant, 1948
Ctenus darlingtoni Bryant, 1948
Ctenus haitiensis Strand, 1909
Ctenus haina Alayón-García, 2004
Ctenus hiemalis Bryant, 1948
Ctenus insulanus Bryant, 1948
Ctenus jaragua Alayón-García, 2004
Ctenus manni Bryant, 1948
Ctenus monticola Bryant, 1948
Ctenus naranjo Alayón-García, 2004
Cupiennius salei (Keyserling, 1877)
*†*Nanoctenus longipes* Wunderlich, 1988
Trujillina hursti (Bryant, 1948)
Trujillina spinipes Bryant, 1948

Family *Desidae*

Paratheuma insulana (Banks, 1902)

Family Dictynidae

Emblyna altamira (Gertsch and Davis, 1942)
*†*Hispaniolyna hirsuta* Wunderlich, 1988
*†*Hispaniolyna magna* Wunderlich, 1988
*†*Palaeodictyna intermedia* Wunderlich, 1988
*†*Palaeodictyna longispina* Wunderlich, 1988
*†*Palaeodictyna? singularis* Wunderlich, 1988
*†*Palaeodictyna spiculum* Wunderlich, 1988
*†*Palaeodictyna termitophila* Wunderlich, 1988
*†*Palaeodictyna unispina* Wunderlich, 1988
*†*Palaeolathys? circumductus* Wunderlich, 1988
*[†]*Palaeolathys copalis* Wunderlich, 1986
*†*Palaeolathys quadruplex* Wunderlich, 1988
*†*Palaeolathys similis* Wunderlich, 1988
*†*Palaeolathys spinosa* Wunderlich, 1986
*†*Succinya longembolus* Wunderlich, 1988
*†*Succinya pulcher* Wunderlich, 1988
*†*Succinya spinipalpus* Wunderlich, 1988

Family *Amaurobiidae*

Neowadotes casabito Alayón-García, 1995
Tugana crassa (Bryant, 1948)
Tugana infumata (Bryant, 1948)
Retiro gratus (Bryant, 1948)

Family *Miturgidae*

Cheiracanthium inclusum (Hentz, 1847)
†*Strotarchus heidti* Wunderlich, 1988
Teminius hirsutus (Petrunkevitch, 1925)
Teminius insularis (Lucas, 1857)
Teminius monticola (Bryant, 1948)

Family *Anyphaenidae*

Anyphaena dominicana Roewer, 1951
Anyphaena modesta Bryant, 1948
Anyphaena pusilla Bryant, 1948
†*Anyphaenoides bulla* (Wunderlich, 1988)
Hibana tenuis (L. Koch, 1866)
Hibana velox (Becker, 1879)
Lupettiana levii Brescovit, 1999
†*Lupettiana ligula* (Wunderlich, 1988)
Lupettiana parvula (Banks, 1903)
Lupettiana spinosa (Bryant, 1948)
Thaloe ennery Brescovit, 1993
Thaloe remotus (Bryant, 1948)
Wulfila fasciculus (Bryant, 1948)
Wulfila fragilis (Bryant, 1948)
Wulfila gracilipes (Banks, 1903)
†*Wulfila spinipes* Wunderlich, 1988

Family *Liocranidae*?

Lausus pulchellus Bryant, 1948 (*I. sedis*)

Family *Clubionidae*

Elaver implicata (Gertsch, 1941)
†*Elaver nutua* (Wunderlich, 1988)

Family *Corinnidae*

†*Castianeira tenebricosa* Wunderlich, 1988
*†*Chemmisomma dubia* Wunderlich, 1988
†*Corinna flageliformis* Wunderlich, 1988
Corinna parvula Bryant, 1940
Corinna toussainti Bryant, 1948
Creugas gulosus Thorell, 1878
†*Megalostrata grandis* Wunderlich, 1988

Phrurolithus spinosus Bryant, 1948
Trachelas bicolor Keyserling, 1887
Trachelas dilatus Platnick and Shadab, 1974
Trachelas erectus Platnick and Shadab, 1974
†*Trachelas poinari* Penney, 2001
Trachelas tomaculus Platnick and Shadab, 1974
Xeropigo tridentiger (O.P.-Cambridge, 1869)

Family *Trochanteriidae*

*†*Veterator angustus* Wunderlich, 1988
*†*Veterator ascutum* Wunderlich, 1988
*†*Veterator incompletus* Wunderlich, 1982
*†*Veterator longipes* Wunderlich, 1988
*†*Veterator loricatus* Wunderlich, 1988
*†*Veterator porrectus* Wunderlich, 1988
*†*Veterator viduus* Wunderlich, 1988

Family *Prodidomidae*

Zimiris doriai Simon, 1882

Family *Gnaphosidae*

Cesonia nadleri Platnick and Shadab, 1980
*†*Drassyllinus aliter* Wunderlich, 1988
Gnaphosa sericata (L. Koch, 1866)
Sergiolus magnus (Bryant, 1948)

Family *Selenopidae*

Selenops bani Alayón-García, 1992
†*Selenops beynai* Schawaller, 1984
†*Selenops dominicanus* Wunderlich, 2004
Selenops insularis Keyserling, 1881
Selenops marcanoi Alayón-García, 1992
Selenops pensilis Muma, 1953
Selenops phaselus Muma, 1953

Family *Sparassidae*

Heteropoda venatoria (Linnaeus, 1767)
Olios antiguensis (Keyserling, 1880)
†*Pseudosparianthis pfeifferi* (Wunderlich, 1988)
Stasina saetosa Bryant, 1948

Family *Philodromidae*

Apollophanes punctatus (Bryant, 1948)

Family *Thomisidae*

*†*Heterotmarus altus* Wunderlich, 1988

Isaloides toussainti Banks, 1903
*†*Komisumena rosae* Ono, 1981
Misumenops asperatus (Hentz, 1847)
Misumenops bellulus (Banks, 1896)
Misumenops californicus (Banks, 1896)
Misumenops celer (Hentz, 1847)
Rejanellus pallescens (Bryant, 1940)
Rejanellus venustus (Bryant, 1948)
Tobias taczanowskii Roewer, 1951

Family *Salticidae*

Agobardus anormalis Keyserling, 1885
Agobardus anormalis montanus Bryant, 1943
Agobardus brevitarsus Bryant, 1943
Agobardus obscurus Bryant, 1943
Agobardus perpilosus Bryant, 1943
Anasaitis morgani (Peckham and Peckham, 1901)
Antillattus gracilis Bryant, 1943
Antillattus placidus Bryant, 1943
Bythocrotus cephalotes (Simon, 1888)
Cobanus cambridgei (Bryant, 1943)
Commoris modesta Bryant, 1943
Compsodecta haytiensis (Banks, 1903)
Compsodecta peckhami Bryant, 1943
Corythalia banksi Roewer, 1951
Corythalia elegantissima (Simon, 1888)
Corythalia locuples (Simon, 1888)
†*Corythalia ocululiter* Wunderlich, 1988
†*Corythalia pilosa* Wunderlich, 1982
†*Corythalia scissa* Wunderlich, 1988
*†*Descangeles pygmaeus* Wunderlich, 1988
Descanso formosus Bryant, 1943
Descanso magnus Bryant, 1943
Descanso montanus Bryant, 1943
Dinattus erebus Bryant, 1943
Dinattus heros Bryant, 1943
Dinattus minor Bryant, 1943
Eris flava (Peckham and Peckham, 1888)
Habronattus brunneus (Peckham and Peckham, 1901)
Habronattus paratus (Peckham and Peckham, 1896)
Hasarius adansoni (Audouin, 1826)
Hentzia antillana Bryant, 1940
Hentzia mandibularis (Bryant, 1943)
Hentzia vittata (Keyserling, 1885)
Hentzia zombia Richman, 1989
Icius separatus Banks, 1903

Jollas armatus (Bryant, 1943)
Jollas crassus (Bryant, 1943)
Lyssomanes antillanus Peckham and Wheeler, 1889
Lyssomanes perplexa Peckham and Peckham, 1901
†*Lyssomanes pristinus* Wunderlich, 1986
†*Lyssomanes pulcher* Wunderlich, 1988
Lyssomanes viridis (Walckenaer, 1837)?
Maeotella perplexa (Peckham and Peckham, 1901)
Menemerus bivittatus (Dufour, 1831)
Metacyrba pictipes Banks, 1903
Metacyrba taeniola (Hentz, 1846)
Myrmarachne parallela (Fabricius, 1798)
Nebridia manni Bryant, 1943
Nebridia mendica Bryant, 1943
Paraphidippus aurantius (Lucas, 1833)
Parathiodina compta Bryant, 1943
Pelegrina proxima (Peckham and Peckham, 1901)
Pensacola darlingtoni Bryant, 1943
Pensacola electa Bryant, 1943
Pensacola maxillosa Bryant, 1943
Pensacola montana Bryant, 1943
Pensacola peckhami Bryant, 1943
*†*Pensacolatus coxalis* Wunderlich, 1988
*†*Pensacolatus spinipes* Wunderlich, 1988
*†*Pensacolatus tibialis* Wunderlich, 2004
Phidippus regius C.L. Koch, 1846
*†*Phlegrata pala* Wunderlich, 1988

Acknowledgements

I thank F. Columbus (Nova Science Publishers) for the invitation to submit to this volume, D. E. Pérez-Gelabert (Smithsonian Institution, Washington DC), M. Solórzano Kraemer (Institut für Paläontologie, Bonn) for information, N. Platnick (American Museum of Natural History, New York), H. Levi (Museum of Comparative Zoology, Harvard) and G. Hormiga (George Washington University, Washington DC) for personal communications.

References

Alayón-García, G. (1992). La familia Selenopidae (Arachnida, Araneae) en República Dominicana. *Poeyana*, 419, 1–10.

Alayón-García, G. (1995). Nuevo género de Agelenidae (Arachnida, Araneae) de República Dominicana. *Poeyana*, 450, 1–8.

Alayón-García, G. (2000). Las arañas endémicas de Cuba (Arachnida, Araneae). *Rev. Ibér. Aracnol.*, 2, 1–48.

Alayón-García, G. (2001). Lista preliminar de las arañas (Araneae) de la isla de Navassa. *Cocuyo*, 10, 18–22.

Alayón-García, G. (2002). Nueva espècie de *Odo* Keyserling (Araneae, Zoridae) de República Dominicana. *Rev. Ibérica Aracnol.*, 5, 29–32.

Alayón-García, G. (2004). Notas sobre la familia Ctenidae (Arachnida, Araneae) en la Hispaniola, con la descripción de tres nuevas especies. *Rev. Ibér. Aracnol.*, 9, 277–283.

Alayón-García, G., Armas, L. F. de and Antún, A. J. A. (2001). Presencia de *Cyrtophora citricola* Forskål, 1775) (Araneae, Araneidae) en las Antillas. *Rev. Ibér. Aracnol.*, 4, 9–10.

Archer, A. F. (1958). Studies in the orb-weaving spiders. 4. Amer. Mus. Novit., 1922, 1–21.

Banks, N. (1903). A list of the Arachnida from Haiti, with descriptions of new species. *Proc. Acad. Nat. Sci. Philadelphia,* 55, 340–345.

Baroni-Urbani, C. and Saunders, J. B. (1980). The fauna of the Dominican Republic amber, the present status of knowledge. In *Proceedings of the Ninth Caribbean Geological Conference* (Santo Domingo; August 1980) (pp. 213–223).

Bell, J. R., Bohan, D. A., Shaw, E. M. and Weyman, G. S. (2005). Ballooning dispersal using silk, world fauna, phylogenies, genetics and models. *Bull. Entomol. Res.*, 95, 69–114.

Berman, J. D. and Levi, H. W. (1971). The orb–weaver genus *Neoscona* (Araneidae). *Bull. Mus. Comp. Zool.*, 141, 465–500.

Boucot, A. J. (1989) *Evolutionary palaeobiology of behaviour and coevolution.* Amsterdam: Elsevier.

Brescovit, A. D. (1993). *Thaloe* and *Bromelina*, novos géneros de aranhas neotropicais da la familia Anyphaenidae (Arachnida, Araneae). *Rev. Bras. Entomol.*, 37, 693–703.

Brescovit, A. D. (1999). Revisao das aranhas do género *Lupettiana* Brescovit (Araneae, Anyphaenidae, Anyphaeninae). *Rev. Bras. Zool.*, 16 (Supl. 2), 63–76.

Briggs, J. C. (1991). A Cretaceous–Tertiary mass extinction? Were most of Earth's species killed off? *BioScience,* 41, 619–624.

Bryant, E. B. (1943). The salticid spiders of Hispaniola. *Bull. Mus. Comp. Zool.*, 92, 445–522.

Bryant, E. B. (1945). The Argiopidae of Hispaniola. *Bull. Mus. Comp. Zool.*, 95, 357–418.

Bryant, E. B. (1948). The spiders of Hispaniola. *Bull. Mus. Comp. Zool.*, 100, 332–447.

Chickering, A. M. (1967). The genus *Nops* (Araneae, Caponiidae) in Panama and the West Indies. *Breviora*, 274, 1–19.

Coyle, F. A. (1995). A revision of the funnelweb mygalomorph spider subfamily Ischnothelinae (Araneae, Dipluridae). *Bull. Amer. Mus. Nat. Hist.*, 226, 1–133.

Coyle, F. A. and Meigs, T. E. (1989). Two new species of kleptoparasitic *Mysmenopsis* (Araneae, Mysmenidae) from Jamaica. *J. Arachnol.*, 17, 59–70.

Coyle, F. A., O'Shields, T. C. and Perlmutter, D. G. (1991). Observations on the behaviour of the kleptoparasitic spider, *Mysmenopsis furtiva* (Araneae, Mysmenidae). *J. Arachnol.*, 19, 62–66.

Crame, J. A. and Rosen, B. R. (2002). Cenozoic palaeogeography and the rise of modern biodiversity patterns. In J. A. Crame and A. W. Owen (Eds.), *Palaeobiogeography and Biodiversity Change, the Ordovician and Mesozoic–Cenozoic Radiations* (Geological Society Special Publication No. 194, pp. 153–168). London: The Geological Society.

Curtis, J. H., Brenner, M. and Hodell, D. A. (2001). Climate change in the circum-Caribbean (Late Pleistocene to present) and implications for regional biogeography. In C. A. Woods and F. E. Sergile (Eds.). *Biogeography of the West Indies: patterns and perspectives* (second edition, pp. 35–54). Boca Raton: CRC Press.

Cutler, B. (1984). *Late Oligocene amber salticids from the Dominican Republic.* Peckhamia, 2, 45–46.

Dondale, C. D. and Redner, J. H. (1984). Revision of the *milvina* group of the wolf spider genus *Pardosa* (Araneae, Lycosidae). *Psyche*, 91, 67–117.

Edwards, G. B. and Wolff, R. J. (1995). A list of the jumping spiders (Salticidae) of the islands of the Caribbean region. *Peckhamia*, 3, 27–60.

Eskov, K. Y. (1987). A new archaeid spider (Chelicerata, Araneae) from the Jurassic of Kazakhstan, with notes on the so-called "Gondwanan" ranges of Recent taxa. *N. Jb. Geol. Paläont., Abh.*, 175, 81–106.

Eskov, K. Y. (1990). Spider palaeontology, present trends and future expectations. *Acta Zool. Fennica*, 190, 123–127.

Eskov, K. Y. (1992). Archaeid spiders from Eocene Baltic amber (Chelicerata, Araneida, Archaeidae) with remarks on the so-called "Gondwanan" ranges of Recent taxa. *N. Jb. Geol. Paläont., Abh.*, 185, 311–328.

Eskov, K. Y. and Golovatch, S. I., (1986). On the origin of trans-Pacific disjunctions. *Zool. Jb. Syst.*, 113, 265–285.

Exline, H. and Levi, H. W. (1962). American spiders of the genus *Argyrodes* (Araneae, Theridiidae). *Bull. Mus. Comp. Zool.*, 127, 75–204.

Flint, O. S. Jr. and Pérez-Gelabert, D. E. (1999). Checklist of the Caddisflies (Trichoptera) of Hispaniola. *Nov. Carib.*, 1, 33–46.

García-Villafuerte, M. A. and Vera, V. F. (2002). Arañas fosiles incluidas en ambar del Oligoceno-Mioceno de Simojovel, Chiapas, México. In Entomología Mexicana 2002 Vol. 1. 50 aniversario de la Sociedad Mexicana de Entomología (pp. 20–25). Guanajuato, Gto, México: E.G.V. a. A.E.M. e. J.R. Napoles.

García-Villafuerte, M. A. and Penney, D. (2003). *Lyssomanes* (Araneae, Salticidae) in Oligocene–Miocene Chiapas amber. *J. Arachnol.*, 33, 400–404.

Genaro, J. A. and Tejuca, A. E. (2001). Patterns of endemism and biogeography of Cuban insects. In C. A. Woods and F. E. Sergile (Eds.). *Biogeography of the West Indies: patterns and perspectives* (second edition, pp. 77–83). Boca Raton: CRC Press.

Gertsch, W. J. (1958). The spider genus *Loxosceles* in North America, Central America, and the West Indies. *Amer. Mus. Novit.*, 1907, 1–46.

Gertsch, W. J. and Ennik, F. (1983). The spider genus *Loxosceles* in North America, Central America and the West Indies (Araneae, Loxoscelidae). *Bull. Amer. Mus. Nat. Hist.*, 175, 264–360.

Gillespie, R. G., Rivera, M. and Garb, J. (1998). Sun, surf and spiders, taxonomy and phylogeography of Hawaiian Araneae. In P. A. Selden (Ed.) *Proceedings of the 17th European Colloquium of Arachnology*, Edinburgh 1997 (pp. 41–51). Burham Beeches: British Arachnological Society.

Grimaldi, D. A. (1992). Vicariance biogeography, geographic extinctions, and the North American Oligocene tsetse flies. In M. J. Novacek and Q. D. Wheeler (Eds.) *Extinction and phylogeny* (pp. 178–204). New York: Columbia University Press.

Grimaldi, D. and Engel, M. S. (2005). *Evolution of the insects.* New York: Cambridge University Press.

Gutiérrez, E. and Pérez-Gelabert, D. E. (2000). Annotated checklist of Hispaniolan cockroaches. *Trans. Amer. Entomol. Soc.,* 126, 423–445.

Hall, J. P. W., Robbins, R. K. and Harvey, D. J. (2004). Extinction and biogeography in the Caribbean, new evidence from a fossil riodinid butterfly in Dominican amber. *Proc. R. Soc. Lond. B,* 271, 789–801.

Hedges, S. B. (2001). Biogeography of the West Indies: an overview. In C. A. Woods and F. E. Sergile (Eds.). *Biogeography of the West Indies: patterns and perspectives* (second edition, pp. 15–33). Boca Raton: CRC Press.

Henwood, A. (1993). *Ecology and taphonomy of Dominican Republic amber and its inclusions. Lethaia.* 26, 237–245.

Huber, B. A. (2000). New World pholcid spiders (Araneae, Pholcidae), A revision at generic level. *Bull. Amer. Mus. Nat. Hist.,* 254, 1–348.

Hueber, F. M. and Langenheim, J. H. (1986). Dominican amber tree had African ancestors. *Geotimes*, 31, 8–10.

Iturralde-Vinent, M. A. (2001). Geology of the amber-bearing deposits of the Greater Antilles. *Carib. J. Sci.,* 37, 141–167.

Iturralde-Vinent, M. A. and Lidiak, E. G. (Eds.) (2006). Caribbean Plate Tectonics: Stratigraphic, Magmatic, Metamorphic and Tectonic Events (UNESCO/IUGS IGCP Project 433). *Geol. Acta,* 4 (1–2), 1–341.

Iturralde-Vinent, M. A. and MacPhee, R. D. E. (1996). Age and palaeogeographical origin of Dominican amber. *Science*, 273, 1850–1852.

Iturralde-Vinent, M. A. and MacPhee, R. D. E. (1999). Paleogeography of the Caribbean region, implications for Cenozoic biogeography. Bull. Amer. Mus. Nat. Hist., 238, 1–95.

Kraus, O. (1978). Zoogeography and plate tectonics, introduction to a general discussion. *Abh. Vehr. Naturwiss. Ver. Hamburg,* 21/22, 33–41.

Labandeira, C. C. (2005). The fossil record of insect extinction, new approaches and future directions. *Amer. Entomol.* 51, 14–29.

Langenheim, J. H. (1995). Biology of amber-producing trees, focus on case studies of *Hymenaea* and *Agathis*. *Proc. ACS Symp. Ser.,* 617, 1–31.

Levi, H. W. (1955). The spider genera *Episinus* and *Spintharus* from North America, Central America and the West Indies (Araneae, Theridiidae). *J. New York Entomol. Soc.,* 62, 65–90.

Levi, H. W. (1957). The spider genera *Chrysso* and *Tidarren* in America (Araneae, Theridiidae). *J. New York Entomol. Soc.*, 63, 59–81.

Levi, H. W. (1959). The spider genera *Achaearanea, Theridion* and *Sphyrotinus* from Mexico, Central America, and the West Indies (Araneae, Theridiidae). *Bull. Mus. Comp. Zool.,* 121, 55–163.

Levi, H. W. (1962). More American spiders of the genus *Chrysso* (Araneae, Theridiidae). *Psyche*, 69, 209–237.

Levi, H. W. (1963a). The American spider genera *Spintharus* and *Thwaitesia*. *Psyche*, 70, 223–234.

Levi, H. W. (1963b). The American spiders of the genus *Theridion* (Araneae, Theridiidae). *Bull. Mus. Comp. Zool.,* 129, 483–592.

Levi, H. W. (1964). The spider genus *Thymoites* in America (Araneae, Theridiidae). *Bull. Mus. Comp. Zool.,* 130, 445–471.

Levi, H. W. (1977). The American orb-weaver genera *Cyclosa*, *Metazygia* and *Eustala* North of Mexico (Araneae, Araneidae). *Bull. Mus. Comp. Zool.,* 148, 61–127.

Levi, H. W. (1986). The Neotropical orb-weaver genera *Chrysometa* and *Homalometa* (Araneae, Tetragnathidae). *Bull. Mus. Comp. Zool.,* 151, 91–215.

Levi, H. W. (1991). The Neotropical orb-weaver genera *Edricus* and *Wagneriana* (Araneae, Araneidae). *Bull. Mus. Comp. Zool.,* 152, 363–415.

Levi, H. W. (1993). American *Neoscona* and corrections to previous revisions of Neotropical orb-weavers (Araneae, Araneidae). *Psyche*, 99, 221–239.

Levi, H. W. (1995). The Neotropical orb-weaver genus *Metazygia* (Araneae, Araneidae). *Bull. Mus. Comp. Zool.,* 154, 63–151.

Levi, H. W. (1999). The Neotropical and Mexican orb weavers of the genera *Cyclosa* and *Allocyclosa* (Araneae, Araneidae). *Bull. Mus. Comp. Zool.,* 155, 299–379.

Levi, H. W. (2001). The orb weavers of the genera *Molinaranea* and *Nicolepeira*, a new species of *Parawixia*, and comments on orb weavers of temperate South America (Araneae, Araneidae). *Bull. Mus. Comp. Zool.,* 155, 445–475.

MacPhee, R. D. E and Iturralde-Vinent, M. A. (2000). A short history of Greater Antillean land mammals: biogeography, paleogeography, radiations, and extinctions. *Tropics*, 10, 145–154.

MacPhee, R. D. E and Iturralde-Vinent, M. A. (2005). The interpretation of Caribbean palaeogeography, reply to Hedges. In J. A. Alcover and P. Bover (Eds.) Proceedings of the international symposium "insular vertebrate evolution, the palaeontological approach". Monograf. *Soc. d'Hist. Nat. Balears,* 12, 175–184.

Maddison, W. (1986). Distinguishing the jumping spiders *Eris militaris* and *Eris flava* in North America (Araneae, Salticidae). *Psyche*, 93, 141–149.

Martínez-Delclòs, X., Briggs, D. E. G., Peñalver, E. (2004). Taphonomy of insects in carbonates and amber. Palaeogeog. Palaeoclimatol. *Palaeoecol.*, 203, 19–64.

Morrone, J. J. and Crisci, J. V. (1995). Historical biogeography, introduction to methods. *Ann. Rev. Ecol. Syst.*, 26, 371–401.

Muma, M. H. (1953). A study of the spider family Selenopidae in North America, Central America, and the West Indies. *Amer. Mus. Novit.,* 1629, 1–55.

Nagel, P. (1997). New fossil paussids from Dominican amber with notes on the phylogenetic systematics of the paussine complex (Coleoptera, Carabidae). *Syst. Entomol.,* 22, 345–362.

Ogden, R., and Thorpe, R. S. (2002). Molecular evidence for ecological speciation in tropical habitats. *Proc. Nat. Acad. Sci.,* USA, 99, 13612–13615.

Ono, H. (1981). First record of a crab spider (Thomisidae) from Dominican amber (Amber Collection Stuttgart, Arachnida, Araneae). *Stuttgarter Beitr. Naturk.* (B), 73, 1–13.

Opell, B. D. (1979). Revision of the genera and tropical American species of the spider family Uloboridae. *Bull. Mus. Comp. Zool.,* 148, 443–549.

Penney, D. (1999). Hypotheses for the Recent Hispaniolan spider fauna based on the Dominican Republic amber spider fauna. *J. Arachnol.,* 27, 64–70.

Penney, D. (2000a). Miocene spiders in Dominican amber (Oonopidae, Mysmenidae). *Palaeontology,* 43, 343–357.

Penney, D. (2000b). Anyphaenidae in Miocene Dominican Republic amber (Arthropoda, Araneae). *J. Arachnol.*, 28, 223–226.

Penney, D. (2001). Advances in the taxonomy of spiders in Miocene amber from the Dominican Republic (Arthropoda, Araneae). *Palaeontology*, 44, 987–1009.

Penney, D. (2002). Paleoecology of Dominican amber preservation—spider (Araneae) inclusions demonstrate a bias for active, trunk–dwelling faunas. *Paleobiology*, 28, 389–398.

Penney, D. (2003). *Afrarchaea grimaldii*, a new species of Archaeidae (Araneae) in Cretaceous Burmese amber. *J. Arachnol.*, 31, 122–130.

Penney, D. (2004a). New extant and fossil Dominican Republic spider records, with two new synonymies and comments on taphonomic bias of amber preservation. *Rev. Ibér. Aracnol.*, 9, 183–190.

Penney, D. (2004b). Does the fossil record of spiders track that of their principal prey, the insects? Trans. *Royal Soc. Edinb., Earth Sci.*, 94, 275–281.

Penney, D. (2005a). Fossil blood droplets in Miocene Dominican amber yield clues to speed and direction of resin secretion. *Palaeontology*, 48, 925–927.

Penney, D. (2005b). Importance of Dominican Republic Amber for determining taxonomic bias of fossil resin preservation—A case study of spiders. *Palaeogeog. Palaeoclimatol. Palaeoecol.*, 223, 1–8.

Penney, D. (2005c). First Caribbean *Floricomus* (Araneae, Linyphiidae), a new species in Miocene Dominican Republic amber. A new synonymy for the extant American fauna. *Geol. Acta*, 3, 59–64.

Penney, D. (2005d). First fossil Filistatidae, A new species of *Misionella* in Miocene amber from the Dominican Republic. *J. Arachnol.*, 33, 93–100.

Penney, D. (2006a, for 2005). An annotated systematic catalogue, including synonymics and transfers, of Miocene Dominican Republic amber spiders described up until 2005. *Rev. Ibér. Aracnol.*, 12, in press.

Penney, D. (2006b) A new fossil oonopid spider, in lowermost Eocene amber from the Paris Basin, with comments on the fossil spider assemblage. *Afr. Invert.*, 47, in press.

Penney, D. (in press). Tertiary Neotropical Hersiliidae (Arthropoda, Araneae) with new combinations for the extant fauna and comments on historical biogeography of the family. *Palaeontology*.

Penney, D. and Langan, A. M. L. (2006). Comparing amber fossils across the Cenozoic. *Biol. Lett.*, doi:10.1098/rsbl.2006.0442

Penney, D. and Pérez-Gelabert, D. E. (2002). Comparison of the Recent and Miocene Hispaniolan spider faunas. *Rev. Ibér. Aracnol.*, 6, 203–223.

Penney, D. and Selden, P. A. (2006). First fossil Huttoniidae (Araneae), in Late Cretaceous Canadian Cedar and Grassy Lake ambers. *Cret. Res.*, 27, 442-446.

Penney, D., Wheater, C. P. and Selden, P. A. (2003). Resistance of spiders to Cretaceous–Tertiary extinction events. *Evolution*, 57, 2599–2607.

Pérez-Asso, A. R. and Pérez-Gelabert, D. E. (2001). Checklist of the millipedes (Diplopoda) of Hispaniola. *Bol. S.E.A.*, 28, 67–80.

Pérez-Gelabert, D. E. (1999). Catálogo systemático y bibliografía de la biota fósil en ámbar de la República Dominicana. *Hispaniolana*, 1, 1–65.

Pérez-Gelabert, D. E. (2001). Preliminary checklist of the Orthoptera (Saltatoria) of Hispaniola. *J. Orthoptera Res.*, 10, 63–74.

Pérez-Gelabert, D. E. and Flint, O. S. Jr. (2000). Annotated list of the Neuroptera of Hispaniola, with new faunistic records of some species. *J. Neuropterol.*, 3, 9–23.

Petrunkevitch, A. (1928). The Antillean spider fauna—a study in geographic isolation. *Science*, 68, 650.

Petrunkevitch, A. (1963). Chiapas amber spiders. *Uni. Calif. Pubs. Entomol.*, 31, 1–40.

Petrunkevitch, A. (1971). Chiapas amber spiders, 2. *Uni. Calif. Pubs. Entomol.*, 63, 1–44.

Piel, W. H. (2001). The systematics of Neotropical orb-weaving spiders in the genus *Metepeira* (Araneae, Araneidae). *Bull. Mus. Comp. Zool.*, 157, 1–92.

Platnick, N. I. (2006). The world spider catalogue (version 6.5). Online at, http://research.amnh.org/entomology/spiders/catalog/INTRO2.html.

Platnick, N. I. and Nelson, G. (1978). A method of analysis for historical biogeography. *Syst. Zool.*, 27, 1–16.

Platnick, N. I. and Penney, D. (2004). A revision of the widespread spider genus *Zimiris* (Araneae, Prodidomidae). *Amer. Mus. Novitat.*, 3450, 1–12.

Platnick, N. I. and Shadab, M. U. (1974a). A revision of the *bispinosus* and *bicolor* groups of the spider genus *Trachelas* in North and Central America and the West Indies. *Amer. Mus. Novit.*, 2560, 1–34.

Platnick, N. I. and Shadab, M. U. (1974b). A revision of the *tranquillus* and *speciosus* groups of the spider genus *Trachelas* in North and Central America and the West Indies. *Amer. Mus. Novit.*, 2553, 1–34.

Platnick, N. I. and Shadab, M. U. (1978). A review of the spider genus *Mysmenopsis* (Araneae, Mysmenidae). *Amer. Mus. Novit.*, 2661, 1–22.

Platnick, N. I. and Shadab, M. U. (1980). A revision of the spider genus *Cesonia* (Araneae, Gnaphosidae). *Bull. Amer. Mus. Nat. Hist.*, 165, 335–386.

Platnick, N. I. and Shadab, M. U. (1989). A review of the spider genus *Teminius* (Araneae, Miturgidae). *Amer. Mus. Novit.*, 2963, 1–12.

Poinar, G. O. Jr. (1991). *Hymenaea protera* sp.n. (Leguminosae, Caesalpinioideae) from Dominican amber has African affinities. *Experientia,* 47,1075–1082.

Poinar, G. O. Jr. (1992). Life in amber. California: Stanford University Press.

Poinar, G. O. Jr. and Hess, R. (1982). Ultrastructure of 40-million-year-old insect tissue. *Science*, 215, 1241–1242.

Poinar, G. O. Jr. (1999). Extinction of tropical insect lineages in Dominican amber from Plio-Pleistocene cooling events. *Russian Entomol. J.*, 8, 1–4.

Poinar, G. O. Jr. and Poinar, R. (1999). *The amber forest: a reconstruction of a vanished world.* New Jersey: Princeton University Press.

Raven, R. J. (2000). Taxonomica Araneae I, Barychelidae, Theraphosidae, Nemesiidae and Dipluridae (Araneae). *Mem. Qld Mus.*, 45, 569–575.

Reiskind, J. (1989). The potential of amber fossils in the study of the biogeography of spiders in the Caribbean with the description of a new species of *Lyssomanes* from Dominican amber (Araneae, Salticidae). In C. A. Woods (Ed.) *Biogeography of the West Indies: past, present and future* (pp. 217–227). Gainseville: Sandhill Crane Press.

Rheims, C. A. and Brescovit, A. D. (2004). Revision and cladistic analysis of the spider family Hersiliidae (Arachnida, Araneae) with emphasis on Neotropical and Nearctic species. *Insect Syst. Evol.*, 35, 189–239.

Richman, D. B. (1989). A revision of the genus *Hentzia* (Araneae, Salticidae). *J. Arachnol.*, 17, 285–344.

Robertson, D. S., McKenna, M. C., Toon, O. B., Hope, S., and Lillegraven, J. A. (2004). Survival in the first hours of the Cenozoic. *GSA Bull.,* 116, 760–768.

Ross, M. I. and Scotese, C. R. (1988). A hierarchical tectonic model of the Gulf of Mexico and Caribbean region. *Tectonophysics*, 155, 139–168.

Sánchez-Ruíz, A. (2005). Una nueva especie de *Nops* MacLeay, 1839 (Araneae, Caponiidae) de República Dominicana. *Rev. Ibér. Aracnol.,* 11, 23–27.

Sanderson, M. W. and Farr, T. H. (1960). Amber with insect and plant inclusions from the Dominican Republic. *Science,* 131, 1313.

Santos, A. J. and Gonzaga, M. O. (2003). On the spider genus *Oecobius* Lucas, 1846 in South America (Araneae, Oecobiidae). *J. Nat. Hist.,* 37, 239–252.

Schawaller, W. (1981a). The spider family Hersiliidae in Dominican amber (Amber Collection Stuttgart, Arachnida, Araneae). *Stuttgarter Beitr. Naturk.* (B), 79, 1–10.

Schawaller, W. (1981b). Survey of the spider–families in Dominican amber and other Tertiary resins (Amber Collection Stuttgart, Arachnidae, Araneae). *Stuttgarter Beitr. Naturk.* (B), 77, 1–10.

Schawaller, W. (1982). Spinnen der Familien Tetragnathidae, Uloboridae und Dipluridae in Dominikanischem Bernstein und allgemeine Gesichtspunkte (Arachnida, Araneae). Stuttgarter Beitr. *Naturk.* (B), 89, 1–19.

Schawaller, W. (1984). The family Selenopidae in Dominican amber (Arachnida, Araneae). Stuttgarter Beitr. *Naturk.* (B), 103, 1–8.

Schubert, C. (1988). Climatic changes during the last glacial maximum in northern South America and the Caribbean, a review. *Interciencia*, 13, 128–137.

Smith, A. M. (1986). *The tarantula classification and identification guide.* London: Fitzgerald Publishing.

Wilson, E. O. (1985). Invasion and extinction in the West Indian ant fauna, evidence from Dominican amber. *Science*, 229, 265–267.

Wolff, R. J. (1990). A new species of *Thiodina* (Araneae, Salticidae) from Dominican amber. *Acta Zool. Fennica,* 190, 405–408.

Woods, C. A. and Sergile, F. E. (Eds.) (2001). Biogeography of the West Indies: patterns and perspectives (second edition). *Boca Raton:* CRC Press.

Wunderlich, J. (1981). Fossile Zwergsechsaugenspinnen (Oonopidae) der Gattung *Orchestina* Simon, 1882 in Bernstein mit Ammerkungen zur Sexual-biologie (Arachnidae, Araneae). *Mitt. Geol.-Paläont. Inst. Univ. Hamburg,* 51, 83–113.

Wunderlich, J. (1982). Die häufigsten Spinnen (Araneae) des Dominikanischen Bernsteins. *Neue Entomol. Nachr.,* 1, 26–45.

Wunderlich, J. (1986). Spinnenfauna Gestern und Heute, 1: Fossile Spinnen in Bernstein und Ihre Heute Lebenden Verwandten. Wiesbaden: Erich Bauer Verlag bei Quelle and Meyer.

Wunderlich, J. (1987). *Tama minor*, n. sp., eine fossile Spinnerart der Familie Hersiliidae in Dominikanischem Bernstein (Arachnida, Araneae). *Entomol.* Z., 97, 93–96.

Wunderlich, J. (1988). Die Fossilen Spinnen im Dominikanischem Bernstein. *Beitr. Araneol.,* 2, 1–378.

Wunderlich, J. (1998). Beschreibung der ersten fossilen Spinnen der Unterfamilien Mysmeninae (Anapidae) und Erigoninae (Linyphiidae) im Dominikanischen Bernstein (Arachnida, Araneae). *Entomol. Z.,* 108, 363–367.

Wunderlich, J. (1999). Two subfamilies of spiders (Araneae, Linyphiidae, Erigoninae and Anapidae, Mysmeninae) new to Dominican amber – or falsificated amber? *Est. Mus. Cienc. Nat. Àlava,* 14 (Num. Espec. 2), 167–172.

Wunderlich, J. (Ed.) (2004). Fossil spiders in amber and copal. *Beitr. Araneol.,* 3ab, 1–1908.

In: Focus on Biodiversity Research
Editor: Jan Schwartz, pp. 101-114

ISBN: 1-60021-372-3

Chapter 5

ON THE STRUCTURE OF BIODIVERSITY: SOME METAPHYSICAL ESSAYS

Igor Ya. Pavlinov*

Zoological Museum of Moscow Lomonosov State University, Moscow, Russia

ABSTRACT

Under consideration are some metaphysical backgrounds of the concept of biodiversity (BD) as an immanent property of the evolving biota. Such a consideration makes it possible to deduce an idea of hierarchy as a principal property of BD, as well as to analyze the latter's structure in a general form.

It is stated that the recent "species-oriented" BD concept, though operational, is outdated, as it does not correspond to the current predominating taxonomic concept. Representation of taxic diversity (as a component of overall BD) in the form of hierarchical phylogenetic pattern seems to be more adequate to the recent developments in taxonomy. Its operational interpretation implies that different species could be "weighted" in respect to their position in the taxonomic hierarchy. This presumes a certain system of fixed taxonomic ranks and does not enjoy an idea of "rankless classifications" declared by neo-cladists.

As far as the species notion is concerned, the major challenge of recent taxonomic theory to the BD researches includes pluralistic treatment of the species which put certain limitations on counting and direct comparison of number of species in different groups of organisms. In this respect, the so called "Hull's principle" is reconsidered to stress that various species definitions are not just mutually exclusive but complementary. Some consideration is given to the focal point of the species problem: if any ontological model exists which might ensure there is *the* species as a universal fundamental unit of the structure of the entire biosphere.

It is stressed that one more component of overall BD structure, in addition to ecological and taxic diversity, could be recognized: it is the diversity of biomorphs (life

* e-mail: igor_pavlinov@zmmu.msu.ru

forms). Some questions of their definition and classification are considered, and ideas of urgent need for a special classificatory discipline (biomorphics) dealing especially with that diversity is declared.

INTRODUCTION

Biology has been long dealing with two natural phenomena – with the diversity of structures and with the structure of diversity. The last third of the 20th century gave birth to quite an important idea, according to which it is this diversity which constitutes one of the principal results of biological evolution. One can anticipate that understanding of the structure and dynamics of the diversity of the living beings will become one of the central themes of biology of the 21st century. That is why it is hardly accidental that a natural history science emerged recently for studying biological diversity (BD) that was offered to entitle dyatropics (Chaikovsky, 1990).

Activities related to BD has won a great stimuli of two kinds during the last several decades. From one side, the so called "Rio Declaration" of 1992 has made BD one of the key notions in the modern nature conservation strategies, thus turning it out to be one of the most important research fields in biology. From another side, a new world picture having been developed by synergetics that, unlike classic paradigm, accentuates attention not on laws of nature as such but on the diversity of the world (Barantsev, 2003), including diversity of organisms, thus stimulating exploration of the diversity proper.

It so occurs that the above two sources of rise of interest to the BD matter are in mutual conflict, in a sense. Including concept of BD in the world picture elaborated by synergetics makes it most important to understand its essential properties, which belongs to the scope of metaphysics. Conservation biology, contrary to this, concentrates itself on practical aspects of applications of that concept appealing to its operationalizing and thus leaving no room for the metaphysics (Sarkar, Margules, 2001; Sarkar, 2002). The latter is aggravated by participation of managing officers in elaboration and, especially, application of bioconservation strategies which are much far from the metaphysical problems.

It is clearly enough that concepts and notions of modern, basically rational, science are to be operational which is prerequisite of their being applicable. However, it seems hardly correct to reduce all the problematics of the BD concept to its operationalization. As matter of fact, it is needed first to define what are we dealing with before to try to shroud it into any operational form. In other words, one has to answer first the question "what is it?" before to search an answer to the question "how to deal with it?" For this, it is necessary to be able to put the first question as much correctly as possible.

The present paper considers, quite briefly and hence (unfortunately) quite superficially, some questions of metaphysics of biodiversity: what and why is it, how and why is it structured, what and why are its basic elements, etc. Each question deserves special attention and detailed consideration at much more length than it is done here. So, my task (and hope) is just to grasp possible correct forms of putting those questions about biodiversity rather than to answer them.

What does Biodiversity Mean: Or, A Property of What Is It?

The above "title" question presumes some understanding (at best, definition) of "what is the biodiversity?" Within the positivism tradition still quite influential among conservation biologists (Roebuck, Phifer, 1999), such an understanding is basically "elementaristic". Actually, within this tradition, to understand (to learn) an object under exploration means in a large part an ability to measure it; respectively, two objects are usually considered the "same" in a sense if they are commensurable (on application of this idea to BD, see Sarkar 2002). On the other hand, the "elementarism" exhibits itself in understanding (and definition) of a complex object through its constituent elements which are being measured some or other way. Such an approach applied to BD means its understanding as a set of species, or of populations, or of organisms (Norton, 1986; Heywood, Watson, 1995). It is evident that such kind of understanding is quite operational allowing for an easy assessment of BD at both local and global scales which makes it more than attractive to conservation biologists and managers.

Such a definition (called extensional) is at the minimum insufficient for understanding and expression of the essence of BD. Additional to it is the intensional way of definition (Il'in, 2003) which allows to define an object, in the terms of formal logic, by "genus proximus and species differentia". In less formal words, it means that, in order to define BD, it is requested to set first up some natural phenomenon (a "genus"), a partial subphenomenon (property, aspect etc., that is a "species") of which the BD being defined could be considered (Pavlinov, Rossolimo, 2004).

This principle elaborated by the formal logic is quite consistent with a metaphysical approach adopted by non-classical scientific epistemology. Accordingly to it, BD just as any natural history object could be properly understood and outlined only in the context of a certain non-formal model (theory) which indicates a more general phenomenon to contain BD as its property. From such a viewpoint, it is evident that any "elementaristic" definition of BD is not only insufficient, but it simply is incorrect. To my opinion, there is no more sense in defining BD by reference to species than to define an organism by reference to its cells. As a matter of fact, the species is a "part" of BD and not a more general (inclusive) phenomenon. Instead, a more correct form of putting the question "what?" about BD seems to be the following: *a property of what is it*?

It follows from the above that the first step toward intentional definition of BD would be implication of that natural phenomenon, a property which the BD could be thought about. One of the most general and quite correct (because of tautology) understanding of BD is treating it as "diversity of all forms of life" (Schulze, Mooney, 1994; Faith, 2003). But it seems to be too general and thus not certain enough, as it requests prior understanding of what is life (and who knows that?). Close to it by sense but more "local" and so less ambiguous might be taking the Earth biota as a whole relative to which the BD could be considered. It implies understanding of the BD as a *property of biota* of the Earth (and of any local fragment of it). It is certainly not a definition in any strict sense but it is at least a possible background, or prerequisite for it.

What does such a background give? Before all, it allows the consideration of basic properties of BD through the reference to basic properties of the biota as a whole. In particular, it concerns the structure of BD: what is it, how it comes to existence, what are its principal aspects and elements, etc. There are two principal approaches to consider this subject, structuralistic and functionalistic (Faith, 2003). The first one implies that the BD structure is analyzed as such: but there seems to be a problem of establishing a kind of close argumentation circle, as we have to define a structure by itself. The second approach is free of such a danger, as a structure is considered as a result of a process generating it. Such an approach looks more promising: it presumes analysis of mechanisms structuring the biota (Pavlinov, 2001; Pavlinov, Rossolimo, 2004).

By understanding the causes of structuring of the biota, the background model of the latter as of open non-equilibrium system looks most attractive. From this viewpoint, existence of the biota is not being but becoming, that is, it is a process of irreversible development. It is evident that the latter is an historical development (evolution in modern sense) which, unlike an individual one (evolution of scholasticists), it is not programmed toward any final state. It is this process that causes, at a very general level, structuring the biota thus generating a variety of its elements and states. This structuring involves two flows in the biota (Brooks, Wiley, 1986), one of which orders matter and energy transformations, while another orders information transmissions. The first flow is responsible for ordering the trophic chains, so it structures the biota into ecosystems. The second flow is responsible for the chains of genetic information by phyletic lines, so it structures the biota into monophyletic groups.

One might conclude from the above that, among two general interpretations of BD, functional and compositional ones (Faith, 2003), the first one is to be considered as a primary relative to the second. Indeed, as it follows from what was just stated, it is reasonable to think, more or less roughly, that it is the processes that generate structures and not vice versa. However, such a straightforward causal opposition is rarely found to be true. Actually, biotic diversification processes do run on the background of some basic structure of biogeochemical relations to "build into" it the emerging diversity of organisms to assure proper fulfillment of their ecological functions (Barskov et al., 1996). Such a view of interrelations between structural and functional parameters of BD seems to have certain concern to assessment of the place of biomorphs into overall BD structure (see below).

At any rate, these two processes of structuring the biota allow to designate at first glance quite naturally two principal BD components generally acknowledged, ecological (biocenotical) diversity (ED) and taxonomic (phylogenetic) diversity (TD) (Schulze, Mooney, 1994; Faith, 2003). It is to be stressed that they are not "parts" of the overall BD and they do not exist by themselves: they are simply different manifestations of the same phenomenon. So, they could be characterized as interconnected: organisms belong to each and to all of them simultaneously.

From an epistemological standpoint, such a type of "existence" of ED and TD could be interpreted as their aspect setness. Each of them corresponds to a certain consideration aspect which is fixed up by a researcher on the basis of certain background assumption. It is evident that contents of a recognized BD component depends on the way in which the respective consideration aspect is defined by one or another assumption. So, it means that any fixation of BD components is not a trivial task with a unique solution.

The latter leads unavoidably to pluralism in particular interpretations of BD structure and its components. Such a position fits well into ideas of modern nonclassic scientific epistemology and means impossibility of universal understanding and description (assessment) of BD (Faith, 2003).

The Hierarchy of Biodiversity: Where does it Come from?

The currently dominating BD concept seems to be one-level, as it considers species as its key element. Respectively, above-species hierarchy is generally treated as just an auxiliary means of describing BD structure. This method of consideration of BD was rooted in ideology of the New Systematics dominated in perception of scientists formulated the current concept of BD. According to that ideology, the basic problem of taxonomy is the one of the species (Mayr, 1969).

This taxonomic doctrine is now outdated and has been replaced by the phylogenetic paradigm in which the basic problem of taxonomy is one of hierarchy of monophyletic groups (Eldredge, Cracraft, 1980). It evidentially leads to a hierarchical view of BR, or at least of its taxonomic component (TD) (Faith, 1992, 1994).

From such a standpoint rises an important question of ontological substantiation of taxonomic hierarchy. The matter is that the latter's objective status is, generally speaking, not evident. As a matter of fact, that hierarchy has appeared in taxonomy initially as an application of so called genus–species classificatory algorithm elaborated by scholasticism (Panchen, 1992; Pavlinov, 1992). That algorithm is a formal logical one applicable to any kind of objects being classified; that is, it has not initially presumed any specific ontology of their hierarchy. Instead, it is the hierarchical algorithm which leads to hierarchical classification of any kind of diversity, even if it is arranged non-hierarchically. So, the basic question is: where does the TD hierarchy come from – from its own nature or from an algorithm we describe.

In recent phylogenetics, ontological substantiation of taxonomic hierarchy is given by reference to the phylogenesis (Eldredge, Cracraft, 1980; Pavlinov, 2005). The latter is usually defined as a process of historical development of groups of organisms producing their inclusive hierarchy. So, a taxonomist does not "create" an hierarchy by his/her own will but tries to discover it in Nature.

In order to provide that substantiation with a more general background, let us go back to treating the biota as an open non-equilibrium developing system. Development of such a system structures it, and while a certain level of complexity is achieved, its diversity tends to become hierarchically structured (Chernavsky, 2004). Phylogenesis is one of the aspects of historical development of the biota, and hierarchically arranged array of phylogenetic groups is one of the aspects of the structure of the biota. One may conclude (or assume, or suppose) that both phylogenesis and phylogenetic pattern are immanent to the developing biota. So, not just a set of species but the entire hierarchy of monophyletic groups resulted from the phylogenesis constitute a basic property of TD and hence of BD in general.

Such an understanding of the hierarchy of BD (at least of its taxonomic component) allows for the to consideration of a problem of the "Linnaean hierarchy" recently debated actively by taxonomists. It is the problem of existence and number of fixed taxonomic ranks (for review, see Ereshefsky, 2001). These ranks have been initially borrowed from the abovementioned genus–species logical scheme; their number was increasing steadily with the time. Some taxonomists believe there are "principal" ("Linnaean") and "auxiliary" ("post-Linnaean") ranks and even try to substantiate them "experimentally" (e.g. Mednikov, 1974). According to others, there are no fixed ranks of biological taxa in the nature, so the entire classification is to be "rankless" (Ereshefsky, 2001). Unfortunately, the latter term is misleading, as it actually means abolishing not the ranks proper but just the fixed ranks with the fixed nomenclatorial designation.

One of the possible approaches to resolve the above problem could be the following. Hierarchically arranged diversity produced by historical development of the biota could be treated as a fractal (Chernavsky, 2004). The latter formalism allows to stress the following important properties of TD. Firstly, any fractal constitutes a unity of both the hierarchical structure and algorithm generating it: one cannot exist without other. It seems to remove contradiction between functional and compositional interpretations of BD: the fractal consideration unites them. Secondly, this formalism allows, at least at first approximation, to suppose that there are no "foreground" taxonomic ranks, all of them are probably equivalent in a sense.

Treating hierarchy of BD as a consequence of the biota's development provides a key for understanding some of its important characters. Before all, one may suppose that the BD hierarchy becomes more "perfect" as the biota was developing and becoming more structured. This means that hierarchical structure of BD was probably more fuzzy at earlier stages of the history of biota and less so in more recent times. If it is so, than hierarchy of BD could be different in the groups that have emerged at different stages of the biota's history.

From a pragmatic standpoint, hierarchical treatment of the taxonomic aspect of BD should have certain impact on elaboration of priorities in developing conservation strategies. Providing that the real taxonomic hierarchy is not symmetrical (phylogenetic tree is not strictly dichotomous), various species may have different input in the overall BD dependent on their phylogenetic position: is being a member of a monotypic family is one thing and belonging to a specious genus is quite another. Such inputs could be reflected by respective differential weighting of species depending on that position (Faith, 1992, 1994; Pavlinov 2001), and the entire hierarchy could be modeled as a set of interconnected "phylogenetic weights".

Such an approach would certainly presume the existence of a fixed weighing scale allowing to assign certain commensurable weights to the species belonging to far distant fragments of the "Tree of Life". This scale, in its turn, is to be based on some system of fixed taxonomic ranks. So, as far as the pragmatic aspect of BD problematics is concerned, the idea of "rankless classifications" should be abandoned as non-operational, although metaphysically it is well supported.

However, there are several important problems that does not allow one to be especially optimistic about easy adoption of the above idea. First of them is the need for some theoretical background for such a unified scale of both taxonomic ranks and respective "weights" to be assigned to the species. For such a scale to be unified and to give commensurable species weights, that background is also to be unified. At present, phylogenetic ideology seems to prevail in contemporary taxonomy (as it was stated above), but it would be too bold to hope that it is adopted unanimously by the entire taxonomic community: a lot of typologists, pheneticians, and just "pragmaticians" study the diversity of life. So, although cladists now believe they won the "struggle for existence" and even elaborate the Phylocode to substitute the standard nomenclatorial Codes their belief alone does not warrant any unity of opinion existing or having to exist in taxonomy.

So, a possibility of elaboration of a unified background for elaboration of as unified weighting scale in question is problematic. This produces another problem – the one of dubious direct comparability of various classifications elaborated for various groups of organisms: they could be based on different taxonomic philosophies. But even within the framework of the phylogenetic paradigm, the background definitions of the proper phylogenesis may vary (e.g., Haeckelian vs. Simpsonian vs. Hennigian ones), and the phylogenetic pattern may supposedly be different in various "branches" of the "Tree of Life" (see above).

At last, particular phylogenetic reconstructions do not, generally speaking, belong to well-supported and reliably tested hypotheses: they heavily depend on characters used for reconstructions, and on methods of phylogenetic tree building. These reconstructions used to be made with various degrees of precision and reliability in various groups which certainly aggravate the entire business.

Does all this mean that the idea of taking phylogenetic pattern into consideration in elaboration of BD conservation strategies should be abandoned? Probably not, because this pattern is the fundamental property of the BD and therefore cannot be ignored. All the above (and other) problems just mean that the ranking and weighting scales, formally speaking, are to be formulated in terms of the fuzzy set theory (Kril, Bo, 1995), which would allow one to consider them as linguistic variables with particular values being dependent on the expert estimates.

At any rate, such scales would be of use both for global and local ("place", in sense of Sarkar, 2002) assessments of BD. In the second case, they would allow to estimate uniqueness of local biotas in respect to the "phylogenetic component" of the BD depending of the "phylogenetic weights" of the species composing those biotas.

WHAT IS *THE* SPECIES: OR, HOW DOES IT COME TO EXIST?

As it was indicated above, the species concept still predominates in the modern conservation biology. At the same time, recent discussions indicate that the context of this concept becomes more and more unclear: *the* species seems to disappear behind its various particular interpretations and definitions. At maximum, a principal question is put forward: is the species a unique and universal natural phenomenon, or do we just use this concept without justification to designate unlinked natural phenomena principally different by their essence (Ereshefsky, 2001)?

It is to be reinforced that the notion of species has been introduced to the taxonomy and then to the other branches of biology from the abovementioned genus–species logical scheme. That notion, as the entire scheme, was pure based on the presumption that there was no particular natural history ontology. However, it did not prevent scholasticists from declaring that "would we not know what is the species, nothing would secure us from misunderstandings" (Boetius). Such a standpoint predetermined subsequently most close attention to the species phenomenon in the biology.

One of the first steps toward ontologization of the species became famous from Linnaeus' aphorism: "There are as many species as it was created by God". Later on, evolutionists also concentrated on the species, trying to explain how the species were"created" by evolutionary process. Thus the species became the central theme of evolutionary theory in the synthetic theory of evolution which made it of nearly the same high significance in ecology: the species concept appeared to be basic for the niche concept, as well as for interpretation of a ecosystem as a set of interconnected species niches (Allen, Starr, 1982).

Generally speaking, the question of existence of *the* species as a universal life entity is not even metaphysical but quite naturphilosophical. The latter refers to unconscious belief of biologists that the species actually exist in nature in some form or another. At maximum, such a belief is expressed by a suggestion to recognize "species" as a peculiar ontological entity, along with "sets", "natural kinds", "individuals", etc. (Holsinger, 1984). Even operationalists, when rejecting discussion of non-empirical aspects of the species problem, often appeal to *the* species, as far as they proclaim their need for practically applicable "species criteria". Contrary to this, the so called "species pluralism" rejects universality of that general category but instead acknowledges the reality of various categories of species called "phylospecies", "ecospecies" etc. (Ereshefsky, 2001).

In my opinion, there is nothing non-scientific in a metaphysical (and even "naturphilosophical") standpoint in considering the species problem (Ghiselin, 1997). Indeed, the latter is to be placed at first position, and not concealed, when the species problem is being discussed. For me, it is the naturphilosophy that allows one to approach correctly theelaboration of some general (or "ideal", see Hull, 1997) species concept and to the problem of hierarchy of particular species concepts (Mayden, 1997; Brigandt, 2001). As it follows from the principal position of the present author, such an approach first requests some general contentwise model which would be taken as a starting point in the discussion of what is or might be a "primary" (basic) concept of "the species in general". Such a model is to be means for searching the answer to the most fundamental question: *why and how did anything like species appeared in the structure of biota?*

Consideration of the biota as non-equilibrium developing system was adopted above as such a model. Its immanent property is the hierarchical structure of BD which elements are groups of organisms of various levels of generality. In the way that this hierarchy is universal for the biota, these levels and groups are also universal by definition. Somewhere among them exists a level which is usually referred to as the species level: it is evidentially universal by initial condition of universality of the entire hierarchy. The question is now: is it possible, and if it is then how, to fix and define such a level of hierarchy and the groups belonging to it in a general sense without referring to a particular already existing concept (save using the same term "species").

At least within metaphysical framework free of any operationalization precondition, a representation of development of the biota as a temporal flow may serve as one of the suitable metaphors to answer the above question. This flow is structured in a twofold way, laminarily (phyletic lines) and in forms of local stable vortexes (groups of organisms). It is possible that the least of such vortexes (groups) could be considered as the species – the lowest-level elements of TD (or BD in general) capable of stable self-reproduction (Pavlinov, 1992).

Such a view of "species" does not obviously provide any practical solution of the "species problem" as it is understood in most of modern reviews and discussions. Rather, it brings new fundamental insights regarding ontology of *the* species as a unit of the BD.

The first of these insights concern ontology of the species. The latter, as it is understood here, does not belong to any of the ontological group categories usually considered in this connection – individual, natural kind, class etc. It seems attractive to identify it as belonging to the group category of *kumatoid* (Zuev, 2002). This category could be interpreted as the natural kind with the time dimension added explicitly: the species-as-kumatoid remains more or less stable as a part of the overall "historical flow" of developing biota while its composition changes with each new generation.

The synergetic formula "order out of chaos" (Prigogine, Stengers, 1984) may also have some concern to the species as it is here understood. Indeed, according to the classical Darwinian view of the origin of species, the latter emerges as ordered entities out of stochastic fluctuations routinely interpreted as individual variation. However, an evolutionary scenario may be supposed according to when the entire fluctuation background is being transformed into individualized species at each stage of evolution of a particular phylum, some (not strictly fixed) portion of the "chaos" may persist. This means, in taxonomic terms, a possibility for the given genus to be divided completely into recognized species, contrary to the above logical scheme which is usually adopted by practicizing taxonomists (Pavlinov, 1992, 2006).

The idea of gradual hierarchical structurization of the evolving biota seems to be applicable to the species level, as well. This means that some "perfectioning" of that level was done along the progression of the entire biota, beginning from more fuzzy (less discrete) forms of species in primordial organisms (Hanage et al., 2005) and advancing toward developing more perfect isolating mechanisms providing for more discrete species organization as a part of the biotic structure. The above perfectioning involved, among other things, prohibition of horizontal transfer of genetic materials quite common among prokaryotes and practically lacking among higher eukaryotes (Shestakov, 2003).

All the above concerning the species could hardly be called "the species concept"; it is rather a look at the idea of "general species". Its outcome is twofold. On the one hand, this look agrees with that idea: there might be a contentwise model of the evolving biota suggesting some universal unity of the BD which might be called "species". On the other hand, it presumes that that "general species" so understood is heterogeneous by its own characteristics: particular species in various groups of organisms can differ by degree of their discreteness and by mechanisms supporting it.

This heterogeneity is reflected in the abovementioned "species pluralism" denotating the impossibility of elaboration of a universal operational species concept equally applied to different groups (phyla etc.) of organisms. This impossibility was formalized several years ago as "species delimitation uncertainty principle" by B. Adams (2001). Adams thaught that his principle named after David Hull was analogous to the uncertainty principle of Heisenberg, so it was so called. But he seems to me to be wrong. As a matter of fact, he wrote, following D. Hull (1997), about reciprocal correspondence between operationality and theoretical precision of any particular species concept. So, it occurs to me that both Hull and Adams describe a relation which fits not the Heisenberg's uncertainty principle but rather the logical principle of reciprocal correspondence between strictness and pithiness of a notion (Ghorsky, 1983). Respectively, the "Hull's principle", if I clearly understood B. Adams' idea, is to be reformulated the following way: *the more strictly a species concept is defined, the narrower is the range of its operational application.*

However, Heisenberg's principle in its original sense does have certain concern with the situation of the interpretation of species as an element of the biotic structure. It seems to me that the uncertainty in the given case involves a much more important problem of description of BD than just delimitation of several particular species (which after all is just a matter of their proper diagnostics). It involves a problem of relation between complementary characteristics corresponding to different aspects of consideration of the species as a biological phenomenon.

Taking into consideration the two basic components of BD outlined above, the species as a unit of the latter could be considered in at least two respective aspects – either as a historical or as an ecological entity. These two aspects evidently cannot be reduced to either of them, nor can they be reduced to some third aspect. So, they are complementary and, as such, can be considered as connected by an uncertain relation: *fixing any one of these aspects means impossibility to consider the species in all details from another aspect*. For instance, considering and characterizing the position of a species in the structure of phylogenetic relationships makes it rather problematic to strictly define its position in the network of ecological relations, and vice versa. If one wishes to study a species from both aspects, an iterative procedure should be launched in which this species is considered first, say, as a historical unit and then as an ecological unit (or vice versa), with subsequent alteration of the consideration aspects.

It follows from the immediate above that various species concepts, each fixing some way particular consideration aspect(s), are not mutually "concurring" and exclusive but rather complementary to each other. The question is: might there be any general (or "ideal", after Hull) concept which would unite those consideration aspects and would allow us to treat the species as multi-aspectous but yet unitary entity? One of them could be exemplified by the evolutionary species concept of Simpson which considers the species as an historical entity taking certain place in the structure of ecological relations (Wiley, Mayden, 2000). Such a concept is actually non-operational, as it suggests implicitly consideration of all aspects simultaneously, but this does not follow the "Hull's principle", but from its being of higher "dimensionality" (after K. Popper) than any more narrow concept.

The above position seems to conciliate in a manner so called "species pluralism" of M. Ereshefsky with the classical idea of unitarity of each species to which a particular organism belongs. Indeed, each species, by initial assumption, is unique, but it can nevertheless be considered from various aspects. These aspects can not necessarily define conterminating partitions of the same set of organisms. So, if those different aspects are treated as different species – for instance, ecospecies or phylospecies (Ereshefsky, 2001), then each organism can be considered actually belong to those as if "different species" thus making an impression of "plurality" of species as units of the biota.

However, I wonder if there is any common sense in playing such a therminological game. It seems more appropriate to burden the notion of species with some one but more certain contents (essence) rather then to speak about "essentially different species". Looking back to the roots of the varying notions of species and taking into consideration the nowadays predominating taxonomic paradigm, it looks plausible to treat the species historically (phylogenetically), for instance as a fragment of a phylogenetic lineage (de Queiros, 1999). And to provide an ecological treatment of species with some other therminological designation – for instance, to equate Ereshefsky's "ecospecies" with the elementary biomorph (life form).

The Biomorph Diversity as a Component of Biodiversity

Biomorphs (life forms) are rarely discussed in the context of problematics of BD. One of the causes of such an attitude of the BD explorers is traditional conjuration that the basic discipline for researches on BD is taxonomy. It classifies species and supraspecific taxa thus dealing with the taxonomic component of BD only and does not consider biomorphs (only if they are not equated with taxa). The ecological component of BD is studied by biocenology (and by its subdiscipline, syntaxonomy) which considers classifications of natural communities (ecosystems, syntaxa and taxocens). Biomorphs are also omitted from such consideration; gilds are recognized at best but their members share only ecological peculiarities but not morphological ones.

The neglecting of biomorphs seems to be a consequence of above mentioned absolutization of species and population as elementary components of the ecosystems. Meanwhile, there is a biocenotical concept according to which the biomorphs are actually the basic ecosystem elements (Chernov, 1991).

Two examples are usually given to illustrate that idea. On the one hand, dragonfly imagoes and larvae unambiguously belong to quite different biomorphs regardless of whether or not they belong to the same or to different bio(phylo)species. On the other hand, in certain ecologically saturated communities of tropical rainforests at least a part of species diversity seems to be "redundant". For instance, closely related (congeneric) species of some of the soil arthropods may represent the same biomorph and so are "equivalent" from the functional ecosystemic standpoint.

I am far from intending to expound here a kind of biomorphic approach to the ecological community. However, a need for such an approach to analysis of BD structure can be substantiated, at a very general (metaphysical) level of consideration adopted here, on the basis of synergetic biotic model outlined above schematically.

Each ecological role requests a certain morphological structure to conduct it regardless of origin of the particular elements compounding that structure. Thus, a certain correspondence must exist between the diversity of the ecological roles responsible for optimal functioning of the ecosystems and the diversity of morphological structures responsible for the "material maintenance" of those roles and, hence, of the entire ecosystems. "Regardless of origin" means that the two internested structures, of biomorphic diversity and of species diversity, need not to coincide with each other: evolutionary parallelisms and convergences prove that. So, from the ecosystemic viewpoint, one may consider phylogenetic lineages just as a means for transforming initial fuzzy biomorphic diversity into a more structured one along with structuring and diversification of the entire developing biota (Rautian, Zherikhin, 1997).

From this standpoint, it seems more than desirable to consider the place of biomorphic diversity (BmD) in the overall structure of BD. The main question is the following: should BmD be considered as a part of ED or as one more component of BD additional to ED and TD? In answering this question, it seems reasonable to take into consideration that ED corresponds to the functional aspect of biota, TD corresponds to its structural aspect, while BmD corresponds to a cross-section (overlapping) of these two. So, it might well be that three-partitioning of the entire BD should be acknowledged and reflected some way in elaboration of conservation strategies. This would make them more adequate to the modern view of the structure of the biota in which bimorphs occurs non-accidentally (Krivolutsky, 1998; Pavlinov, Rossolimo, 2004).

Whatever might be an answer to the above question, an unresolved task of exploration of the structure of BmD remains. For this, a special biological discipline is to be developed which would explore this natural phenomenon – diversity of biomorphs (or life forms); it might be called "*biomorphics*" (Pavlinov, 2001). Its first task could be defined as elaboration of a contentwise framework for classification of biomorphs. It would serve as a backgrounds for resolving though particular but quite important problems such as: how to define homology and analogy in the case of biomorphs, what might be optimal forms (hierarchical, periodical etc.) of biomorph classifications, and at last if it is possible to elaborate an overall classification of biomorphs analogous, say, to phylogenetic "Tree of Life".

Conclusion

No one would seriously object to the idea that security of biodiversity is one of the most serious problems mankind faces today. However, it does not abolish another idea – scientific exploration of biodiversity as a natural phenomenon. It is evident that it is such an exploration that may warrant that the conservation strategy incorporating its results would be really "scientific".

Such an exploration is to be aimed at elaboration of a scientific concept (image, model, etc.) of the biodiversity being considered not from elementaristic but rather the holistic point of view. Such a concept should advert metaphysical matters on the nature (essence) of the varying BD. Among them, the following questions briefly considered above seem to merit first consideration:

- What is that general natural phenomenon, a property (aspect) of which is BD?
- What are causes of BD and of its own structure?
- What are the basic components and elements of BD and how do they interrelate?

References

Adams B.J. (2001) The species delimitation uncertainty principle. *J. Nematol.*, 33, 153–160.

Allen T.F., Starr T.B. (1982) *Hierarchy: Perspectives for ecological complexity.* Chicago: Univ. Chicago Press.

Barantsev R.G. (2003) *Synergetics in contemporary natural sciences.* Moscow: Editorial URSS. In Russian.

Barskov I.S. Zherikhin V.V., Rautian A.S. (1996) Problems in evolution of biological diversity. *J. General Biol.*, 57, 14-39. In Russian with English summary.

Brigandt I. (2003) Species pluralism does not imply species eliminativism. *Phil. Sci.,* 70, 1305-1316.

Brooks D.R., Wiley E.O. (1986) Evolution as entropy. Chicago: Univ. Chicago Press.

Chaikovsky Yu.V. (1990) *Elements of evolutionary dyatropics.* Moscow: Science. In Russian.

Chernavsky D.S. (2004) Synergetics and information (a dynamic theory of information). Moscow: Editorial URSS. In Russian.

Chernov Yu.I. (1991) Biological diversity: its essence and problems. *Adv. Contemp. Biol.*, 111, 499-507.

de Queiros K. (1999) The general lineage concept of species and the defining properties of the species category. In: Ed. Wilson R.A. *Species: new interdisciplinary essay.* Cambridge (MA): MIT Press, 40-89

Ghorsky D.P. (1983) Logic. Moscow: Uchpedgiz. In Russian.

Il'in V.V. (2003) *Philosophy of Science*. Moscow: Moscow Univ. Publ.

Eldredge N., Cracraft J. (1980) *Phylogenetic patterns and the evolutionary process.* New York: Columbia Univ. Press.

Ereshefsky M. (2001) *The poverty of the Linnaean hierarchy: a philosophical study of biological taxonomy.* Cambridge: Cambridge Univ. Press.

Faith D.P. (1992) Conservation evaluation and phylogenetic diversity. *Biol. Conserv.*, 61, 1-10.

Faith D.P. (1994) Phylogenetic pattern and the quantification of organismal biodiversity. *Phil. Trans. Roy. Soc. Lond.*, 345, 45-58.

Faith D.P. (2003) Biodiversity. In: Ed. Zalta E. N. The Stanford Encyclopedia of Philosophy (Summer 2003 Edition). URL http://plato.stanford.edu/archives/sum2003/entries/biodiversity/.

Ghiselin M.T. (1997) *Metaphysics and the origin of species.* New York: New York State Univ. Press.

Hanage W.P., Fraser C., Spratt B.G. (2005) Fuzzy species among recombinogenic bacteria. *BioMed Central, Biol.,* 3, 6-11.

Heywood V.H., Watson R.T., eds (1995) *Global biodiversity assessment.* Cambridge: Cambridge Univ. Press.

Holsinger K.E. (1984) The nature of biological species. *Phil. Sci.*, 1984, 51, 293-307.

Hull D.L. (1997). The ideal species concept – and why we can't got it. In: Eds Claridge M.F. et al. Species. The units of biodiversity. New York: Chapman and Hall. 357–380.

Krill G.J., Bo Yuan (1995) Fuzzy sets and fuzzy logic: theory and applications. New York: Prentice Hall.

Krivolutsky D.A. (1998) Life forms and biological diversity of animals. *Bull. Mosc. Soc. Natur., Biol. Sect.*, 104, 61-67. In Russian.

Mayr E. (1969) Principles of animal taxonomy. New York: McGrow Hill.

Mayden R.L. (1997) A hierarchy of the species concepts: the denouement in the saga of the species problem. In: Eds P. M.F. *Claridge et al. Species: the units of biodiversity.* London, New York: Chapman and Hall, 382-424.

Mednikov B.M., 1974. On reality of the higher taxonomic categories of the vertebrates. *J. General Biol.*, 35, 659-655. In Russian, with English summary.

Norton B.G. (1986) *The preservation of species: the value of biological diversity.* Princeton (NJ): Princeton Univ. Press.

Panchen A.L. (1992) *Classification, evolution, and the nature of biology.* Cambridge (MA): Cambridge Univ. Press.

Pavlinov I.Ya. (1992) Is there a biological species, or what is the "harm" of taxonomy. *J. General Biol.*, 53, 757-767. In Russian, with English summary.

Pavlinov I.Ya. (2001) Concepts of taxonomy and concepts of biodiversity: a problem of interaction. *J. General Biol.*, 62, 362-366. In Russian, with English summary.

Pavlinov I.Ya. (2005) *Introduction to contemporary phylogenetics*. Moscow: KMK Sci Press. In Russian.

Pavlinov I.Ya. (2006) Classical and non-classical taxonomy: where does the boundary pass. *J. General Biol.*, 67, 83-106. In Russian, with English summary.

Pavlinov I.Ya., Rossolimo O.L. (2004) The structure of biological diversity. *Agr. Russia*, 4, 21-24. In Russian.

Prigogine I., Stengers I. (1984) Order out of chaos. London: Heinemann.

Rautian A.S., Zherikhin V.V. (1998) Models of community evolution (phylocenogenesis) and lessons from ecological crisises in the geological past of the Earth. *J. General Biol.*, 58, 20-47. In Russian, with English summary.

Roebuck P., Phifer P. (1999) The persistence of positivism in conservation biology. *Conserv. Biol.*, 13, 444-446.

Sarkar S., Margules C. (2001) Operationalizing biodiversity for conservation planning. *J. Biosci.*, 27, Suppl. 2, 299–308

Sarkar S. (2002). Defining "biodiversity"; assessing biodiversity. *The Monist*, 85, 131-155.

Shestakov S.V. (2003) On the earlier stages of biological evolution from the genomic standpoint. *Paleont. J.*, 6, 50-57

Schulze E.-D., Mooney H.A. (1994) Ecosystem function of biodiversity: a summary. In: Eds Schulze E.-D., Mooney H.A. Ecosystem function of biodiversity. New York: Springer, 497-510.

Wiley E.O., Mayden R.L. (2000) The evolutionary species concept. In: Eds Wheeler Q. D., Meier R. *Species concepts and phylogenetic theory: a debate.* New York: Columbia Univ. Press, 70-224.

Zuev V.V. (2002) *The problem of reality in biological taxonomy.* Novosibirsk: Novosibirsk Univ. Publ.

In: Focus on Biodiversity Research
Editor: Jan Schwartz, pp. 115-130

ISBN: 1-60021-372-3

Chapter 6

SPONTANEOUS PATTERN FORMATION AND DIVERSITY IN SPATIALLY STRUCTURED EVOLUTIONARY ECOLOGY

Hiroki Sayama[1*], Les Kaufman[2], and Yaneer Bar-Yam[3]

[1.] Department of Bioengineering, Binghamton University, State University of New York, Binghamton, NY USA

[2.] Department of Biology, Boston University, Boston, MA USA

[3.] New England Complex Systems Institute, Cambridge, MA USA

ABSTRACT

Spontaneous pattern formation, the appearance of inhomogeneities that are not directly imposed by external forces, can be an important but overlooked factor in the origin and maintenance of genetic diversity in populations, and in parapatric speciation. Using agent-based simulation models with multilocus genetics, we show that such patterns form in spatially distributed populations with local mating neighborhoods. Under these conditions, disruptive selection induces spontaneous pattern formation and coarsening in the spatial distribution of genotypes, and thus increases the persistence of genetic diversity. The lateral motion of hybrid zones between different genotypes of similar fitness is affected by their topological shape. Hence, the shape of these boundaries plays a crucial role in determining the rate of genetic variation and the long time behavior of the population. These patterns also interact with boundary and internal obstacles of habitat structure, so that diversity persists the longest in habitats with the most irregular geographical features. Localized obstacles contribute to the maintenance of genetic diversity even if they do not preclude interbreeding, by inhibiting the movement of hybrid zones between areas dominated by different genotypes. Our results imply that the most effective size of such obstacles for maintaining diversity is about the same as the radius of a local deme. These results are relevant to spatial ecology and conservation

---

* To whom correspondence should be addressed. Email: sayama@binghamton.edu

biology, in which both genetic diversity and the size and shape of population reserves have great import for the long-term survival of species in the wild. They also imply significant coupling of population level and ecological properties to evolutionary dynamics. The persistence and amount of diversity, the relationship of diversity to evolutionary change, and the survival of types is controlled by the genetically undetermined spatial distribution of types and the structure of the habitat.

INTRODUCTION

Spontaneous pattern formation describes a process in which heterogeneous structure arises within a system through internal dynamics, without external forces playing any role in either the existence or the shape of the structure. More than five decades ago British mathematician Alan Turing presented a theoretical model of spatially extended chemical reactions that could create static patterns from homogeneous initial conditions (Turing 1952). After a long period of dormancy, the dynamics of such spatial pattern formation has recently been attracting attention in developmental biology to describe various general phenomena such as spontaneous cell differentiation in embryos, and specific phenomena such as the formation of pigment patterns on animal skins. Besides Turing's static patterns, there are several other distinct classes of spontaneous pattern formation that are found theoretically to be generic to large classes of models of interacting systems and are realized in a wide range of experimental physical/chemical systems.

Dynamic domain formation and coarsening behavior was first recognized in statistical physics for systems in which components take either of two (or a few) stable states and local interactions encourage local homogeneity (Lifshitz and Slyozov 1961). Since homogeneity is local, regions of one type or another form, but there are boundaries between these regions since global homogeneity does not immediately take place. This behavior can be seen in many natural and artificial systems, especially near a phase transition in which ordering takes place. These include magnetic ordering, and vapor to liquid, or liquid to solid transitions. For example, when magnets can order in one of two opposite directions, domains of each direction form, with sharp boundaries between them. Growth of the domains occurs by the process of boundary movement, which is very slow compared to the orientation time of individual atoms.

Here we show that this same ordering in local regions and boundary movement can also arise in evolutionary ecology. The local domains are regions of similar genetic type, while diversity is maintained by having regions of different types separated by boundaries that may be mobile or static hybrid zones. The recognition of this natural mechanism of diversity maintenance has useful implications for park planning and conservation.

The spatial dynamics and characteristics of domain formation have been extensively studied using analytical and computational methods (Bray 1994). The central feature of this behavior is the characteristic time dependence of the typical domain size that increases as a function of time (coarsening) as a result of interdomain boundary motion. Spontaneous pattern formation has recently been introduced into ecology to study spatial interspecific interactions (Durrett and Levin 1994; Tilman and Kareiva 1997), where the competition between two or more species may create a spatial pattern of species that changes over time.

However, these studies do not consider the possibility or implications of such pattern dynamics for intraspecific genetic diversity.

Here we consider a theoretical model of spatially distributed mating populations with multilocus genetics and show via agent-based simulations that spontaneously forming spatial genetic patterns increase the persistence of genetic diversity when selection favors two fit genotypes over their genetic intermediates (disruptive selection) (Bar-Yam 1999; Sayama et al. 2000a, 2000b, 2003). Moreover, we show that these patterns interact with habitat structure so that the habitat structure becomes directly relevant to the maintenance of genetic diversity — but not quite in the way that we are used to thinking about spatial heterogeneity. Most ecologists consider genetic diversity as increasing with habitat heterogeneity due to harlequin selection. In this case, however, diversity arises and persists longer in larger habitats due to the time required for pattern coarsening by the motion of boundaries between extreme genotypes (hybrid zones). The internal structure of the habitat also matters. Localized obstacles that do not preclude interbreeding nevertheless contribute to the maintenance of genetic diversity by inhibiting the movement of hybrid zones. The most effective obstacle size to maintain diversity is found to be approximately the radius of a local deme.

These results are relevant to spatial ecology and conservation biology, in which both genetic diversity and the size and shape of population reserves are of concern due to their implications for long-term survival of species (Diamond 1975; Lynch 1996; Margules and Pressey 2000). They are also relevant to understanding the coupling of ecological properties to evolutionary dynamics. The persistence of diversity due to spatial patterns is counter to the usual Fisherian relationships between population diversity as measured by genetic variance and evolutionary change in well mixed populations. Moreover, the survival of types may be dramatically affected by their spatial arrangement. The spontaneous formation of patterns also can be considered a model for parapatric speciation. All of these processes couple evolutionary and ecological dynamics.

MODEL

We consider a spatially structured population of monoecious organisms with local but overlapping demes. Sexual reproduction of the organisms is described by a haploid, agent/individual-based model with multilocus genetics. An organism's genome has n loci, each of which has one of m possible alleles. One mature organism exclusively occupies one spatial site, reflecting the local carrying capacity. Organism reproduction generates P offspring at each site. Each of the offspring genomes is determined by random recombination of the genomes of two parent organisms randomly chosen within a distance R, the radius of a local deme. After reproduction, the fittest individual in the produced offspring at each site is selected to become a parent at the next generation.

The phenotype of each offspring is assumed to be a quantitative trait determined from its genome in such a way that each allele independently and additively contributes to the trait. Selection is applied using a fitness function that is maximal for extreme values of the quantitative trait. This has the effect of disruptive selection by selecting against organisms with intermediate trait values. It is equivalent to concave fitness sets, such as arise when a population evolves to exploit highly disparate resource types (Thoday 1972). Such disruptive selection is considered a common and general cause of trait divergence (Kondrashov and Kondrashov 1999; Dieckmann and Doebeli 1999). The details of the model and the specific parameter settings chosen for the simulations are given in the Appendix.

Unlike many empirical studies in conservation genetics that focus on the spatial patterns of neutral markers, our study focuses on dynamic changes of adaptive genetic variation within populations under disruptive selective pressure against quantitative traits. In such settings we can characterize adaptive genetic variation by trait variation because they are strongly coupled, and in what follows we use the trait variance as a measure of genetic diversity.

In the results presented below we do not include mutation. We have also carried out simulations including mutation and found that results are not affected within the range of biologically realistic mutation rates (e.g. mutation rate per locus $\mu < 10^{-3}$). Even for unreasonably high mutation rates ($10^{-3} < \mu < 10^{-1}$) the results are only weakly affected through the slowed motion of hybrid zones. This illustrates the generic nature of coarsening behavior and its robustness to many model variations that respect the locality of interactions.

Spontaneous Pattern Formation

There is a dramatic contrast between theoretical predictions based on two different assumptions; (1) a panmictic population where *R* is infinite (i.e., mating is possible between all organisms [Figure 1; top]) and (2) a population with local demes where *R* is significantly smaller than the radius of the habitat (Figure 1; bottom). In the former case the population becomes restricted to one type after only a few generations. Under the latter condition the organisms locally tend to assume one or the other of two fittest types. Globally, spatial patterns with domains of two different genotypes form. Because the organisms are not physically isolated from each other, the hybrid zones between genotypes move laterally, in a direction so as to reduce their local curvature. The rate of motion increases as the local curvature increases. Inclusions of one type surrounded by the other type disappear over time, which may result in a final state that contradicts expectations based upon traditional frequency-dependent fitness (Figure 2). The spatial dynamics of the domains seen in this model is equivalent to the process of pattern formation and coarsening that occurs in nonbiological systems (Bray 1994). The robustness of these dynamics is apparent when we study the effects of variations in the model parameters *n*, *m*, *R*, or *P*. As long as there are two fittest genotypes under disruptive selection, a similar pattern formation and coarsening always results. Moreover, the observed dynamics of spatial patterns is also robust to minor fitness differences between favored genotypes (Figure 3). Even with significant fitness differences patterns still form, but they persist for shorter periods of time.

As mentioned above, such pattern forming and coarsening behavior has recently been proposed to occur in spatially distributed populations due to interspecific ecological interaction (Durrett and Levin 1994; Gandhi et al. 1998) and microevolution (Sayama et al. 2000a, 2000b, 2003). The key property common to both is the frequency-dependent dynamics in the local population growth, which is not captured by other purely random genetic drift models. The spatial patterns of genetic diversity shown in Figure 1 (bottom) are spontaneous self-organizing ones; there is no underlying spatial heterogeneity in the environment as a template. These spatial patterns of polymorphism are dynamic (due to the motion of hybrid zones) and thus distinct from the static polymorphism found in subdivided and weakly coupled metapopulation models of ecological systems (Karlin and McGregor 1972; Levin 1974). The distinction is an important one, for the dynamics described here may be superimposed in nature on the more familiar patterns arising from habitat heterogeneity. Moreover, just as traffic jams are a spontaneous feature of heavy traffic flows, but they couple to local perturbations such as on or off ramps, the spontaneous pattern formation described here can also couple to, and thus dramatically enhance the effects of, the underlying habitat heterogeneity.

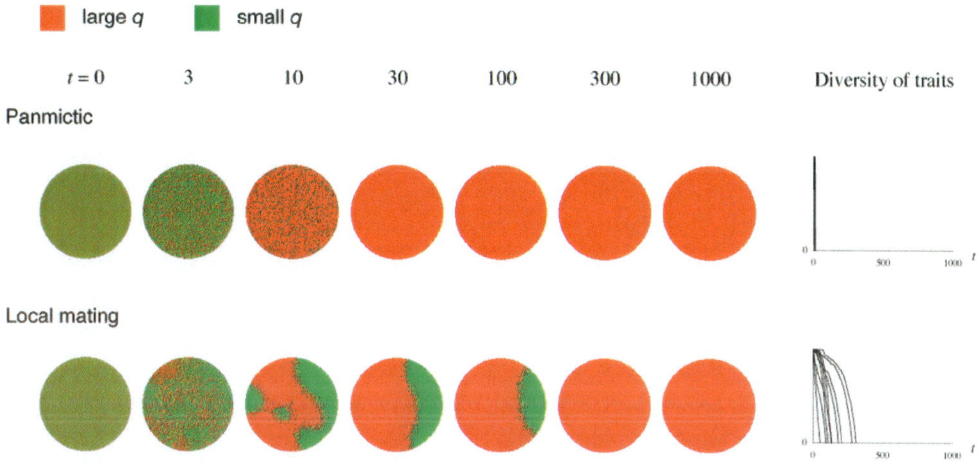

Figure 1. Simulations of pattern formation in a spatially distributed mating population undergoing disruptive selection initiated at t=0. Organisms are characterized by a quantitative trait q (see Appendix). Large and small q are favored over intermediate values of q by disruptive selection. On the right, the time dependence of the diversity is plotted for 10 sample simulation runs. Diversity is characterized by the variance of the trait distribution. These also apply to Figs. 2 to 5. Top: Simulations of a panmictic population, where organisms can mate regardless of spatial distance (i.e. the radius of the local deme R is infinite). In this case diversity vanishes only a few generations after the introduction of disruptive selection. Bottom: Simulations with local mating, where the size of local demes of organisms is smaller than the size of the habitat. Local mating causes pattern formation and thus persistence of genetic diversity.

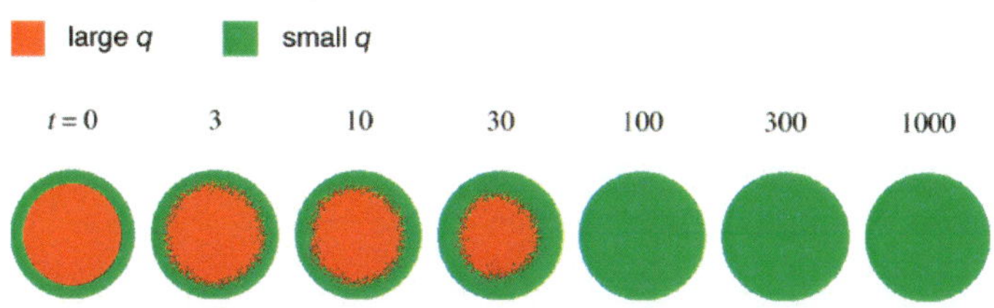

Figure 2. An extreme example of genetic changes determined by topological shape of hybrid zones. Initially, one of the two fittest genotypes (with large *q*) forms a big flock in the middle of the habitat, surrounded by a thin band of the other type (with small *q*). The flock occupies more than 60% of the total population so that frequency-dependent fitness difference predicts their dominance over the other type. Due to the circular shape of the hybrid zones, however, the initially dominant genotype is eventually dominated by the other. This result is robust to small changes in the model. Remarkably, even if the small *q* type has a slightly lower fitness than the high *q* type, it will still dominate over the long term (see also Figure 3).

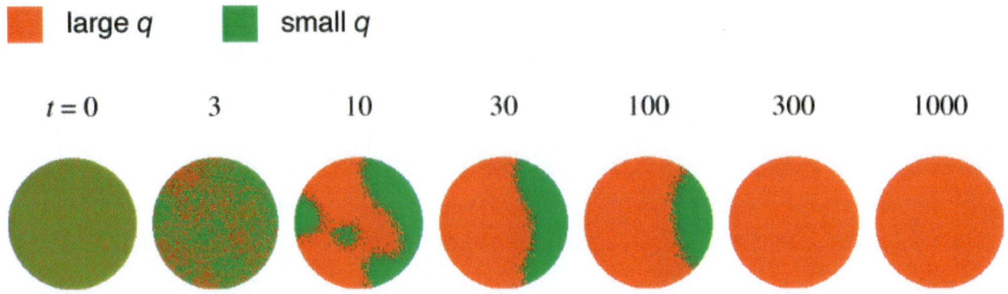

Figure 3. A simulation of pattern formation with small selection bias added to favor species of large q. The fitness of species of large q is increased by 1% compared to species of small q. The seed value of random numbers for this run is the same as that used in Figure 1 (bottom). While the fitness difference slightly promotes the propagation of hybrid zones from large q domains to small q domains, its effect is hard to detect visually. This implies the robustness of spontaneous pattern formation and coarsening dynamics against minor fitness differences between types.

The genetics of hybrid zones has been studied with increasing attention (Endler 1977; Barton 1979; Barton and Hewitt 1981, 1989; Harrison 1986, 1993; Nurnberger et al. 1995; Hewitt 2001), and the motion of hybrid zones has been seen in experimental observations (Pearson 2000; Britch et al. 2001; Rohwer et al. 2001). These studies typically analyze the dynamics of hybrid zones based on the difference between inherent fitness of two competing species where their domains meet. In contrast, our study deals with the spontaneous emergence of spatial separation between two equally favored types and the motion of hybrid zones by a particular kind of context-dependent fitness difference that arises from spatial distribution and dynamics, not simply from inherent biological superiority of one genotype. In our model, the local curvature of hybrid zones gives a local context-dependence to effective fitness. This may have a real-world significance, but has not yet been considered in studies on hybrid zones in the wild.

MAINTENANCE OF GENETIC DIVERSITY

We consider the relevance of the dynamics of such heterogeneous spatial patterns in the genotypes (and their associated hybrid zones) to the persistence of genetic diversity and, particularly, the significant role of habitat size and shape. These issues are typically not considered in theoretical or laboratory experimental studies that assume spatially homogeneous populations and that may specifically exclude consideration of edge and boundary effects.

Our results show that spontaneous pattern formation implies that the genetic diversity persists significantly longer in a population with local demes, as compared to a panmictic population when they are subject to disruptive selection (diversity plots in Figure 1). The time until homogenization increases with the size of the habitat (Figure 4), because the characteristic length scale of spatial patterns showing coarsening behavior grows as a square root of the elapsed time (Sayama et al. 2000a, 2000b) and therefore a longer time is needed to reach homogenization in larger areas. This suggests that protected habitats substantially smaller than natural ones will tend to lose genetic diversity more rapidly when disruptive selection occurs. This is of concern because high genetic diversity is widely considered to be a key to the robustness and resilience of biological systems (Tilman and Downing 1994; Holling et al. 1995). The loss of diversity when only one of two original types is left is not just the loss of the other type, but also of the many possible crosses between them that might be of higher fitness if environmental changes occur.

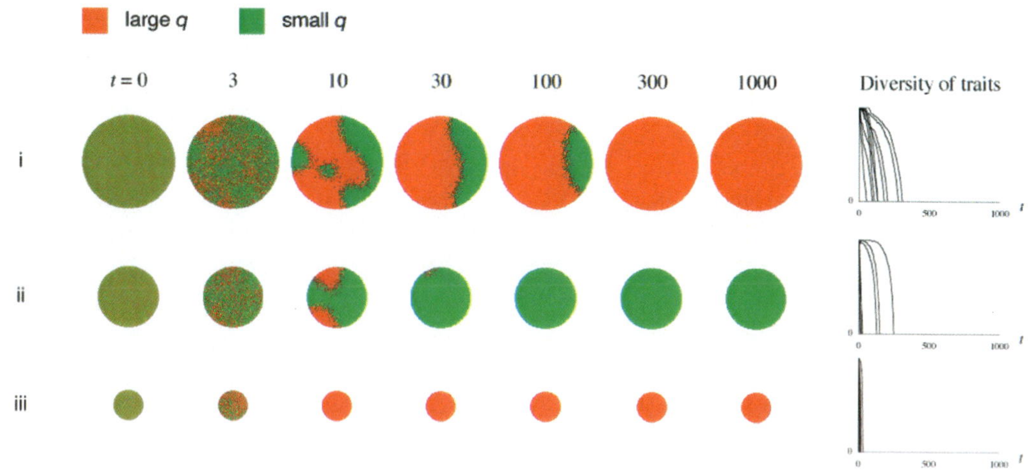

Figure 4. Simulations of pattern formation with local mating for three cases with different habitat sizes. The ratio of the diameter of one local deme to the total habitat diameter is about 16% for i, 25% for ii, and 50% for iii. The results for i is the same as shown in Figure 1. The time of diversity persistence is shorter for smaller habitats.

EFFECTS OF GEOGRAPHICAL IRREGULARITIES

The results in Figs. 1 to 4, and those obtained in other studies (Durrett and Levin 1994; Gandhi et al. 1998; Sayama et al. 2000a, 2000b), all apply only to simple habitat structures. To further study the implications of pattern formation, we consider the irregular geographical features that natural and protected habitats have. The effects of spatial barriers on static clines have been studied using reaction/diffusion equation models of gene frequency (Slatkin 1973; Nagylaki 1976; Fife 1979). Also, the effects of population subdivision on neutral genetic variation and effective population size have been studied using metapopulation models (Gilpin 1991; Gliddon and Goudet 1994; Hedrick and Gilpin 1997). In contrast with these earlier studies that calculated only stationary solutions (final structures) of genetic variation that reflect reduced migration rates between subpopulations in zero- or one-dimensional spaces, we consider the dynamic movement and transformation of curved hybrid zones and their interactions with localized obstacles in two dimensions.

Figure 5 shows examples of simulations representing habitats with internal obstacles of different sizes. Such irregular features inhibit and sometimes eventually stop the motion of hybrid zones and thus freeze coarsening patterns, preventing the reduction of genetic diversity through homogenization by reproductive mixing. We systematically evaluate the effectiveness of the internal obstacles in freezing the coarsening patterns by measuring the probability of persistence of polymorphism over time. Two parameters are varied: the size of an obstacle and the obstacle area fraction inside the habitat (the ratio of the area of the obstacles to the total area of the habitat). We have carried out over 10,000 simulation runs to explore the parameter space. We define an *effectiveness index*, which measures the efficiency per unit area of internal obstacles, as the slope of a hyperbolic tangent fitted to $p(x)$, where p is the probability of maintenance of polymorphism and x is the obstacle fraction. Generally, the probability of polymorphism increases as the density of obstacles increases (Figure 6; embedded graphs). This is due to a decrease in the rate of local patch extinction through the pinning of hybrid zone motion. More significantly, the effectiveness index is found to be maximal when the size of obstacles is equal to the radius of a local deme (Figure 6; main plot). This can be understood in that obstacles smaller than the radius of a deme are not large enough to hamper local reproductive mixing and thus do not effectively freeze coarsening patterns. Obstacles at the size of the radius of a deme can freeze the patterns. Obstacles larger than this are not much more effective, but use a larger area and thus have a lower effectiveness. This result weakly depends on the circular shape of the obstacles we use. It is expected that if one used star-shaped obstacles, polymorphism would be retained with smaller covered areas having larger perimeters. However, the specific scale (diameter) of the most effective obstacles should be independent of their shape because it corresponds to the range of the mating processes.

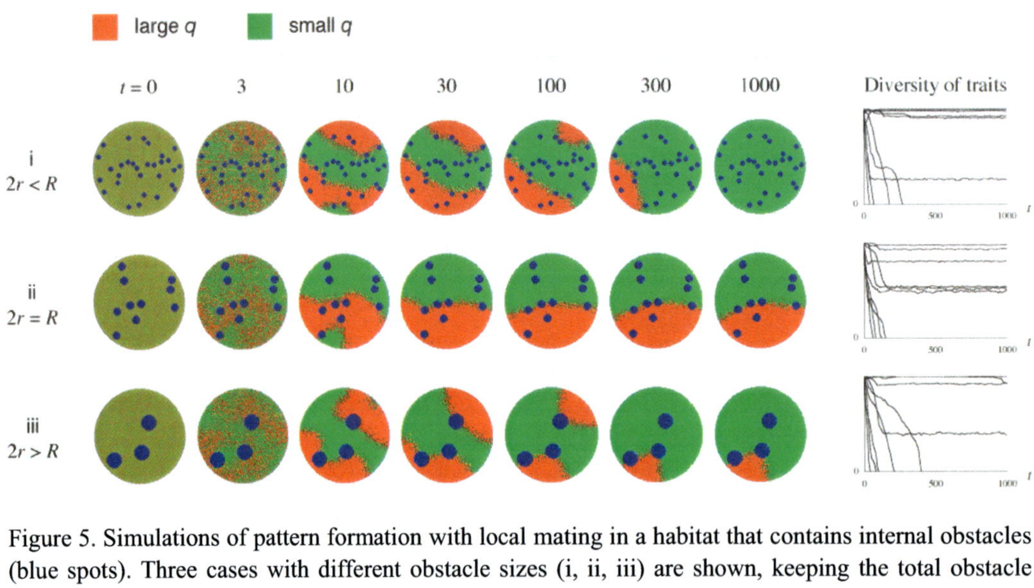

Figure 5. Simulations of pattern formation with local mating in a habitat that contains internal obstacles (blue spots). Three cases with different obstacle sizes (i, ii, iii) are shown, keeping the total obstacle area almost the same (r is the radius of the obstacles). Habitats with obstacles have a longer persistence of diversity due to the pinning of the motion of hybrid zones at the obstacles. The obstacles with a size $2r=R$ are found to be most effective (see Figure 6).

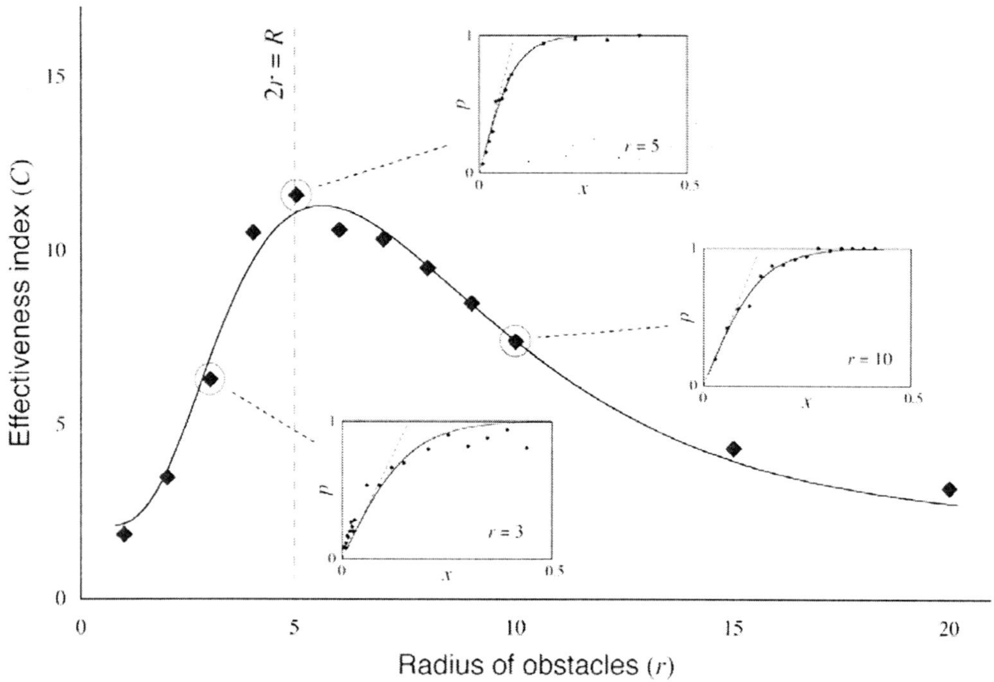

Figure 6. The relationship between obstacle size and effectiveness of obstacles in preserving diversity as measured by the effectiveness index C as a function of the radius of obstacles r. The effectiveness index is the initial slope of a hyperbolic tangent $p=\tanh(Cx)$ fitted to the simulation results $p(x)$, where p is the probability of maintenance of polymorphism and x is the obstacle fraction (the ratio of obstacle area to total area of the habitat). This measures the efficiency per unit area of the obstacles. A lognormal curve is fitted to the results for reference. This plot has a peak near $2r = R$. Thus, obstacles that are about as large as the radius of local demes are the most effective in maintaining polymorphism (Figure 5). The probability of maintenance of polymorphism was measured for each r and x using the final outcomes of multiple simulations for 500 generations, thus without obstacles the result would be zero. Each data point plotted in the embedded $p(x)$ graphs reflects the results of at least 20 runs, with an average of over 50 runs per point. The total number of runs used to obtain these results is over 10,000.

Discussion

Our results on the implications of spatial pattern formation for genetic diversity in natural and artificial habitats can be summarized as follows: In the reference case, when selection occurs in the context of complete reproductive mixing of the population, the population diversity declines rapidly. Local mating results in the preservation of trait variation and genetic diversity over substantially longer times according to our spatially extended population model. Even in the context of local mating, however, the diversity eventually declines, and does so more rapidly for smaller habitats. Local geographical irregularities can further extend the retention of a higher diversity. The effect of localized irregularities should be contrasted with the effect of habitat partitioning. Absolute partitioning is not desirable because it will result in small populations with a high probability of extinction. Spatial obstacles, even though they do not partition the space, inhibit complete reproductive mixing and are effective in maintaining genetic diversity in a geographically contiguous population.

The effect of local obstacles is not restricted to their immediate vicinity. By impeding the motion of hybrid zones, these obstacles can impact on the diversity of the entire habitat, a substantial long-range or non-local effect. Our results imply that where it is possible to design such obstacles, using obstacles whose size approximates the radius of the local demes may lead to maintenance of polymorphism with a minimal sacrificed habitat area. Other factors should also be considered in practical planning (e.g., total availability of suitable habitats, potential influences of obstacle construction, etc.).

In additional studies we found that convoluted habitat boundaries also freeze coarsening patterns (Sayama et al. 2000b). The implications of this result are counter-intuitive, for it implies that genetic diversity could be enhanced by increasing the ratio of habitat edge to area. For actual habitat design, however, taking advantage of this effect would also increase undesirable edge effects, such as microclimate deterioration or invasion by predator and weed species. Thus, from a practical viewpoint, external boundaries of protected habitats should probably be kept as short as possible because they are generally sources of environmental degradation and encroachment.

Our results suggest that internal obstacles in habitats might be introduced to enhance the maintenance of genetic diversity. What would the internal obstacles implied here actually be in any practical application? In an aquatic system they could be islands of a different physical environment, such as rocky reef on a mud or sand bottom. In a terrestrial park they could be fenced-off areas whose inside and outside are both protected and in which, for example, conservation and ecology management experiments could be conducted. In an era where the trend in zoological parks is to become as natural as possible, it might seem abhorrent to unleash interior designers upon the small nature reserves that we fondly but erroneously regard as accessible miniwildernesses (Nairobi National Park comes to mind). With few vast areas of true wilderness remaining in the world, current strategies for endangered species survival call for the maintenance of as many independent, self-sustaining populations as possible to maximize both total population size and genetic diversity. Our work suggests that, at least for some species, careful planning can maximize the conservation value of any one of these highly constrained remnant habitats.

To understand the more general implications for diversity, it is necessary to consider the population of organisms that arise from multiple disruptive selection events occurring at different times, as opposed to a single one. Each of these events may affect different aspects of the organismal traits corresponding to different but possibly overlapping sets of genes. As environmental changes occur, disruptive selection affecting various traits and/or loci introduces new patterns of covariance in the allele population that correspond to patterns of population differences. If the correlation between these distinct disruptive selection events is sufficiently weak, these patterns coarsen over time almost independently. Otherwise they may interfere with each other. Figure 7 shows preliminary studies on spontaneous pattern formation due to sequential introduction of disruptive selection. Two distinct disruptive selection events, which are orthogonal to each other in genome space, create two overlapping and independently coarsening patterns, which divide the population into four different types. In such cases, at a particular time, the history of introduction of disruptive selection events is apparent as a set of overlapping patterns that exist on various spatial scales. Larger scale features correspond to differences that were introduced earlier in time (or had a stronger type dominance).

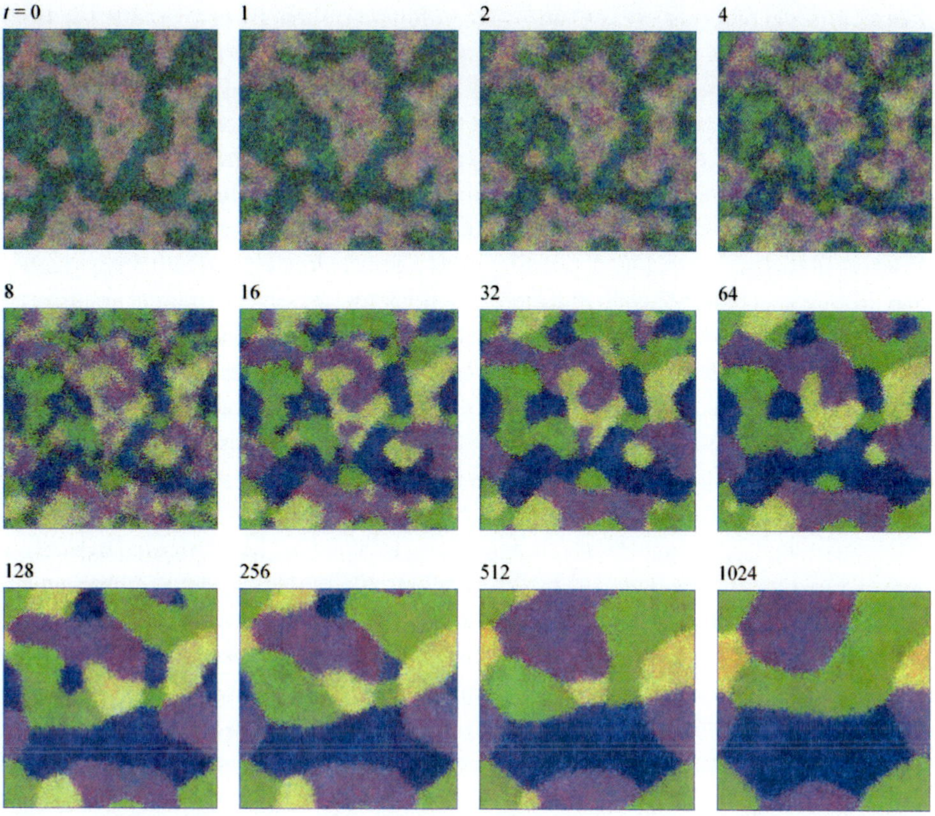

Figure 7 A simulation of pattern formation with local mating under sequential introduction of disruptive selection. The parameters used are n=10, m=10, P=2, R=3, and the space is a square area of 256x256 sites with periodic boundary conditions applied to its edges. These choices are to make the overlapping coarsening behavior more visible. t = 0 is 32 generations after the first disruptive selection was introduced to the population. At this time, the second selection that is orthogonal to the first one in genome space is introduced, leading to overlapping coarsening patterns of four different genotypes. During the coarsening process, the characteristic scale of patterns caused by the first selection (blue-green vs. yellow-purple) is larger than that by the second (blue-purple vs. green-yellow), reflecting its earlier introduction to the population.

This study also provides new insight into the origins of genetic diversity as a possible precursor of speciation. Recent studies have acknowledged the relevance of symmetry breaking to speciation (Kondrashov and Kondrashov 1999; Dieckmann and Doebeli 1999). Our results indicate that a spatially distributed species may spontaneously split into multiple different types that keep physical contact with each other in the absence of inhomogeneous habitat structure. At the hybrid zones between these types, mechanisms for reproductive isolation would be favored since they would increase reproductive success, facilitating trait divergence and formation of multiple species. The first step would be the appearance of a mutation leading to reproductive isolation at a location along the boundary. The benefits of reproductive isolation would cause this trait to propagate along the boundary, but also away from boundaries. The propagation away from boundaries would occur because reproductive isolation at boundaries would enable the types to coexist spatially, leading the boundaries to thicken, increasing the boundary area and consequently the benefit of reproductive isolation would extend to neighboring areas. Thus, the evolutionary benefit of reproductive isolation would spread throughout the population propagating outward from the point of origin both along and away from the original boundaries. This process would give rise to parapatric speciation – speciation that takes place in a spatially distributed population with no geographical barriers. Previous studies on parapatric speciation (Manzo and Peliti 1994; Gavrilets et al. 1998, 2000) used stepping-stone models with several subpopulations and weak migration between them. Our model is unique compared to these earlier models in that it considers a spatially contiguous population in two dimensions with overlapping demes and the behavior of coarsening domains of different genotypes.

Finally, our results demonstrate the existence of powerful linkages between biological processes at three different scales. Spontaneous pattern formation and its interplay with geographical features are emergent traits of ecological populations, caused by attributes of the organismal reproductive process that reflect genetic recombination inside each organism. This is an important example that contradicts the erroneous but commonly held assumption that it is possible to neatly partition evolutionary effects at different spatial scales (i.e., the study of molecule, individual, population, metapopulation, species, or ecosystem). Indeed, when hierarchical structure is considered as a signal attribute of complex systems, it is the dynamic linkages among the levels, rather than the number of levels themselves, that should probably be the focus of attention. We emphasize that the self-organization of dynamic patterns through multiscale interactions is a characteristic of properties of diverse complex systems, including spatially distributed living systems (Bar-Yam 1997), and that recognition of this can lead to deeper insights into community organization and evolution.

APPENDIX: DETAILS OF THE SIMULATION MODEL

The model used for this research is an agent-based model with a two-dimensional spatial grid of sites in which each site can contain one mature monoecious organism with an n-locus genome (also see Figure 8). Haploid and diploid models both yield similar results, and we present the results obtained with haploid models in this chapter. There are m possible alleles for each locus. The genome of an organism determines a quantitative trait q. Each allele is assumed to independently and linearly contribute to the trait (i.e., $q = \sum a_i$, where a_i is an integer ranging from 0 to m-1 that represents the effect of the allele at the ith locus). The possible traits range from 0 to $n(m\text{-}1)$. Modifying the assumption of linearity of allelic effects to traits can change the relative allelic diversity between different loci, but does not significantly change the general results on the dynamics of spatial patterns reported in this chapter. Fitness is a function of the quantitative trait, $f(q)$, which is lowest for intermediate values of the trait and highest for the two extreme values. We have tested several different fitness assignments, including V-shaped, quadratic, and discrete (stepwise) dependences of the fitness as a function of the trait. All these assignments yield the same results. Reproduction in the model assumes random genetic crosses between two parent organisms (i.e., the allele at each locus is inherited from either of two parents independently). The parent organisms are chosen randomly out of a local neighborhood of radius R around each site. Such mating processes are repeated P times per site per generation, then the fittest individual of the P offspring at each site becomes the selected parent at the next generation. We do not consider the effects of mutation for simplicity.

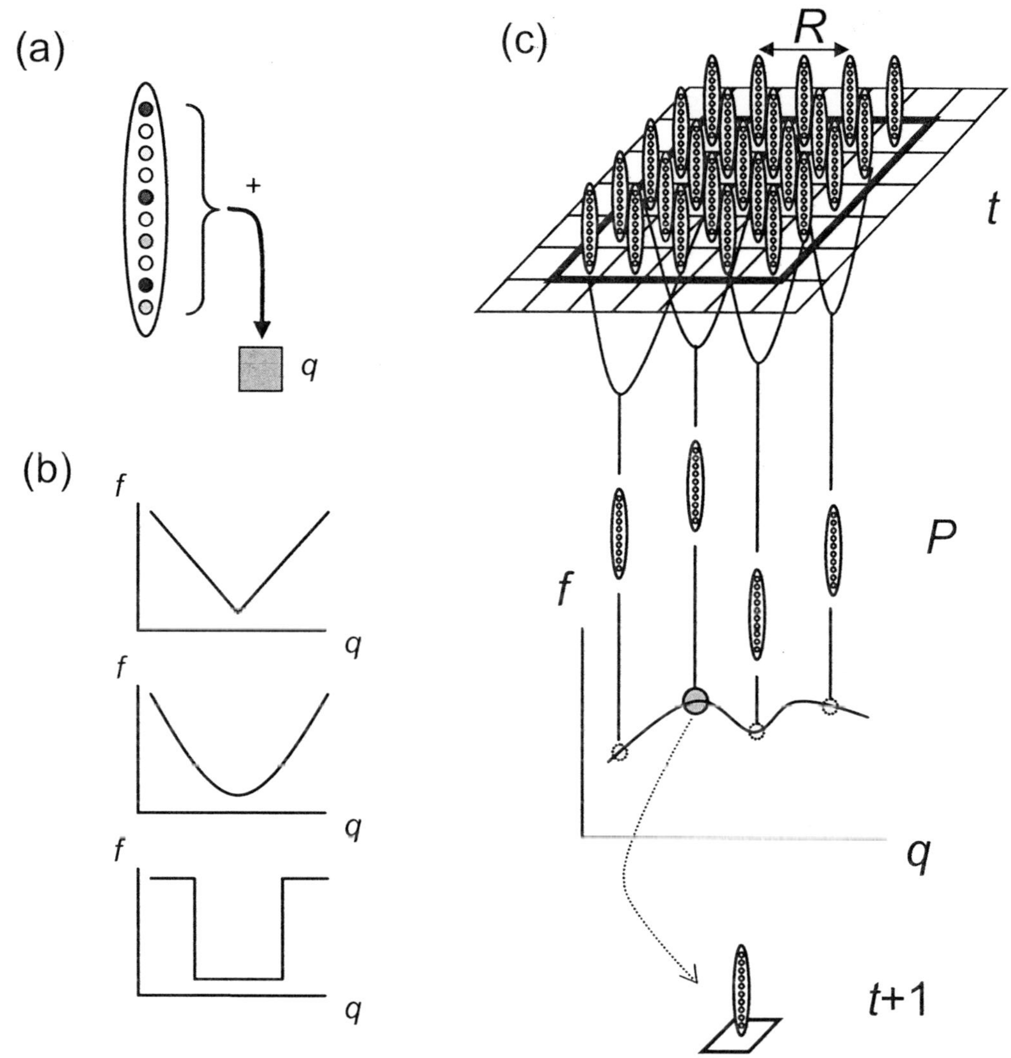

Figure 8. Schematic illustration of the simulation model. (a) An organism has a haploid n-locus genome. Each locus takes one of the m different alleles (shown in grey-scale shades). The quantitative trait of the organism q is determined by the sum of the effects of alleles on all loci. (b) Fitness f is given by a nonlinear function of the quantitative trait q that is lowest for intermediate values of the trait and highest for the two extreme values. (c) Each spatial site in a two-dimensional domain is occupied by one organism. In reproduction, two organisms are randomly picked up within a local neighborhood of radius R (either square or circular) and their genomes are randomly recombined to produce an offspring. Such mating is repeated for P times per each spatial site. The offspring with largest f wins to occupy the site in the next generation.

The parameters used in the figures are n=10, m=10, P=10, R=10 unless otherwise noted, and the fitness of a trait is a V-shaped curve (i.e., the fitness is linearly related to the trait value distance from the median possible value [$f(q) = | q - n(m-1)/2 |$]). The habitat is a circular space with a diameter of 127 site widths long (except for Figs. 4ii, 4iii and 7). For simulations described in Figs. 5 and 6, circular obstacle patterns on which organisms cannot exist are randomly placed inside the habitat in such a way that they neither overlap each other nor stick out from the habitat. The initial conditions consist of organisms with randomly assigned genotypes. This can arise when selection becomes disruptive after being nondisruptive.

Acknowledgments

This work was supported in part by the National Science Foundation under Grant No. 0083885. HS acknowledges additional support by grants from the University of Electro-Communications and the Showa Shell Sekiyu Foundation for Promotion of Environmental Research, and LK acknowledges additional support by a grant from the David and Lucile Packard Foundation.

References

Barton, N. H. 1979. The dynamics of hybrid zones. *Heredity* 43:341-359.

Barton, N. H., and G. M. Hewitt. 1981. Hybrid zones and speciation. Pages 109-145 in W. R. Atchley, and D. S. Woodruff, editors. Evolution and speciation: essays in honor of M. J. D. White. Cambridge University Press, Cambridge, United Kingdom.

Barton, N. H., and G. M. Hewitt. 1989. Adaptation, speciation and hybrid zones. *Nature* (London) 341: 497-503.

Bar-Yam, Y. 1997. Dynamics of complex systems. Perseus Books, Cambridge, Massachusetts.

Bar-Yam, Y. 1999. Formalizing the gene centered view of evolution. Advances in Complex Systems 2:277-281.

Bray, A. J. 1994. Theory of phase-ordering kinetics. *Advances in Physics* 43:357-459.

Britch, S. C., M. L. Cain, and D. J. Howard. 2001. Spatio-temporal dynamics of the Allonemobius fasciatus-A. socius mosaic hybrid zone: a 14-year perspective. *Molecular Ecology* 10:627-638.

Diamond, J. M. 1975. The island dilemma: lessons of modern biogeographic studies for the design of natural reserves. *Biological Conservation* 7:129-146.

Dieckmann, U., and M. Doebeli. 1999. On the origin of species by sympatric speciation. *Nature* (London) 400:354-357.

Durrett, R., and S. Levin. 1994. The importance of being discrete (and spatial). *Theoretical Population Biology* 46:363-394.

Endler, J. A. 1977. Geographic Variation, Speciation, and Clines. Princeton University Press, Princeton, New Jersey.

Fife, P. C. 1979. Mathematical aspects of reacting and diffusing systems. Lecture notes in biomathematics 28. Springer-Verlag, Berlin.

Gandhi, A., S. Levin, and S. Orszag. 1998. "Critical slowing down" in time-to-extinction: an example of critical phenomena in ecology. *Journal of Theoretical Biology* 192:363-376.

Gavrilets, S., H. Li, and M. D. Vose. 1998. Rapid parapatric speciation on holey adaptive landscapes. *Proceedings of the Royal Society of London B Biological Sciences* 265:1483-1489.

Gavrilets, S., H. Li, and M. D. Vose. 2000. Patterns of parapatric speciation. *Evolution* 54:1126-1134.

Gilpin, M. E. 1991. The genetic effective size of a metapopulation. Pages 165-175 in M. E. Gilpin, and I. Hanski, editors. *Metapopulation dynamics: empirical and theoretical investigations*. Academic Press, London, United Kingdom.

Gliddon, C., and J. Goudet. 1994. The genetic structure of metapopulations and conservation biology. Pages 107-114 in V. Loeschcke, J. Tomiuk, and S. K. Jain, editors. *Conservation Genetics*. Birkhauser Verlag, Berlin.

Harrison, R. G. 1986. Pattern and process in a narrow hybrid zone. *Heredity* 56:337-350.

Harrison, R. G. 1993. Hybrid zones and the evolutionary processes. Oxford University Press, New York.

Hedrick, P. W., and M. E. Gilpin. 1997. Genetic effective size of a metapopulation. Pages 166-181 in I. A. Hanski, and M. E. Gilpin, editors. Metapopulation biology. *Ecology, genetics, and evolution*. Academic Press, San Diego, California.

Hewitt, G. M. 2001. Speciation, hybrid zones and phylogeography - or seeing genes in space and time. *Behavioral Ecology*,10:537-549.

Holling, C. S., D. W. Schindler, B. W. Walker, and J. Roughgarden. 1995. Biodiversity in the functioning of ecosystems: an ecological synthesis. Pages 44-83 in C. Perrings, K.-G. Mäler, C. Folke, C. S. Holling, and B.-O. Jansson, editors. *Biodiversity loss: economic and ecological issues*. Cambridge University Press, New York.

Karlin, S., and J. McGregor. 1972. Polymorphisms for genetic and ecological systems with weak coupling. *Theoretical Population Biology* 3:210-238.

Kondrashov, A. S., and F. A. Kondrashov. 1999. Interactions among quantitative traits in the course of sympatric speciation. *Nature* (London) 400:351-354.

Levin, S. A. 1974. Dispersion and population interactions. *The American Naturalist* 108:207-228.

Lifshitz, I. M., and V. V. Slyozov. 1961. The kinetics of precipitation from supersaturated solid solutions. *Journal of Physics and Chemistry of Solids* 19: 35-50.

Lynch, M. 1996. A quantitative-genetic perspective on conservation issues. Pages 471-501 in J. C. Avise, and J. L. Hamrick, editors. *Conservation genetics: case histories from nature*. Chapman and Hall, New York.

Manzo, F., and L. Peliti. 1994. Geographic speciation in the Derrida-Higgs model of species formation. *Journal of Physics A: Mathematical and General* 27:7079-7086.

Margules, C. R., and R. L. Pressey. 2000. Systematic conservation planning. Nature (London) 405:243-253.

Nagylaki, T. 1976. Clines with variable migration. *Genetics* 83:867-886.

Nurnberger, B. D., N. H. Barton, C. MacCallum, J. Gilchrist, and M. Appleby. 1995. Natural selection on quantitative traits in the Bombina hybrid zone. *Evolution 49*:1224-1238.

Pearson, S. F. 2000. Behavioral asymmetries in a moving hybrid zone. Behavioral *Ecology* 11:84-92.

Rohwer, S., E. Bermingham, and C. Wood. 2001. Plumage and mitochondrial DNA haplotype variation across a moving hybrid zone. *Evolution* 55:405-422.

Sayama, H., L. Kaufman, and Y. Bar-Yam. 2000a. Symmetry breaking and coarsening in spatially distributed evolutionary processes including sexual reproduction and disruptive selection. *Physical Review E* 62:7065-7069.

Sayama, H., L. Kaufman, and Y. Bar-Yam. 2000b. The role of spontaneous pattern formation in the creation and maintenance of biological diversity. InterJournal (http://www.interjournal.org/) Brief Article 417.

Sayama, H., L. kaufman, and Y. Bar-Yam. 2003. Spontaneous pattern formation and genetic diversity in habitats with irregular geographical features. *Conservation Biology* 17:893-900.

Slatkin, M. 1973. Gene flow and selection in a cline. *Genetics* 75:733-756.

Thoday, J. M. 1972. Disruptive selection. *Proceedings of the Royal Society of London B* 182:109-143.

Tilman, D., and J. A. Downing. 1994. Biodiversity and stability in grasslands. *Nature* (London) 367:363-365.

Tilman, D., and P. Kareiva, editors. 1997. *Spatial ecology: the role of space in population dynamics and interspecific interactions.* Princeton University Press, Princeton, New Jersey.

Turing, A. 1952. The chemical basis of morphogenesis. *Philosophical Transactions of the Royal Society of London B* 237: 37-72.

In: Focus on Biodiversity Research
Editor: Jan Schwartz, pp. 131-177

ISBN: 1-60021-372-3

Chapter 7

CHEMOTHERAPEUTICS BRAZILIAN PLANT SPECIES AS SOURCES OF ANTITUBERCULAR AND ANTIMALARIAL CHEMOTHERAPEUTICS

L. A. Basso, [1] ***A. G. Fett-Neto,*** [2] ***M. M. Campos,*** [3] ***D. S. Santos*** [1]

[1] Centro de Pesquisas em Biologia Molecular e Funcional, Instituto de Pesquisas Biomédicas, Pontifícia Universidade Católica do Rio Grande do Sul, RS, Brazil

[2] Laboratório de Fisiologia Vegetal, Centro de Biotecnologia, Universidade Federal do Rio Grande do Sul, RS, Brazil

[3] Faculdade de Odontologia, Pontifícia Universidade Católica do Rio Grande do Sul, RS, Brazil

1. ABSTRACT

The elucidation of biological and biochemical mechanisms of natural products with therapeutic action is of great utility for unveiling new chemical structure platforms on which to base the development of new drugs. Various organisms have been valuable sources for natural products with therapeutic activity. Among these, plants occupy a prominent position for the isolation of pharmacologically relevant metabolites. Approximately 25% of prescribed drugs come from plants. Paradoxically, there is a considerable lack of knowledge on plant chemistry compared to the existing diversity, particularly in the tropics. Tuberculosis (TB) is the leading cause of mortality due to a bacterial pathogen, *Mycobacterium tuberculosis*. More than 95% of 1.8 million deaths caused by TB in the year 2000 occurred in developing countries. Malaria, caused by the protozoan parasites of the genus *Plasmodium*, continues to be a major cause of children's morbidity and mortality throughout the world, particularly in the tropics. Even though antitubercular and antimalarial agents are available to treat these infectious diseases, drug resistance has posed a major challenge to public health programs. Accordingly, new drugs are needed to face resistant strains and improve current drug regimens. Brazil has extensive areas regarded as biodiversity hotspots, including the Atlantic Forest in the coast and the Cerrado in central Brazil. Moreover, the Amazon is probably going to be regarded as a hotspot as its species are described and their chemical diversity is investigated. In order to maximize the chances of finding bioactive components amongst a vast number of plant species, ethnobotanical and chemotaxonomical approaches have been employed. In addition, a description of a few biological targets to screen against natural-compound library is presented because this is an important decision that is critical to the long-term success of a drug discovery program. Here we also intend to present an array of plants from the Brazilian biodiversity and related taxa that have been reported as having antitubercular or antimalarial activity.

2. Introduction

In this section we try to convey the rationale for mining natural products as scaffolds for new chemotherapeutic agents (2.1), the need for the development of antitubercular and antimalarial agents (2.2), the use of defined molecular targets for the development of chemotherapeutic agents (2.3), and a brief description of brazilian plant species as possible sources of these agents (2.4).

2.1. Why Search for Natural Molecules?

The first records, written on clay tablets in cuneiform, are from Mesopotamia and date from about 2600 B.C.; among the substances that they used were oils of *Cedrus* species (cedar) and *Cupressus sempevirens* (cypress), *Glycyrrhiza glabra* (licorice), *Commiphora* species (myrth), and *Papaver somniferum* (poppy juice), all of which are still in use today for the treatment of ailments ranging from coughs and colds to parasitic infections and inflammation (Cragg and Newman, 2001). The Egyptian pharmaceutical record "Ebers papyrus" from 1500 B.C. documents some 700 drugs, which mostly are plants. The first record of Chinese Materia Medica dates from about 1100 B.C. and documentation of the Indian Ayurvedic system dates from about 1000 B.C. In the ancient Western world, the Greeks contributed substantially to the rational development of the use of herbal drugs. Theophrasus (in approximately 300 B.C.) describes in his History of Plants the medicinal qualities of herbs, Dioscorides (in approximately 100 A.D.) recorded the collection, storage and use of medicinal herbs. Galen (130-200 A.D.), a practitioner and teacher of pharmacy and medicine in Rome, described in detail his complex prescriptions and formulas used in compounding drugs. The monasteries in England, Ireland, France and Germany preserved the remnants of this Western knowledge during the Dark and Middle Ages. However, the Arabs were responsible for the preservation of much of the Greco-Roman expertise and for expanding it to include the use of their own knowledge, together with Chinese and Indian herbs unknown to the Greco-Roman world. Natural products are an attractive source of structural diversity and complexity of agents that exhibit potent biological activities and desirable pharmacological profiles. Natural products have thus inspired chemists and physicians for millennia. The elucidation of biological and biochemical mechanisms of natural products with therapeutic action have been invaluable to the efforts of organic and medicinal chemists as tools for deciphering the logic of biosynthesis and as platforms for drug development (Newman et al., 2000). Natural products and their derivatives have played a highly significant role in the drug discovery and development processes. Roughly half (49 %) of the New Chemical Entities (NCEs), which are also known as New Active Substances (NAS), introduced between 1981 and 2002 were natural products, semi-synthetic natural product analogues or synthetic compounds based on natural-product pharmacophores, which are an ensemble of steric and electronic features that is necessary to ensure optimal interactions with a specific biological target structure and to trigger or block its biological response (Newman et al., 2003). The vast majority of NCEs with antibacterial activity are natural products, their derivatives, a natural product mimic, or derived from total synthesis on the pharmacophore derived from a natural product. The structural complexity of many natural products and/or the need for renewable supply can limit the scope for making chemical modifications to optimize their therapeutic efficacy and be an obstacle to large-scale production. However, total synthesis in large quantities of discodermolide (a potent anti-cancer natural product) shows that the increasing efficiency of synthetic organic chemistry has reduced the barrier posed by limited natural supply, even for compounds with very complex structures (Gunasekera et al., 2004).

There has been a dwindling emphasis in the pharmaceutical industry on the discovery of natural products with biological activity. A number of factors can be attributed to this trend: the introduction of high-throughput screening (HTS) against defined molecular targets (moving from natural products extracts to "screen-friendly" synthetic chemical library); the development of combinatorial chemistry; advances in molecular biology, cellular biology and genomics (increasing the number of molecular targets and shortening drug discovery timelines); a declining emphasis among major pharmaceutical companies on infectious disease therapy (a traditional area of strength for natural products); and possible uncertainties with regard to collection of biomaterials as a result of the 1992 Rio Convention on Biological Diversity (Koehn and Carter, 2005). Coupled with the reduced emphasis on natural product discovery, large companies have greatly curtailed, wholly eliminated or spun off their antibacterial research. The reason for the Big Pharma stopping antibacterial drug discovery programs is the prevailing view that there are better ways to invest research dollars (Projan, 2003a). The public pressures on the pharmaceutical industry are driving drug discovery towards commercially "safer" products, as it is considered unseemly, by both the public and politicians alike, to make a profit on life-saving medications (Projan, 2003b). Therefore, as pointed out by Projan (2003b), governments believe themselves morally justified to ignore intellectual property rights, with the result that more and more companies will focus on chronic rather than acute diseases. Ironically, as more effective means of prolonging life-span and more effective treatments of ageing are available, infectious diseases are predicted to increase in significance. An additional reason for the weak pipeline of novel anti-infective drugs has been the lack of productivity of target-based drug discovery to identify novel agents (Projan, 2003b). The target-based screening identifies agents with both activity against the target and antimicrobial activity; however, at a later point in the drug discovery process it is determined that the antimicrobial activity is unrelated to the activity against the original molecular target. Notwithstanding, the target-based screening represents a rational approach to identifying promising lead compounds with the distinct advantage that knowledge of the target allows optimization of chemical compounds guided by a defined structural framework as long as the three-dimensional structure of the target is also known.

Nature uses a combinatorial approach in which fragments are attached to a central core by, for instance, methylations, acylations, glycosylation and oxidations. Synthetic molecules are increasingly produced by combinatorial chemistry approaches, in which a common core is elaborated by attaching combinations of fragments to reactive sites on the periphery of the core structure, thereby creating a large collection of synthetic chemical compounds or a synthetic combinatorial library. However, even large libraries of compounds from combinatorial chemical approaches used in screening do not reflect the rich chemical diversity indicated by the much larger chemical space covered by natural products (Dobson, 2004), since combinatorial libraries are designed more on the basis of chemical accessibility and maximum achievable size than on biologically relevant chemical diversity or properties (Martin and Critchlow, 1999). The reason for the lack of lead compounds from synthetic libraries in some therapeutic areas such as anti-infective, immunosuppressant, oncology, and metabolic diseases may be due to the different chemical space occupied by natural products and synthetic compounds. Accordingly, combinatorial libraries based on natural product templates, which have been honed by their evolutionary history for biological activity, are excellent starting points for structural diversification to increase the likelihood that a chemical compound with useful biological action is found especially in therapeutic areas that have a dearth of lead compounds (Brohm et al., 2002). Hits are defined as nonreactive compounds of verified structure and purity, which a confirmed minimal *in vitro* potency (< 20 μM) derived from high-throughput screening. Leads are compounds that, in addition to their potency, show some degree of selectivity for the molecular target, along with confirmed binding specificity, emerging structure-activity relationships, biophysical and absorption, distribution, metabolism, excretion and toxicity (ADME-Tox) properties, which are promising for their eventual optimization into small-molecule drugs (Wunberg et al., 2006). Chemical synthesis accessibility (e.g. number of synthetic steps) and ease in the synthesis of analogs may play an important role when considering a set of different leads for optimization. As compared to synthetic drugs or synthetic drug candidates, natural products typically have more stereogenic centres and more architectural complexity, they contain relatively more carbon, hydrogen and oxygen, and less nitrogen and other elements (Clardy and Walsh, 2004). Natural products have chemical functionalities that are compatible with the aqueous milieu of biological microenvironments and built-in conformational constraints that reflect the importance of reducing the entropy loss upon ligand binding to biological targets and thus increasing their potency. In addition, the functional-group array used by nature has instructed synthetic- and medicinal-chemists about bio mimetic strategies and isostere (shape-conserving) replacements. Computational chemistry has been used to measure those desirable chemical features that distinguish natural products from other sources of drug leads. Cristopher A. Lipinski's experimental and computational analyses of the World Drug Index led to the "rule-of-five" that identifies several key properties that should be considered for small molecules that are intended to be orally administered (Lipinski et al., 1997). These properties are: molecular mass <500 Daltons, number of hydrogen-bond donors <5, number of hydrogen-bond acceptors <10, calculated octanol-water partition coefficient <5 (using lipohilicity as an indicator of the ability of a molecule to cross biological membranes). Lead optimization of low-potency leads into drugs is often accompanied by increase in molecular weight and lipophilicity (Teague et al., 1999). Drug discovery should thus start with lead-like structures not only fulfilling the "rule-of-five" but also with even lower molecular weights and lipophilicity. Representative combinatorial, synthetic and natural-product compound libraries have been examined on the basis of molecular mass, number of chiral centres, molecular flexibility as measured by number of rotatable bonds and ring topology, distribution of heavy atoms, and Lipinsky-type descriptors (Feher and Schmidt, 2003). Natural products and drugs have similar molecular weight distribution, whereas combinatorial compounds appear to have higher molecular weights. The mean number of chiral centers among natural compounds is 6.2, in contrast to 2.3 in drugs and 0.4 in combinatorial molecules. Natural products have fewer rotable bonds than combinatorial compounds suggesting that natural compounds exploit the thermodynamic advantage conferred by rigidity to achieve superior binding properties, since a rigid ligand will exhibit stronger binding due to lower entropic losses if a flexible and a rigid ligand can form the same pattern of hydrogen bonds and hydrophobic interactions with the target. With respect to elemental composition, drug and combinatorial molecules tend to have a significantly higher number of nitrogen-, sulphur- and halogen-containing groups, whereas natural products bear a higher number of oxygen atoms. As compared to synthetic drugs and combinatorial libraries, natural products have lower ratio of aromatic ring atoms to total heavy atoms, and higher number of solvated hydrogen-bond donors and acceptors and greater molecular rigidity. Natural-product libraries also have a broader distribution of molecular properties such as molecular mass, octanol-water partition coefficient and diversity of ring systems compared with synthetic and combinatorial counterparts. Interestingly, the fraction of natural product structures with two or more "rule-of-five" violations is quite low (approximately 10 %) and equal to that of trade drugs (Lee and Schneider, 2001). A statistical analysis on the physical properties and structure fragments of over 1000 marketed orally administered drugs concluded that they tend to have lower molecular mass and have fewer hydrogen-bond donors, acceptors, and rotable bonds than drugs that have other routes of administration (Vieth et al., 2004). It has more recently been shown that there is also a dependence of molecular properties on proteomic family for marketed oral drugs (Vieth et al., 2006). Hence these rules can be used to guide selection of natural products derived from plant extracts with drug-like properties to increase the likelihood of discovering a potential therapeutic agent and subsequent market launch of a New Molecular Entity (NME is a medication containing an active ingredient that has not been previously approved for marketing in any form). In addition, natural products have been described as a population of privileged structures selected by evolutionary pressures to interact with a wide variety of proteins and biological targets for specific purposes which is supported by the fact that natural products have become effective drugs in a wide variety of therapeutic indication (Koehn and Carter, 2005). Privileged scaffolds are structures commonly found in a wide variety of bioactive natural products. Accordingly, the emerging area of diversity-oriented synthesis is beginning to provide natural product-like chemical compounds that will hopefully be effective as drugs. In addition, libraries of natural-product derivatives, natural-product-like compounds prepared by total synthesis, and libraries derived from natural-products are different approaches to constructing screening libraries (Boldi, 2004). These approaches may also benefit from an elegant description of a strategy to increase the likelihood of success of the hit-to-lead process, including analysis of solubility, molecular volume, lipophilicity, polar surface area, rotable bonds, metabolic clearance, human ether-a-go-go related gene activity, cytochrome P450 3A4 inhibition, stability in human microsomes, and undesirable chemical properties (Wunberg et al., 2006).

For thousands of years medicine and natural products have been closely linked through the use of traditional medicines and natural poisons (Newman et al., 2000). Although there are a number of sources for natural products that have proved extremely valuable as therapeutic agents, in this section we will focus on a few examples of natural products derived from plants. In addition, the focus in this section will be placed on a few examples of antibacterial drugs derived from plants. Clinical, pharmacological, and chemical studies of traditional medicines, which were derived predominantly from plants, were the basis of most early medicines such as aspirin, digitoxin, morphine, quinine, emetine and pilocarpine. Quinoline antimalarials and related aryl alcohols are based on the chemical structure of quinine, an active ingredient of *Chinchona* bark. In addtion, the sesquiterpene lactone endoperoxide Artemisinin - the active ingredient of the Chinese herb “qinghao” (*Artemisia annua*), and its derivatives have been used increasingly over the past two decades as antimalarials (Haynes, 2001). Screening of plant extracts against the antimalarial target plasmepsin II identified active compounds from the stem bark and leaves of *Albizia adinocephala* collected from Panama, and fractionation of the stem bark extract led to isolation of two new macrocyclic spermine alkaloids as the active components (Butler, 2004). There is no plant-derived compound in the current chemotherapy of tuberculosis. Notwithstanding, there have been reports on antitubercular activity of plant-derived compounds, for instance, chemical compounds found in hexane and dichloromethane extracts of air-dried roots of *Polyalthia evecta*, a small tree found in Thailand (Kanokmedhakul et al., 2006).

As mentioned above, plants have formed the basis for traditional medicine systems, which have been used for thousands of years in countries such as China and India. Clearly, plants hold a prominent position in the available sources of natural bioactive molecules. The sessile character of plants required the development of a large array of molecules to effectively respond to biotic and abiotic stimuli *in situ*. The chemical diversity of plants has made them the source of choice for the isolation of pharmacologically relevant metabolites. Approximately 80% of the world’s inhabitants rely mainly on traditional plant-based medicines for their primary health care (Cragg and Newman, 2001). Approximately 25% of the drugs prescribed worldwide come from plants, whereas 11% of the 252 drugs considered as basic and essential by the World Health Organization (WHO) derive exclusively from plants (Rates, 2001). Developing countries continue to rely on ethnobotanical remedies as their primary medicines, leaving almost 75% of the world population without access to the modern healthcare products taken for granted in the West (Raskin et al., 2002). Prescription drugs containing phytochemicals were valued at more than US$ 30 billion in 2002 in the USA alone (Raskin et al., 2002). Paradoxically, in contrast to this extreme relevance of plants as sources for new drugs, there is a considerable lack of knowledge on plant chemistry compared to the existing diversity, particularly in the tropics. Of the estimated 400,000 higher plant species in the world, only about 10% have been characterized chemically to some extent, whereas of the approximately 100,000 secondary metabolites that have been discovered from the plant kingdom, only half have been fully elucidated (Oksman-Caldentey and Inzé, 2004). Therefore, the unexplored potential of plants as a source of novel bioactive chemicals is enormous.

The typical process of discovering natural-product hits, and their progression towards development involves a number of steps (Koehn and Carter, 2005). The natural product is extracted from the source, concentrated, bio-assay-guided fractionated and purified, yielding a single biologically active compound. Efforts are made to rapidly identify known compounds by HPLC-mass spectrometry to avoid duplication. The correlation of molecular mass and UV absorption data with known compounds by database searching is ordinarily sufficient to classify sets of compouns. The *de novo* structure determination of compounds that are New

Molecular Entities is currently carried out by high-resolution NMR technologies. Preliminary structure-activity relationship studies are carried out for the candidates meeting criteria for potency and selectivity and the purification process is scaled up. Once the feasibility of mudulating biological response through synthetic modification is established, the hit is declared a lead and proceeds onward for additional optimization by traditional medicinal chemistry. In addition to being a proven and important source of drug leads, natural products-derived drugs also contribute significantly to the profitability of many pharmaceutical companies. Despite the many advances in technology, the cost of generating new drugs is inexorably rising, leading to ever greater pressure on pharmaceutical companies to focus on developing therapies primarily for the common diseases of wealthy countries (Service, 2004). Many disciplines and resources have to come together to deliver success to drug discovery and development efforts. Genomics can provide many potential molecular targets, but only those targets that can be readily manipulated and tested are likely to provide an opportunity to discover a lead molecule. After selecting a molecule for further development, other factors have to be considered such as ease to manufacture (low cost is crucial for antimycobacterials and antimalarials), stability, ready formulation, bioavailability (that is, extensively adsorbed from the gut and avoiding first pass metabolism in the liver to achieve effective concentrations in the systemic circulation), an appropriate half-life, and low toxicity. Good drugs are only useful if they are made available to the people who need them and are used properly. Hence the availability of inexpensive compounds, sound policies on their use, strong healthcare infrastructure, and sufficient funds to purchase the drugs that are needed are all vital (Ridley, 2002). In addition, many effective and affordable antimycobacterial and antimalarial drugs should be in reserve to meet the challenge of resistance as it arises.

2.2. Why Chemotherapeutic Agents to Treat Tuberculosis and/or Malaria?

Tuberculosis

Despite the availability of drugs to treat tuberculosis (TB) since the 1940s, TB remains an important cause of mortality due to a bacterial pathogen, *Mycobacterium tuberculosis*, second only by the human immunodeficiency virus (HIV) (WHO, 2004). It is estimated that 8.9 million new cases of TB occurred worldwide in 2004 based on results of surveys of the prevalence of infection and disease, assessments of the effectiveness of surveillance systems and death registrations (Dye, 2006). Approximately 3.9 million cases were sputum-smear positive, which is the most infectious form of the disease (WHO, 2006), and approximately 80% of individuals newly diagnosed with the disease every year live in the 22 most populous countries (Dye, 2006). Of the 1.8 million deaths estimated to have occurred in the year 2000, more than 95 % of these were in developing countries (Corbett et al., 2003). In addition, approximately 12 % (226,000) of total deaths from TB was attributed to co-infection with *M. tuberculosis* and human immunodeficiency virus (TB-HIV). Immune deficient patients with HIV are at increased risk of latent *M. tuberculosis* infections progressing to active disease, and being transmitted to others (Morens et al., 2004). TB is a leading cause of dedath among people who are HIV-positive (WHO, 2006). A total of 39 million people were estimated to be infected with HIV in 2004 (UNAIDS, 2004). Approximately 2 billion individuals are believed to harbor latent TB based on tuberculin skin test surveys (Dye et al., 1999). The TB-HIV association can thus have catastrophic effects on human health worldwide demanding control policies to be revised and investments in drug development to combat TB to be increased.

Approximately 3.2 % of new cases of TB worldwide in 2000 were multidrug-resistant tuberculosis (MDR-TB), defined as resistant to at least isoniazid and rifampicin (Espinal, 2003). According to the 2004 Global TB Control Report of the World Health Organization, there are 300,000 new cases per year of MDR-TB worldwide, and 79 % of MDR-TB cases are now "super strains", resistant to at least three of the four main drugs used to treat TB (WHO, 2004). Localized high incidence rates of MDR-TB have been found in particular regions, which have been defined as hot zones based on either areas where >5 % of current cases are MDR-TB (prevalence) or areas where >5 % of new cases are MDR-TB (incidence) (Espinal, 2003). MDR-TB is an airborne bacterium that is spread just as easily as drug-sensitive TB. An individual who is sick with any strain of TB will infect between 10 and 20 people each year with that same strain (Nunn, 1997), thereby making the hot zones of particular concern to public health officials. MDR-TB treatment requires the use of second-line drugs (SLDs). However, the bacteriostatic SLDs (amikacin, kanamycin, capreomycin, cycloserine, *para*-aminosalicylic acid, ethionamide, and fluoroquinolones) are less effective, more toxic, and are given for at least three times as long and at 100 times the cost of basic short-course chemotherapy regiments (Pablos-Méndez et al., 2002). In 2000, the Stop TB Partnership's Green Light Committee was created to increase access to SLDs worldwide while ensuring their proper use to prevent increased drug resistance. While assisting MDR-TB treatment programs worldwide, the committee encountered reports of multiple cases of TB with resistance to virtually all SLDs. The Centers for Disease Control and Prevention (CDC) and the World Health Organization surveyed the WHO/International Union Against Tuberculosis and Lund Disease Global Supranational TB Reference Laboratory Network to assess the frequency and distribution of extensively drug-resistant (XDR) TB cases, which are defined as cases in persons with TB whose isolates were resistant to isoniazid and rifampicin and at least three of the six main classes of SLDs (aminoglycosides, polypeptides, fluoroquinolones, thioamides, cycloserine, and para-aminosalicylic acid). This survey determined that, during 2000-2004, of 17,690 TB isolates, 20% were MDR and 2% were XDR (CDC, 2006). XDR-TB has a wide geographic distribution, including the USA where TB control has been effective for many years, and is associated with worse treatment outcomes than MDR-TB. This is the first report on data regarding the occurrence of SDR-TB worldwide. The emergence of XDR-TB worldwide raises the bleak prospect of a future epidemic of virtyually untreatable TB. The factors that most influence the emergence of drug-resistant strains include inappropriate treatment regimens, and patient noncompliance in completing the prescribed courses of therapy due to the lengthy standard "short-course" treatment or when the side effects become unbearable (Duncan, 2003). Improper treatment of patients with drug-resistant TB due to the use of fewer drugs or drug for too short a time or relying on limited access to poor quality second-line drugs might lead to an increase in SDR-TB cases.

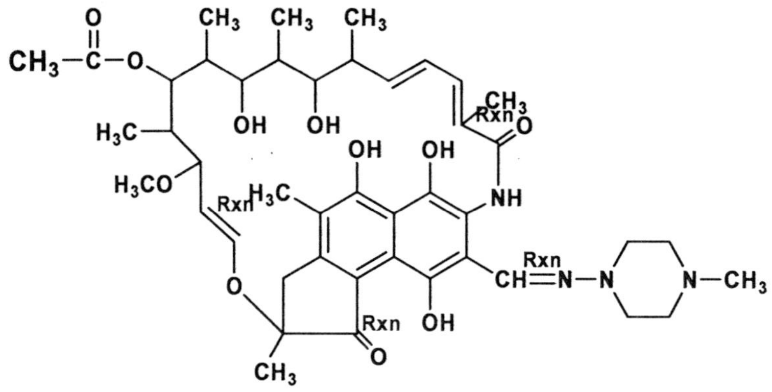

Rifampicin

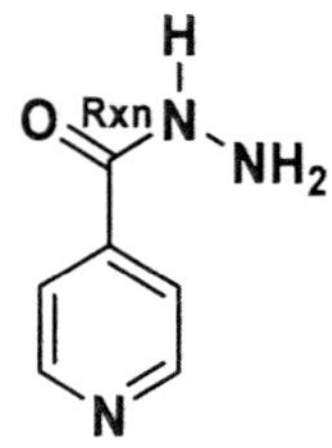

Isoniazid

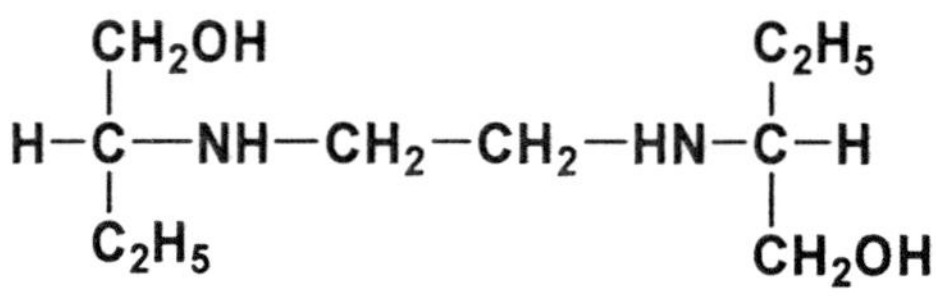

Ethambutol

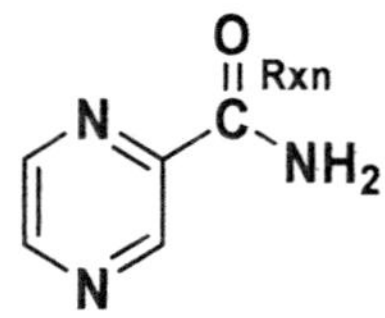

Pyrazinamide

Figure 1. First-line TB drugs.

TB is transmitted by *M. tuberculosis*-containing aerosolized droplets generated by the cough of a person infected with bacilli. The primary infection involves replication of *M. tuberculosis* within alveolar macrophages of the terminal air spaces of the lung, spread to local lymph nodes within the lung, and eventual dissemination of infection to remote sites in the body. However, only 5-10 % of infected individuals develop the disease because the host immune response against *M. tuberculosis* is highly effective in controlling bacterial replication. Nevertheless, *M. tuberculosis* is almost never eradicated due to its ability to establish and maintain latency, a period during which the infected person does not have clinically apparent TB. Post-primary TB, which is predominantly a pulmonary disease, develops later in life, and can be caused either by reactivation of bacteria remaining from the initial infection or by failure to control a subsequent infection. Post-primary TB involves extensive damage to the lungs and efficient aerosol transmission of bacilli. Effective chemotherapies that decreased TB mortality rate worldwide were introduced in the 1940s and 1950s, including streptomycin in 1944 (Schatz et al., 1944), and both isoniazid and pyrazinamide in 1952 (Middlebrook, 1952; Kushner et al., 1952). The later introduction of ethionamide, rifampicin, ethambutol and ciprofloxacin to the arsenal used to treat tuberculosis seemed to provide an adequate number of effective antimicrobial agents (Basso and Blanchard, 1988). The modern, standard "short-course" therapy (Mitchison, 1985) for TB is based on a four-drug regimen of isoniazid, rifampicin, pyrazinamide, and ethambutol or streptomycin for two months, followed by treatment with a combination of isoniazid and rifampicin for an additional four months (**Figure 1**). This combination therapy must be strictly followed to prevent drug resistance and relapse, and direct observation of patient compliance is the most reliable way to ensure effective treatment and prevent the acquisition of resistance. The DNA mutations identified in drug-resistant strains of *M. tuberculosis* have been reviewed elsewhere (Basso and Blanchard, 1998; Ramaswamy and Musser, 1998; Rattan et al., 1998; Glickman and Jacobs, 2001; Schroeder et al., 2002). Interestingly, it has recently been proposed that the error-prone DnaE2 Polymerase protein may be an important mechanism for generating mutations that result in the establishment of drug-resistant strains of *M. tuberculosis in vivo* (Boshoff et al., 2003). The bacteriostatic second-line drugs are reserved to strengthen the treatment of drug-resistant disease or when bactericidal drugs are prohibited because of toxicity. *M. tuberculosis* has been considered the world's most successful pathogen and this is largely due to the ability of the bacillum to persist in host tissues, where drugs that are rapidly bactericidal *in vitro* require prolonged administration to achieve comparable effects (Hingley-Wilson et al., 2003; Gomez and McKinney, 2004). Hence more effective and less toxic anti-tubercular agents are urgently needed to shorten the duration of current treatment, to provide effective treatment of latent tuberculosis infection, and to improve the treatment of both MDR-TB and, importantly, XDR-TB.

Malaria

Approximately 3 million deaths due to malaria and close to five billion episodes of clinical illness possibly meriting antimalarial therapy occur throughout the world, with more than 90 % of cases occurring in Africa (Breman et al., 2004). Malaria continues to be a major cause of children's morbidity and mortality throughout the world. Morbidity, mortality, and disability from malaria (and other diseases) is summarized in disability-adjusted life years (DALYs) metric, an aggregate measure of premature mortality, morbidity, and disability, which can also be used for analyzing cost-effectiveness of interventions. Malaria deaths are responsible for almost 3 % of the world's DALYs. More than 50 % of the world's population is exposed to malaria, despite advances in understanding malaria ecology and development of interventions. An empirical approach combining epidemiological, geographical and demographic data used to estimate the number of clinical events caused by *Plasmodium falciparum* worldwide deduced that there were 515 (range 300-600) million episodes of clinical *P. falciparum* malaria in 2002 (Snow et al., 2005). These global estimates are up to 50 % higher than those reported by the World Health Organization. A recent report on global burden of diseases in 2001 showed that the malaria mortality rate increased during the 1990s with the proportion of deaths in children younger than 5 years old increasing from 15% to 22% (Lopez et al., 2006). This study also reported that there were 1.21 million deaths from malaria in low-and-middle-income countries. Malaria is caused by protozoan parasites of genus *Plasmodium* (four species affecting humans) and transmitted by the *Anopheles* mosquito. *Plasmodium falciparum* is responsible for the most severe forms of malaria that frequently can be deadly disease if not conveniently treated. Malaria starts by the bite of infected female *Anopheles* mosquitoes that inject sporozoites of *Plasmodium* parasite. Sporozoites invade hepatocytes, transform, multiply, and develop into tissue schizonts. Tissue schizonts rupture and release thousands of pathogenic merozoites that invade erythrocytes in the blood. In erythrocytes, most parasites undergo asexual development from young ring forms to trophozoites and finally to mature schizonts. Erythrocytes containing schizonts rupture and release merozoites. The released merozoites invade more erythocytes to continue the cycle. These repeated cycles of asexual reproduction of the parasite is responsible for the disease while the sequestration of mature forms of the *P. falciparum* parasites inside the erythocytes at the endothelial cells of capillary vessels is responsible for severe malaria forms like cerebral and pulmonary malaria (Kappe et al., 2003). Some erythrocytic parasites differentiate into sexual forms known as gametocytes. After infected human blood is ingested by a female mosquito, exflagellation of the male gametocyte is followed by male gametogenesis and fertilization of the female gametocyte in the gut of the insect. The resulting zygote, which develops as an oocyst in the gut wall, eventually gives rise to the infective sporozoite, which invades the salivary gland of the mosquito. The insect can then infect another human host by taking a blood meal. The invasive merozoite stage of the parasite is important immunologically because it is briefly extracellular and thus exposed to host antibodies between leaving one red blood cell and entering the next. An excellent description of the ultrastructure of *Plasmodium falciparum* asexual blood stages has been published by Bannister et al. (2000).

The traditional mainstay of treatment of *Plasmodium falciparum* malaria is quinine. During World War II, owing to shortage in supply of quinine, synthetic alternative therapeutic drugs replaced quinine for prophylaxis and routine treatment. Knowledge of quinine structure inspired the development of hundreds of synthetic quinoline-based antimalarial drugs, including chloroquine (the classic hemozoin-targeted agent), mefloquine, primaquine, and amodiaquine (**Figure 2**). For decades people in underdeveloped regions of the world have relied upon chloroquine for treatment of malaria due to its easy access, wide distribution, and minimal side effects. Nevertheless, the appearance of widespread chloroquine-resistant malaria has compromised this therapeutic course of action (Arav-Boger and Shapiro, 2005). Clinical failure of mefloquine and atovaquone (a hydroxy-naphthoquinone synthetic derivative) was reported soon after their introduction (Edwards and Biagini, 2006). The antifolate combination of pyrimethamine, a dihydrofolate reductase inhibitor related to trimethoprim, and sulfadoxine, a long-acting sulfonamide, has been used for prophylaxis and human malaria, in particular for those cases of chloroquine-resistant strains of *P. falciparum*. However, resistance to this formulation has rapidly developed and some fatal reactions have been associated with this drug combination. This combination is thus no longer recommended for long-term prophylaxis. Proguanil (a biguanide derivative) is metabolized to cycloguanil which, in turn, inhibits the bifunctional plasmodial dihydrofolate reductase-thymidylate synthetase. However, overlapping resistance to cycloguanil and pyrimethamine may cast doubts on the long-term efficacy of its therapeutic use. New combination therapies are in the pipeline to minimize the spread of resistance and it is hoped that they can curb the relentless march of drug resistance (Edwards and Biagini, 2006). The emergence of resistant strains made it mandatory to broaden efforts to look into alternative effective antimalarial drugs. Artemisinin (a sesquiterpene lactone endoperoxide) and and semi-synthetic derivatives have been used increasingly over the past two decades as antimalarials. These derivatives include artemether (an oil-soluble methyl ester), arteether and artesunate (the water-soluble hemisuccinate salt of dihydroartemisinin), which are all metabolized to dihydroartemisinin – the main active agent in the body, and are fast acting and act against gametocytes, the sexual stages of *Plasmodium* that infect mosquitoes. Although artemisinin is highly effective against multidrug-resistant *Plasmodium* spp., it is in short supply and unaffordable to most malaria sufferers because artemisinin total synthesis is difficult and costly (Enserink, 2005). The semi-synthesis of artemisinin using artemisinic acid, its immediate precursor, derived from a microbial source could be a cost-effective, environmentally friendly, high-quality and reliable strategy.

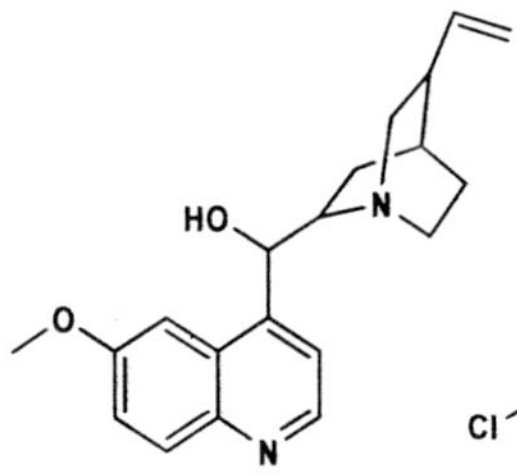

Quinine

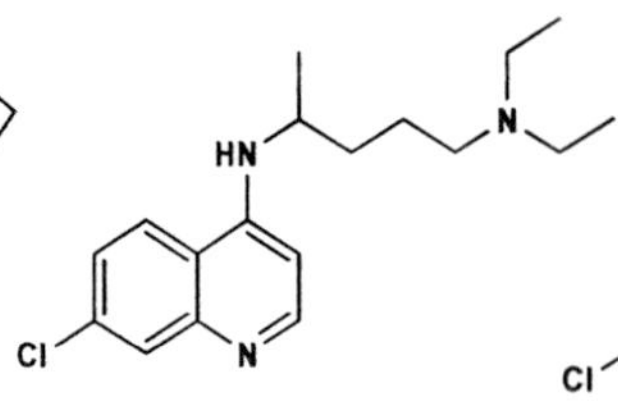

Chloroquine

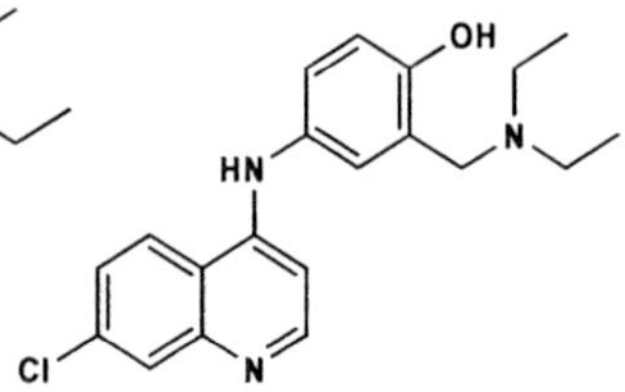

Amodiaquine

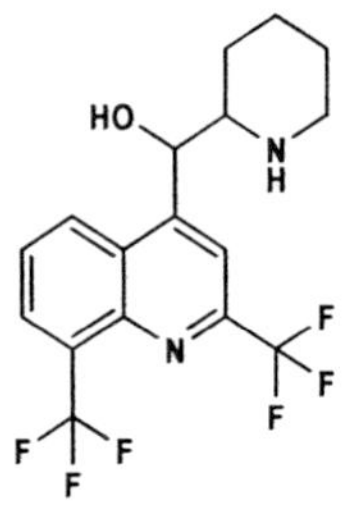

Mefloquine

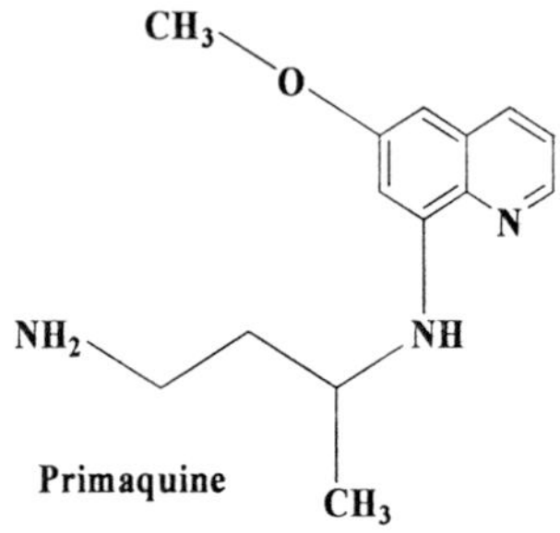

Primaquine

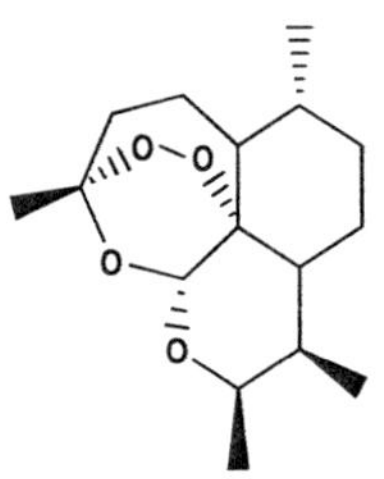

Artemisinin

R=H	Dihydroartemisinin
R=Me	Artemether
R=Et	Arteether
R=CO(CH2)2CO2H	Artesunate

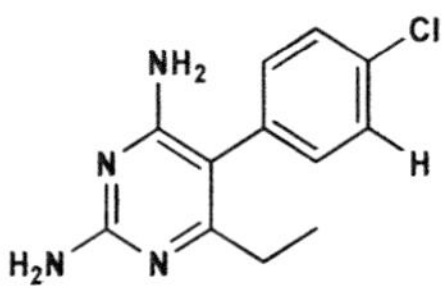

Pyrimethamine

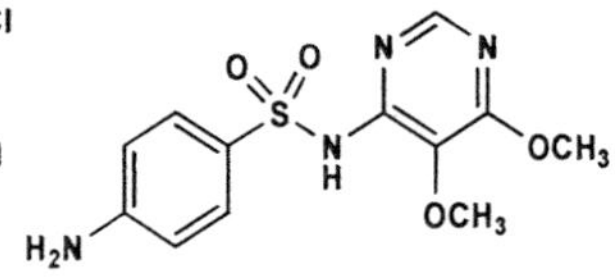

Sulphadoxine

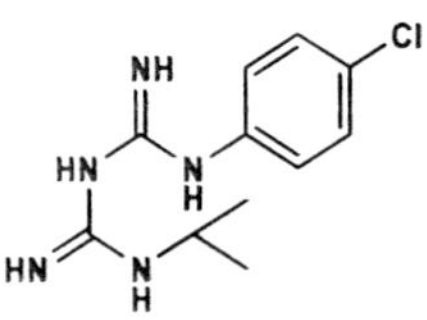

Proguanil

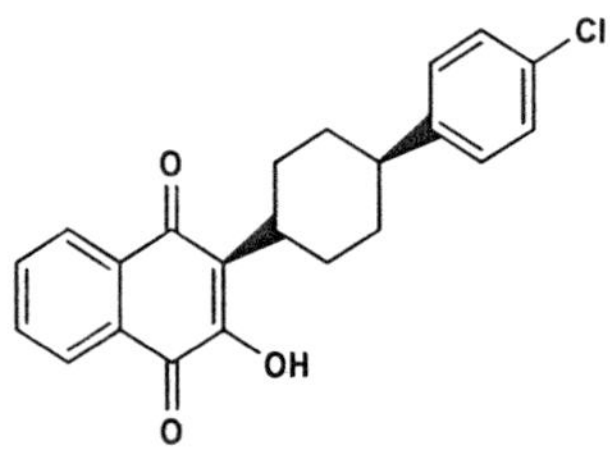

Atovaquone

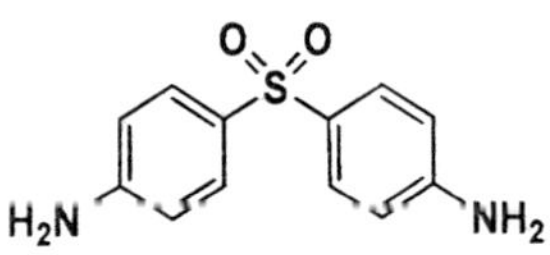

Dapsone

Atovaquone

Figure 2. Structures of antimalarial drugs.

It has recently been reported the engineering of *Saccharomyces cerevisiae* to produce high titres of artemisinic acid using an engineered mevalonate pathway, amorphadiene synthase, and a novel cytochrome P450 monooxygenase (CYP71AV1) from *Artemisia annua* that performs a three-step oxidation of amorpha-4,11-diene to artemisinic acid (Ro et al., 2006). Coupled with relatively high-yield chemistry for the conversion of artemisinic acid to artemisinin (Acton and Roth, 1992), the microbially produced artemisinic acid is a promising low-cost source of artemisinin. Artemisinin inhibits the sarcoendoplasmic reticulum calcium-dependent ATPase (SERCA) *PfATPase6* thereby disabling the calcium pump in malaria parasites (Uhlemann et al., 2005). DNA mutations resulting in resistance to artemether, an artemisinin derivative, have been identified in clinical isolates of *P. falciparum* from areas with uncontrolled use of artemisinin derivatives (Jambou et al., 2005). As has recently been pointed out the prospect of plentiful artemisinin is encouraging, but if the parasite becomes resistant, increased drug production will be worthless (Towie, 2006). Continued and sustainable improvements in antimalarial medicines through focused research and development are essential for the future ability to treat and control malaria worldwide. Unfortunately, malaria is a disease of poverty, and despite a wealth of scientific knowledge there is insufficient market incentive to generate the competitive, business-driven industrial antimalarial drug research and development that is normally needed to deliver new products. Mechanisms of partnering with industry should be established to overcome this obstacle and to open up and build on scientific opportunities for improved chemotherapy in the future. In the meantime, the best prospect for drug development and use in the short term is the government commitment to mechanisms and projects to ensure that improved drugs are sustainably discovered and developed and new targets for antimalarial agents are unveiled.

2.3. Defined Targets for Antituberculosis and Antimalarial Drug Development

The sequencing of the genomes of pathogens should allow identification of genes essential for the survival of the pathogens, and their encoded proteins may serve as molecular targets for new drug discovery efforts. The choice of what biological targets to screen against a natural-product library is an important decision that is critical to the long-term success of a drug discovery program. In addition, as history has already shown, development of chemotherapeutic agents aimed at a single target has failed to stem the tide of drug resistance and spread of a particular infectious disease. Target identification is an important step in the drug discovery process that can provide the foundation for years of dedicated research. Drug targets must satisfy a variety of criteria, such as low homology between target and host to prevent toxicity, activity of the target in the disease state, and the essentiality of the target to the pathogen's growth and survival. The availability of the *M. tuberculosis* genome sequence (Cole et al., 1998) and mycobacterial genetic tools, such as transposon mutagenesis, gene knockout and gene transfer should greatly facilitate target identification and validation. There have been a number of reviews describing molecular targets for antitubercular agent development, including proteins involved in lipid metabolism, carbohydrate and amino acid transport and metabolism, inorganic ion transport and metabolism, nucleotide transport and metabolism, RNA synthesis, protein synthesis, folate pathway and mycobactin synthesis, membrane energy metabolism, cell wall biogenesis, DNA replication, recombination and repair, post-translational modification, coenzyme metabolism and signal transduction (Schroeder et al., 2002; Sassetti and Rubin, 2003; Anishetty et al., 2005; Basso et al., 2005; Raman et al., 2005; Hasan et al., 2006; Zhang et al., 2006). Completion of genome sequence of *P. falciparum* (Gardner et al., 2002), and growing sequence databases for other *Plasmodium* species (Bahl et al., 2002), should provide valuable tools for whole-genome analysis of Aplicomplexa to allow identification of new targets for drug development. Targets for antimalarial drug development include the folate metabolism (dihydrofolate reductase and dihydropteroate synthase), electron transport, host hemoglobin catabolism by plasmodial proteases (plasmepsins, falcipains, and falcilysin), hemozoin formation, purine salvage pathway (malaria parasites cannot synthetize purines), pyrimidine synthesis (malaria parasites cannot salvage pyrimidines), parasite phospholipid metabolism, mitochondrial electron transport (ubiquinol-cytochrome *c* oxidoreductase), and surface anion channels (Gornick, 2003; Rosenthal, 2003; Biagini et al., 2005; Sharma, 2005). The search for new and effective drugs against these pathogens has been boosted during the last years by an unexpected finding. Through molecular and cell biological analysis it was realized that probably most members of this phylum harbor a plastid-like organelle, called the apicoplast, which probably is derived from the engulfment of a red algae in ancient times. Although the apicoplast itself contains a small circular genome, most of the proteome of this organelle is encoded in the nuclear genome, and the proteins are subsequently transported to the apicoplast. The latter is assumed to contain a number of unique metabolic pathways not found in the vertebrate host, making it an ideal source of drug targets (reviewed by Seeber, 2003). A detailed description of the protein targets of the metabolic pathways mentioned above is beyond the scope of this chapter. Emphasis will be placed on the metabolic pathways from *M. tuberculosis* and *P. falciparum* that our research group has chosen for antitubercular and antimalarial drug development efforts.

2.3.1. Tuberculosis

2.3.1.1. Biosynthesis of Mycolic Acids (FAS-II System)

The readers are referred to a number of reviews that describe in more detail the structure, function, and biogenesis of the mycobacterial cell wall, including the FAS-II system (Brennan and Nikaido, 1995; Kolattukudy et al., 1997; Barry et al., 1998; Asselineau et al, 2002; Schroeder et al., 2002; Brennan, 2003; Basso and Santos, 2005; Takayama et al., 2005). Here a brief description of the mycobacterial cell wall structure and biosynthesis will be given to avoid repeating information given in previous manuscripts except when deemed necessary for clarity. The mycobacterial cell wall is comprised of three covalently linked macromolecules: peptidoglycan, arabinogalactan, and mycolic acid, which is often described as mycolyl-arabinogalactan-peptidoglycan complex (mAGP; **Figure 3**). Mycolic acids are high-molecular-weight α-alkyl, β-hydroxy fatty acids (**Figure 4**), which appear mostly as bound esters in tetramycolylpentaarabinosyl cluster in the mycobacterial cell wall. In the pyrolytic cleavage of mycolic acids the intact fatty acid released is often referred to as the α branch since it occupies the alpha position with respect to the carboxylic acid group. The aldehyde released is referred to as the meroaldehyde and the corresponding segment of the intact mycolate is often referred to as the meromycolate branch.

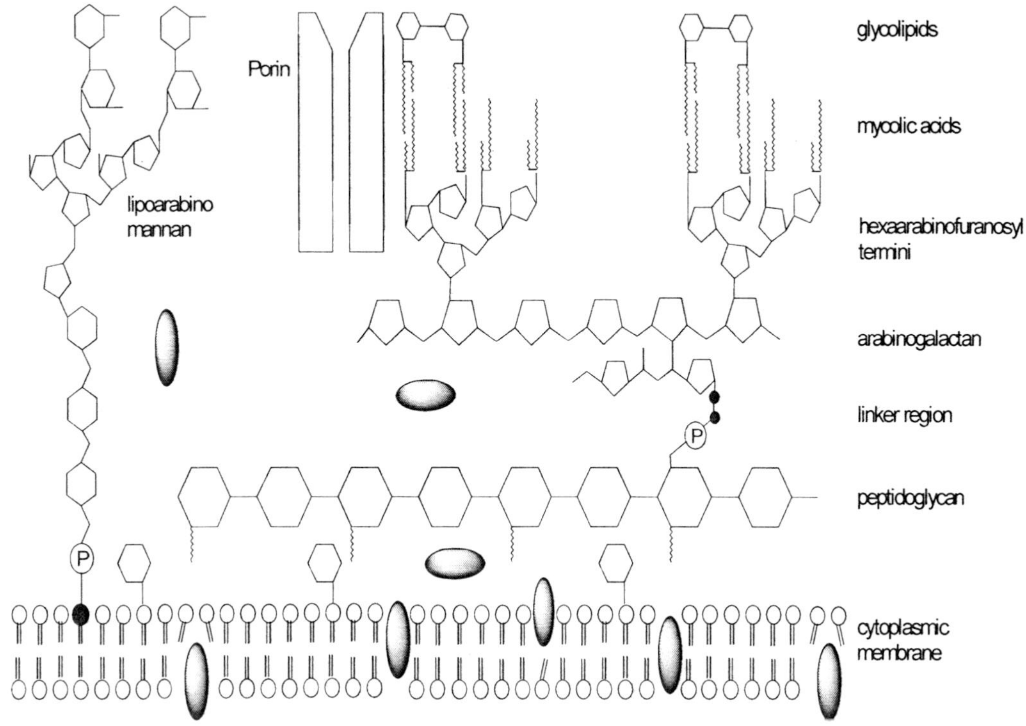

Figure 3. Schematic representation of the mycobacterial cell wall (not drawn to scale). The cytoplasmic membrane is encapsulated by a layer of peptidoglycan. The peptidoglycan backbone is attached to arabinogalactan through an unusual disaccharide phosphate linker region. The arabinogalactan is a branched-chain polysaccharide consisting of a proximal galactose chain linked to a distal arabinose chain. The hexaarabinofuranosyl termini of arabinogalactan are esterified to mycolic acids. The mycolic acid chains are shown perpendicular to the cytoplasmic membrane with the exposed chains interacting with the mycolic chains of trehalose dimycolate. Another major component non-covalently associated to the mycobacterial cell wall is the immunogenic lipoarabinomannan, which is attached to the cytoplasmic membrane by a phosphatidylinositol anchor. Small and hydrophilic solutes diffuse through water-filled protein channels, porins, whereas hydrophobic compounds use the lipid pathway.

Fatty acid elongation occurs through repetitive cycles of condensation, β-keto reduction, dehydration, and enoyl reduction which are catalyzed by, respectively, β-ketoacyl synthase (KAS, condensing enzyme), β-ketoacyl reductase (KAR; MabA in *M. tuberculosis*), β-hydroxyacyl dehydrase (DE), and enoyl reductase (ENR; InhA in *M. tuberculosis*). These chemical reactions are catalyzed by two types of fatty acid synthase systems (FAS). The FAS-I system is a multidomain polyprotein that encodes all the enzymes necessary for fatty acid synthesis in one large polypeptide and is generally present in most eukaryotes, except in plants. FAS-II systems, which are present in bacteria and plants, catalyze the individual reactions by separate proteins readily purified independently of the other enzymes of the pathway and are encoded by unique genes. Mycobacteria, unlike most organisms, have both FAS-I and FAS-II systems (Brindley, 1969). The mycobacterial FAS-I system (**Figure 5**) catalyzes the synthesis of C_{16} and C_{18} fatty acids, the normal products of *de novo* synthesis, and elongation to produce C_{24} and C_{26} fatty acids (Bloch, 1975). The mycobacterial FAS-II (**Figure 6**) is analogous to other bacterial FAS-II systems, with the notable exception of primer specificity. The mycobacterial FAS-II is not capable of *de novo* synthesis from acetate but instead elongates the "short-chain" acyl-ACP intermediates to fatty acids ranging from 24 to 56 carbons in length (Mdluli et al., 1998; Slayden and Barry, 2002). In summary, the mycobacterial FAS-I produces a bimodal ($C_{14:0}$-$C_{16:0}$ to $C_{24:0}$-$C_{26:0}$) distribution of acyl-CoA fatty acids. The mycobacterial FAS-I system would provide the "shorter" acyl-CoA fatty acid precursors ($C_{14:0}$-$C_{16:0}$) for condensation with malonyl-ACP by mtFabH enzyme activity whose products, in turn, would be elongated by the FAS-II system, yielding the long carbon chain of the meromycolate branch (50-60 carbons) of mycolic acids. The longer chain acyl-CoA products ($C_{24:0}$-$C_{26:0}$) of FAS-I would be excluded from chain elongation and remain available to be utilized, presumably in the CoA form, as substrates for formation of the α-alkyl branch (20-26 carbons) of mycolic acids.

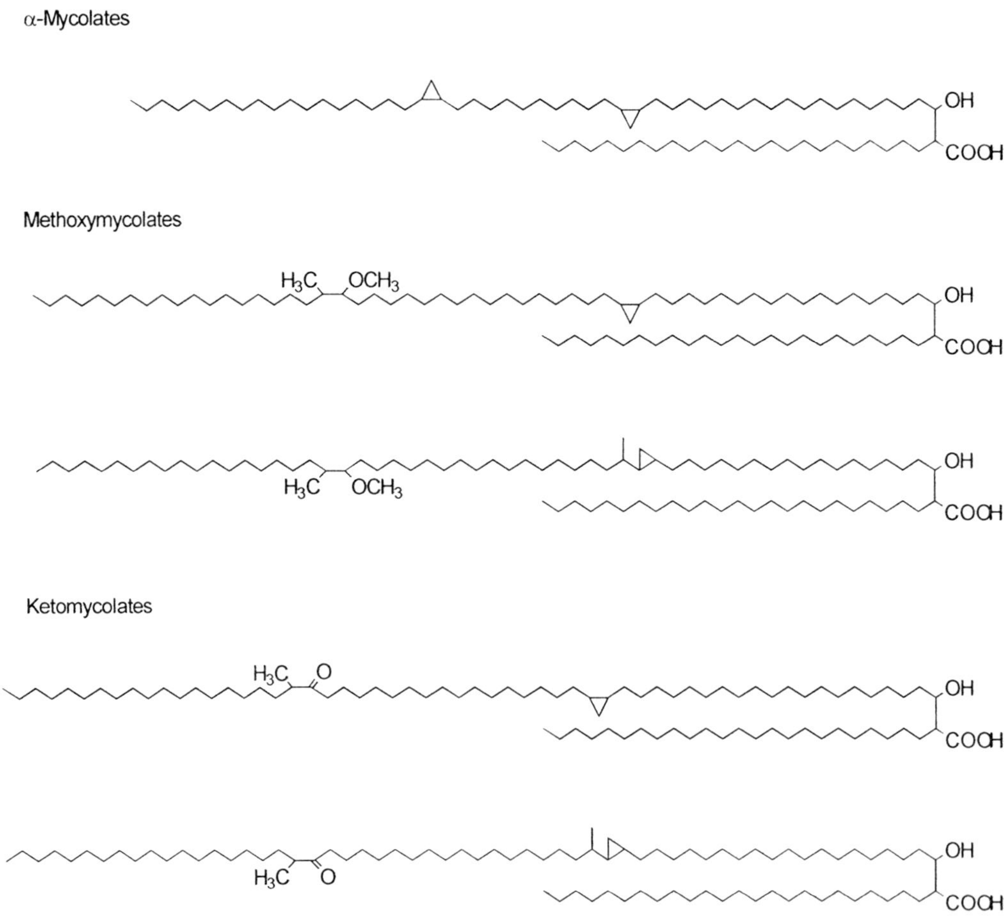

Figure 4. The structures of mycolic acids identified in *M. tuberculosis*. α-Mycolates: its meromycolate chain contains two *cis*-cyclopropanes. Methoxymycolates: its meromycolate chain contains an α-methyl methyl-ether moiety in the distal position and a *cis*-cyclopropane or an α-methyl *trans*-cyclopropane in the proximal position. Ketomycolates: its meromycolate chain contains an α-methyl ketone moiety in the distal position and proximal functionalities as in the methoxy series.

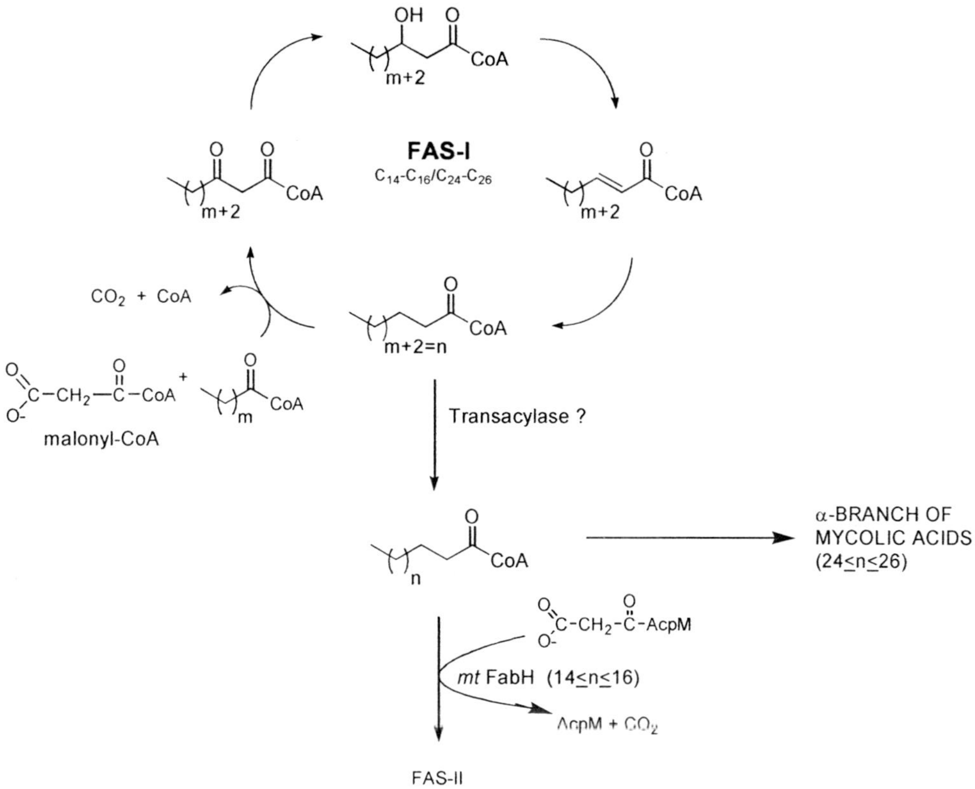

Figure 5. The mycobacterial Type I fatty acid synthase system (FAS-I).

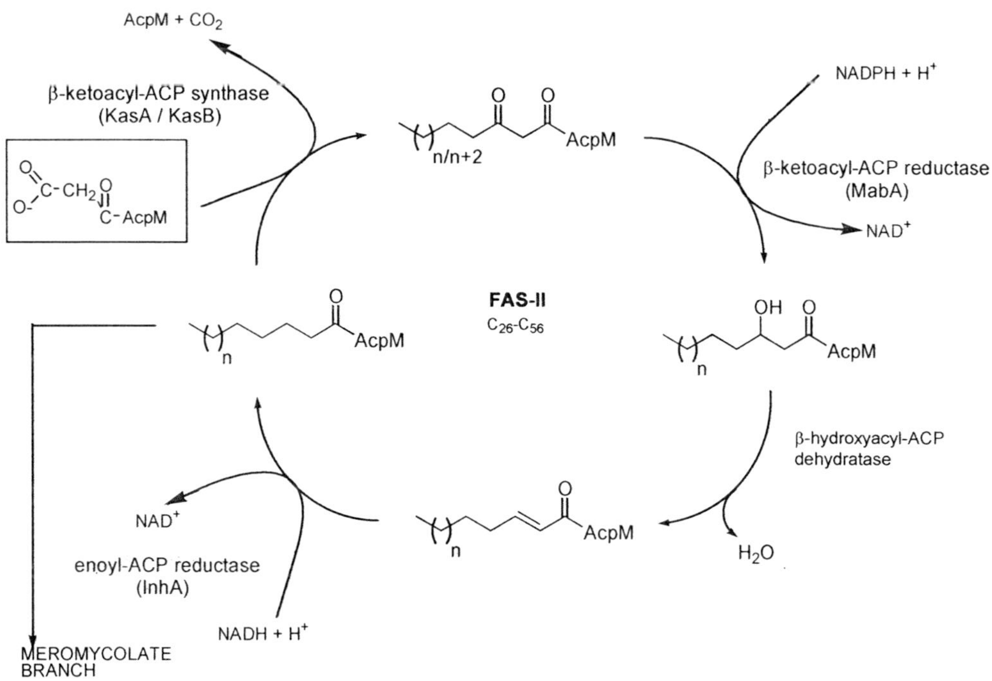

Figure 6. The mycobacterial Type II fatty acid synthase system (FAS-II).

Isoniazid (INH, isonicotinic acid hydrazide) is a pivotal first-line synthetic antitubercular agent, and the most prescribed drug for active infection and prophylaxis. The product of the *M. tuberculosis inhA* structural gene (InhA or ENR), which is located downstream of an ORF coding for a β-ketoacyl reductase (MabA or KAR) in the *inhA* operon, has been shown to be a major target for isoniazid (Banerjee et al., 1994). InhA has been shown to be an NADH-dependent enoyl-ACP (acyl carrier protein) reductase enzyme, which exhibits specificity for long-chain enoyl thioester substrates (Quémard et al., 1995). InhA is a member of the mycobacterial FAS-II system. Inactivation of the *M. tuberculosis inhA*-encoded enoyl reductase and INH treatment resulted in similar morphological changes to the mycobacterial cell wall leading to cell lysis (Vilchèze et al., 2000). Overexpression of *inhA* has been shown to confer resistance to INH and ethionamide (ETH) in *M. smegmatis*, *M. bovis* BCG, and *M. tuberculosis* (Larsen et al., 2002). Further biochemical and genetic evidence has been given likewise showing that InhA is the primary target of INH (Kremer et al., 2003). Consistent with these results, mutations in the *inhA* structural gene and in the *inhA* locus promoter region have been associated with isoniazid resistance (Basso et al., 1998). Molecular dynamics simulations of WT and INH-resistant clinical isolates of *M. tuberculosis* showed that mutations of the glycine-rich loop residues I21V and I16T resulted in a change in the pattern of direct hydrogen bond contacts with the pyrophosphate moiety of NADH, in which the pyrophosphate moiety of NADH moves apart from its binding site (Schroeder et al., 2005). More recently, a comparison of the crystal structures of binary complexes of WT and INH-resistant mutant enoyl reductases with NADH and pre-steady state kinetics showed that disruption of hydrogen bonds results in increased values for the dissociation rate constants of NADH from the binary complexes that could account for higher dissociation constant values for the coenzyme in INH-resistant enzymes (Oliveira et al, 2006). These results provide a mechanism of resistance at molecular level for this clinically important drug and point to InhA as the bona fide target for INH in *M. tuberculosis*. The mechanism of action of isoniazid is complex, as mutations in at least five different genes (*katG*, *inhA*, *ahpC*, *kasA*, and *ndh*) have been found to correlate with isoniazid resistance (Schroeder et al., 2002). INH is a pro-drug that is activated by the mycobacterial *katG*-encoded catalase-peroxidase enzyme in the presence of manganese ions, NAD(H) and oxygen (Johnsson and Schultz, 1994; Johnsson et al., 1995; Basso et al., 1996; Zabinski and Blanchard, 1997). The KatG-produced acylpyridine fragment of isoniazid is covalently attached to the C4 position of NADH, and forms a binary complex with the wild-type enoyl reductase of *M. tuberculosis* (Rozwarski et al., 1998). This isonicotinyl-NAD^+ adduct has been characterized spectroscopically and shown to bind to InhA with a dissociation constant value lower than 0.4 nM (Lei et al., 2000), and to act as a slow, tight-binding competitive inhibitor of WT InhA with an overall dissociation constant value of 0.75 nM (Rawat et al., 2003). Based on the mechanism of activation proposed for isoniazid, via electron transfer reaction (DeBarber et al., 2000), an alternative self-activation route has recently been proposed by our group for designing new drugs for the treatment of WT and INH-resistant tuberculosis. Accordingly, we have recently shown that a pentacyano(isoniazid)ferrateII complex inhibits enzyme activity of both wild-type InhA and I21V mutant InhA identified in isoniazid-resistant clinical isolates of *M. tuberculosis* (Oliveira et al., 2004). An MIC value of 0.2 $\mu g\ mL^{-1}$ for this inorganic complex was determined by the radiometric BACTEC AFB system for *M. tuberculosis* H37Rv strain, and toxicity assays in HL60 leukemia and MCS-7 breast cancer cells yielded an IC_{50} value >

25 μg mL^{-1}; thereby indicating a good selectivity index (SI = IC_{50}/MIC > 125; as suggested by the Tuberculosis Antimicrobial Acquisition and Coordinating Facility of USA for a compound to move forward through screening programs SI should be larger than 10). More recently, we have shown that the pentacyano(isoniazid)ferrateII complex is a slow-onset inhibitor of *M. tuberculosis* InhA enzyme activity, with a true overall dissociation constant value of 70 nM (Oliveira et al., 2006). In this mechanism of action an initial enzyme-inhibitor complex is rapidly formed, which then undergoes a slow isomerization reaction of to an enzyme-inhibitor binary complex in which the inhibitor is more tightly bound to enzyme. The weakness in the use of classical enzyme inhibitors as drugs for clinical conditions is that inhibition results in the upstream accumulation of the substrate for the enzyme, which may overcome the inhibition. By contrast, the build up of substrate cannot have any effect on the isomerization of enzyme-inhibitor complex typical of the slow-onset mechanism and hence reversal of the inhibition (Morrison and Walsh, 1988).

As mentioned above, in the *inhA* operon, the *inhA* gene is located immediately downstream of the *mabA* (also *fabG1* in *M. tuberculosis* or *fabG* in *E. coli*) gene that codes for a β-ketoacyl reductase (MabA) (Banerjee et al., 1998). MabA catalyzes the NADPH-dependent reduction of *β*-ketoacyl-ACP, yielding β-hydroxyl-acyl-ACP (**Figure 6**). MabA has recently been shown to preferentially metabolize long-chain fatty acids, consistent with its role in elongation of mycolic acid precursors (Marrakchi et al., 2002). More recently, determination of MabA crystal structure revealed a conserved Rossman-fold and the Ser-Tyr-Lys catalytic triad conserved amongst the family of short-chain dehydrogenases/reductases SDR members to which it belongs (Cohen-Gonsaud et al., 2002). We have recently determined the enzyme and chemical mechanism of MabA (unpublished results) and site-directed mutagenesis studies are currently underway to assess the role of the catalytic triad amino acid residues. Interestingly, MabA has been shown to be a target of isoniazid (Ducasse-Cabanot et al., 2004). Thus the *M. tuberculosis mabA*-encoded NADPH-dependent β-ketoacyl reductase enzyme also represents a target for anti-TB agent development. Indeed, any enzyme belonging to mycobacterial FAS-II system may represent an attractive target for drug development.

2.3.1.2. Shikimate Pathway

The shikimate pathway is an attractive target for the development of herbicides and antimicrobial agents because it is essential in algae, higher plants, bacteria and fungi, but absent from mammals (Bentley, 1990; Herrmann and Weaver, 1999; Coggins et al., 2003; Basso et al., 2005). The shikimate pathway (**Figure 7**) links metabolism of carbohydrates to biosynthesis of aromatic compounds. The first committed step in the shikimate pathway is the stereospecific condensation of phosphoenolpyruvate (PEP) and D-erythrose 4-phosphate (E4P) forming 3-deoxy-D-arabino-heptulosonate 7-phosphate (DAHP) and inorganic phosphate catalyzed by DAHP synthase (DAHPS; EC 4.1.2.15) (**Figure 7**). The second reaction is the elimination of phosphate from DAHP to generate 3-dehydroquinate (DHQ). The reaction is catalyzed by DHQ synthase (DHQS; EC 4.6.1.3) and requires catalytic amounts of NAD^+ for activity, even though the enzyme catalyzed reaction is redox neutral. In the third enzyme-catalyzed reaction, DHQ is dehydrated to 3-dehydroshikimate (DHS) by 3-dehydroquinate dehydratase (often referred to as 3-dehydroquinase; DHQ dehydratase; EC 4.2.1.10), thereby initiating the process of aromatization by introducing the first of three double bonds. The type II enzyme catalyzes a *trans*-dehydration via an enolate intermediate as in *M. tuberculosis*, whereas the type I catalyzes a *cis*-dehydration of 3-dehydroquinate via a covalent imine intermediate as in *Escherichia coli*. The fourth is the reduction of 3-dehydroshikimate to shikimate catalyzed by shikimate dehydrogenase (*aroE*-encoded SHK dehydrogenase; EC 1.1.1.25). Shikimate kinase (SK; EC 2.7.1.71), the fifth enzyme of the pathway, catalyzes a phosphate transfer from ATP to the carbon-3 hydroxyl group of shikimate forming shikimate 3-phosphate (S3P). The sixth reaction is an unusual transfer of an enolpyruvyl moiety from phosphoenolpyruvate (PEP) to the 5-hydroxyl group of S3P with the elimination of inorganic phosphate forming 5-enolpyruvylshikimate 3-phosphate (EPSP) catalyzed EPSP synthase (*aroA*-encoded EPSPS; EC 2.5.1.19). The seventh and final step in the main trunk of the shikimate pathway is the *trans*-1,4 elimination of phosphate from EPSP to yield chorismate catalyzed by chorismate synthase (CS; EC 4.6.1.4).

The mycobacterial shikimate pathway (the main trunk) leads to the biosynthesis of chorismic acid, which is converted by five distinct enzymes to prephenate (precursor of phenylalanine and tyrosine), anthranilate (precursor of tryptophan), aminodeoxychorismate (precursor of *para*-aminobenzoic acid -PABA – which, in turn, leads to tetrahydrofolate synthesis), *para*-hydroxybenzoic acid (precursor of ubiquinone or Coenzyme Q), and isochorismate (common precursor of naphtoquinones, menaquinones and mycobactins) (Figure 7). Analysis of the complete genome sequence of *M tuberculosis* shows the presence of seven *aro* genes predicted to be involved in the shikimate pathway (Cole et al., 1998). Four homologues to the shikimate pathway enzymes were located in a cluster containing the *aroD*-encoded type II DHQ dehydratase (Rv2537c), *aroB*-encoded DHQ synthase (Rv2538c), *aroK*-encoded type I shikimate kinase (Rv2539c), and *aroF*-encoded chorismate synthase (Rv2540c). The remaining homologues to shikimate pathway enzymes were annotated as follows: *aroG*-encoded class II phenylalanine-regulated DAHPS (Rv2178c), *aroE*-encoded shikimate dehydrogenase (Rv2552c), and *aroA*-encoded EPSP synthase (Rv3227). The salicylate-derived mycobactin siderophores have been shown to be essential for *M. tuberculosis* growth in macrophages (de Voos et al., 2000). In addition, the shikimate pathway has more recently been shown to be essential for the viability of *M. tuberculosis* (Parish and Stoker, 2002). Incidentally, EPSP synthase enzyme is the site of action of glyphosate [*N*-(phosphonomethyl)glycine], which is a widely used-broad-spectrum herbicide (Steinrücken and Amrhein, 1980). The essentiality of mycobacterial shikimate pathway and its absence from human host indicate that any of its enzyme components represents a promising target for the development of potentially non-toxic antimycobacterial agents. Accordingly, our research group has made a concentrated effort to obtain and characterize the enzymes of *M. tuberculosis* shikimate pathway aiming at rational-based drug design as well as at having homogeneous proteins to be immobilized on solid support to screen against libraries of natural compounds. The shikimate kinase and EPSP synthase have been cloned and expressed in functional and soluble form (Oliveira et al., 2001), and a one-step purification of EPSP synthase yielded homogeneous protein (Oliveira et al., 2003). Molecular model (Azevedo et al., 2002) and crystal structure (Pereira et al., 2004) of *M. tuberculosis* shikimate kinase with shikimic acid bound to the active site should provide a structural framework to guide enzyme inhibitor design. Even though we have not yet been able to obtain the crystal structure of EPSP synthase, structural bioinformatics studies on this protein have been carried out (Pereira et al., 2003).

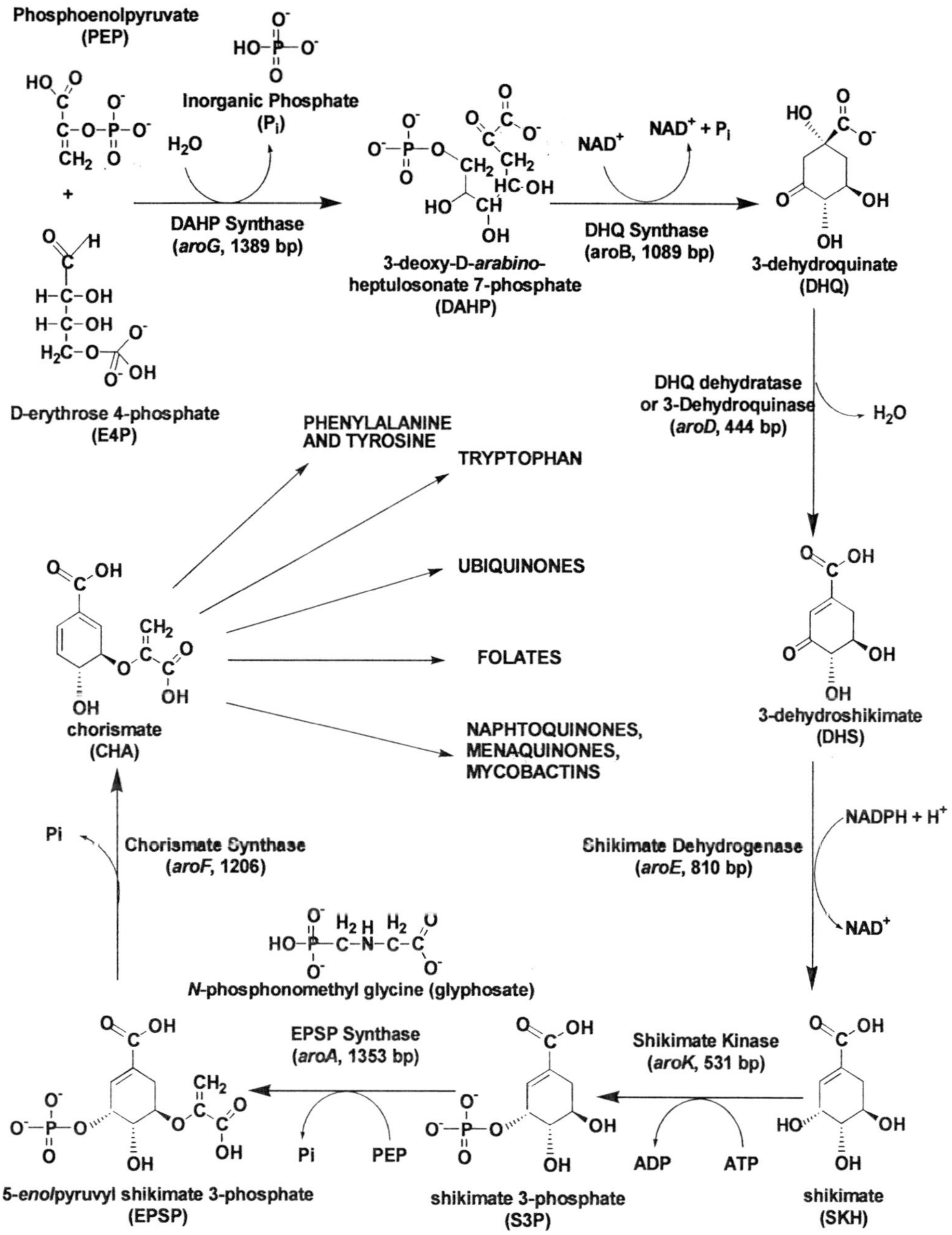

Figure 7. The shikimate pathway (main trunk) leads to the biosynthesis of chorismic acid, which is a key precursor of mycobacterial biosynthesis of aromatic aminoacids, folate, ubiquinone, naphthoquinones, menaquinones, and mycobactines.

We have also described cloning and expression of functional *M. tuberculosis* shikimate dehydrogenase (Magalhães et al., 2002), and purification and enzyme kinetics studies (Fonseca et al., 2006). More recently, studies of primary-kinetic isotope effects and solvent isotope effects and pH-rate profiles provide a thorough description of the enzyme and chemical mechanism of *M. tuberculosis* shikimate dehydrogenase (manuscript in preparation). Crystallization and preliminary X-ray crystallographic analysis of *M. tuberculosis* chorismate synthase (Dias et al., 2004) prompted determination of the three-dimensional structure of this enzyme (Dias et al., 2006). We also reported a structural model for *M. tuberculosis* chorismate synthase in complex with coenzyme and 5-*enol*pyruvyl shikimate 3-phosphate (Fernandes et al., 2006). Cloning, expression and purification of functional *M. tuberculosis* DAHP synthase (Rizzi et al., 2005) should also provide an additional protein target to be immobilized on solid support for antimycobacterial agent screening. Preliminary crystal structure analysis of *M. tuberculosis* DAHP synthase has recently been reported and should guide chemical optimization efforts (Webby et al., 2005).

2.3.2. Malaria

2.3.2.1. FAS-II System

Apicomplexan parasites are a large phylum of unicellular and obligate intracellular organisms of great medical importance. They include the human pathogens *Plasmodium* spp., the causative agent of malaria, and *Toxoplasma gondii*, an opportunistic parasite of immunosuppressed individuals and a common cause of congenital disease. Completion of genome sequence of *P. falciparum* should help identification of new targets for drug development against this parasite. As mentioned above (section 2.3), the finding that probably most members of this phylum Apicomplexa harbor a plastid-like organelle, called the apicoplast (a relict plastid or chloroplast) holds great potential for drug discovery. It has been speculated that the apicoplast is a potential "Achilles' Heel" of the malaria parasite (Seeber, 2003). Although the apicoplast itself contains a small circular genome, most of the proteome of this organelle is encoded in the nuclear genome. Proteins that function in the apicoplast must be targeted from the cytoplasm to the organelle, which is accomplished by means of bipartite leader sequence that is comprised of a hydrophobic signal peptide followed by a basic transit peptide (Ralph et al., 2004). The apicoplast contains a number of unique metabolic pathways not found in the vertebrate host, making it an ideal source of drug targets (Seeber, 2003). The complete fatty acid biosynthetic pathway is located in the apicoplast (Ralph et al., 2004). As mentioned above (section 2.3.1.1), fatty acid elongation occurs through repetitive cycles of condensation, β-keto reduction, dehydration, and enoyl reduction. These chemical reactions are catalyzed by two types of fatty acid synthase systems: FAS-I (a multidomain polyprotein that encodes all the enzyme reactions) and FAS-II (separate proteins catalyze the enzyme reactions and are encoded by unique genes). Interestingly, the malaria parasite only contains the FAS-II system (Ralph et al., 2004). Most apicoplast-targeted FAS-II enzymes have been characterized in *P. falciparum*, for example ACP,malonyl-CoA transacylase (FabD), β-ketoacyl-ACP synthase III (FabH; Prigge et al., 2003), enoyl-ACP reductase (FabI; Surolia and Surolia, 2001), β-ketoacyl-ACP reductase (FabG; Pillai et al., 2003), β-hydroxyacyl-ACP dehydratase (FabZ; Sharma et al., 2003), and, more recently, β-ketoacyl-ACP synthase I/II (FabB/F; Lack et al., 2006). Inhibition of *P. falciparum* enoyl-ACP reductase by triclosan (a chlorinated bisphenol that is widely used in many contemporary consumer and professional health care products) has been reported to inhibit the *in vitro* growth of *P. falciparum* (McLeod et al., 2001). Triclosan has also been shown to offer protection against blood stages of malaria and that the drug target is an enoyl-ACP reductase from *P. falciparum* (Surolia and Surolia, 2001). This enoyl reductase was shown to be NADPH-dependent enzyme and accepts acetoacetyl-CoA as co-substrate (Pillai et al., 2003). Triclosan has been shown to act as a slow-tight binding inhibitor of the *P. falciparum* enoyl-ACP reductase enzyme activity with an overall inhibition constant value of 96 pM (Kapoor et al., 2004). Determination of the three-dimensional structure of malarial enoyl reductase-triclosan NAD^+ ternary complex has provided a structural framework that sheds light on the mode of binding of triclosan (Perozzo et al., 2002). Kinetic and structural analysis showing the features of *P. falciparum* enoyl reductase responsible for the high affinity of the enzyme for triclosan in the presence of NAD^+ has only recently been reported (Kapoor et al., 2004a). It has shown that the enhancement of TCL affinity for *P. falciparum* enoyl reductase in the presence of NAD^+ can be explained by increased van der Waals contacts in the ternary complex, which is facilitated by the movement of residues 318-324 of the substrate-binding loop and the nicotinamide ring of NAD^+. Accordingly, *P. falciparum* enoyl-ACP reductase (FabI) is a validated target for anti-malarial agent development and to be used in drug screening efforts. The reduction step catalyzed by β-ketoacyl-ACP reductase (FabG) is an essential and validated antimicrobial target (Zhang et al., 2004). Kinetic, inhibition and structural studies on *P. falciparum* β-ketoacyl-ACP reductase have recently been reported, showing the antibacterial and anthelminthic agent hexachlorophene is a potent inhibitor of malaria parasite FabG and inhibits *in vitro* growth of *P. falciparum* (Wickramasinghe et al., 2006). However, it is unlikely that hexachlorophene will be useful as an antimalarial agent due to its neurotoxic side effects (Evangelista de Duffard and Duffard, 1996). Screening of a natural product library of flavonoids against *P. falciparum* FabG, FabZ and FabI enzymes identified the polyphenol (-)-catechin gallate as a strong inhibitor of all three enzymes that displayed *in vitro* inhibition growth against chloroquine-sensitive (NF54) and chloroquine-resistant (K1) *P. falciparum* strains in the low to submicromolar range (Tasdemir et al., 2006). The *P. falciparum* FAS-II system enzymes are thus validated targets for the development of antimalarial agents.

2.3.2.2. Purine Salvage Pathway

The life cycle of *Plasmodium spp.* is extraordinarily complex. The three main stages of *P. falciparum* development take place in the liver cells and red blood cells of the human host and the tissues of the mosquito vector. The parasite bloodstage has been used as the target of the majority of vaccine and drug development programs because these forms of the parasite cause malaria-associated pathology (Waters, 2003). An essential step in the life cycle of protozoan parasites is cellular replication of the trophozoite in the human erythrocyte – asexual blood stage of *P. falciparum* – that creates a need for large quantities of purines for RNA and DNA synthesis. However, *P. falciparum* lacks *de novo* purine synthesis (purine auxotroph) and depends exclusively on purine salvage pathway for RNA and DNA synthesis (Sherman, 1979; Reyes et al., 1982). Erythrocytes also lack enzymes for *de novo* purine synthesis. Adenosine is a major purine nucleoside of human blood, but the adenosine kinate activity of *P. falciparum* is very low. Adenosine deaminase is very active in the erythocyte and the parasite. Deamination of adenosine followed by a nucleotidase activity yields inosine, whose phosphorolysis catalyzed by purine nuleoside phosphorylase (PNP) yields hypoxanthine, a major purine precursor for purine salvage. There is no adenosine phosphorylase activity in humans, and adenine is present at very low concentrations (Stramentinoli, 1987). Hypoxanthine is converted to inosine monophosphate by hypoxanthine phosphoribosyltransferase activity, which is present in large amounts in *P. falciparum* (Reyes et al., 1982). This path makes hypoxanthine the common precursor for all purine nucleotides in the parasite (Asahi et al., 1996). Culturing *P. falciparum* in human erythrocytes in medium supplemented with xanthine oxidase, which causes depletion of hypoxanthine, prevents parasite growth (Berman et al., 1991), thereby confirming that a supply of host-derived hypoxanthine is critical for nucleic acid synthesis in *P. falciparum*. The sole pathway of hypoxanthine production in *P. falciparum* and in human erythrocytes is through the phosphorolysis of inosine to hypoxanthine, a reaction catalyzed by PNP (Sherman, 1979). Enzymes of the purine salvage pathway were detected in *P. falciparum* (Reyes et al., 1982), including PNP. Immucillin-H, a transition state analogue that inhibits PNP enzyme activity, has been shown to both target *P. falciparum* PNP reducing the incorporation of inosine into nucleic acids and kill *P. falciparum* cultured in human erythrocytes (Kicska et al., 2002a). Starvation of purines has thus been shown to cause purine-less *P. falciparum* death in cultured cells induced by a transition state analogue, immucillin-H, that inhibits PNP enzyme activity (Kicska et al., 2002a). Immucillin-H has been shown to be a slow-onset tight binding inhibitor of recombinant *P. falciparum* PNP (Kicska et al., 2002b). Determination of the crystal structure of *P. falciparum* PNP in complex with immucillin-H and sulphate ion reveals that its active site presents structural differences from human PNP (Shi et al., 2004). A transition-state analogue that is inhibitor more specific for *P. falciparum* PNP as compared to human PNP, 5'-methylthio-immucillin-H, has been described showing that malaria-specific PNP inhibitors can be designed (Shi et al., 2004). It has also been demonstrated that 5'-methylthio-immucillin-H kills *P. falciparum* strain 3D7 cultures and that inhibition of parasite growth occurs at concentrations where *P. falciparum* PNP but not human PNP is strongly inhibited with a discrimination factor (ratio of Kd values for *P. falciparum* PNP and human PNP) of 112 (Ting et al., 2005). Since *P. falciparum* PNP differs from human PNP by lower Michaelis-Menten constant for inosine, decreased preference for deoxy-guanosine, and reduced affinity for immucillins (with exception of 5'-deoxy-immucillin-H), it has been proposed that specific inhibitors of *P. falciparum* PNP activity is a feasible approach to antimalarial drug development (Shi et al., 2004). Interestingly, it has been proposed that the parasite PNP enzyme may lie at an essential metabolic branchpoint, generating hypoxanthine from both the purine nucleoside pathways and from polyamine synthesis based on its unusual substrate specificity (Shi et al., 2004; Ting et al., 2005). The polyamine pathway forms two molecules of methylthioadenosine for the synthesis of each spermine molecule. In most organisms, methyladenosine is recycled into adenine and methionine pools. Since human erythrocytes cannot synthesize polyamines and synthesis of polyamines is likely essential for malaria parasites, the polyamine pathway enzymes have also been investigated as antimalarial targets (Muller et al., 2001). However, targeting of polyamine synthetic pathway will not be discussed further here. *P. falciparum* is remarkable because it its small number of purine salvage pathway enzymes despite the complete reliance on this pathway. The purine salvage pathway enzymes adenosine deaminase (ADA) and purine nucleoside fosphorylase from *P. falciparum* each have two roles in the parasite and replace the functions of purine nucleoside phosphorylase, adenosine deaminase, 5'-methylthioadenosine phosphorylase, adenine phosphorybosyltransferase, and adenosine kinase in mammals. The action of ADA and PNP permits the parasite to form hypoxanthine from erythrocyte purine pools and to recycle hypoxanthine from polyamine synthesis within the parasite. Hypoxanthine is a precursor for all purines and is a central metabolite for nucleic acid synthesis in *P. falciparum* (Ting et al.,

2005). The enzymes ADA, PNP, and hypoxanthine-guanine-xanthine phosphoribosyltransferase form the major path for purine salvage in *P. falciparum*. It has been proposed that inhibition of *P. falciparum* PNP enzyme activity permits metabolic disruption of two pathways by targeting a single enzyme (Ting et al., 2005). Differences in catalytic features and crystal structures between parasite and host PNP enzymes can be exploited for the synthesis of malaria-specific PNP enzyme inhibitors. The studies described above warrant screening efforts against natural compounds aiming at discovering and developing malaria-specific compounds targeting the parasite PNP enzyme. More recently, the phenotype of the conditionally lethal mutant of *P. falciparum* plasma membrane nucleoside transporter 1 demonstrated that this transporter is a critical component of purine salvage pathway (Bissati et al., 2006), and mutant parasites were not viable at physiological concentrations of hypoxanthine, adenosine, or inosine. This finding has unveiled a new target for antimalarial drug development and supported the purine salvage pathway as an attractive drug target.

2.4. Diversity of Plant Species in Brazil

Biodiversity distribution across the planet is heterogeneous. It has been estimated that 70% of the world's species occur in only 12 countries: Australia, Brazil, China, Colombia, Ecuador, India, Indonesia, Madagascar, Mexico, Peru, and Zaire. Tropical rainforests are one of the richest biomes, containing more than half the number of species on Earth. Brazil is quite significant in the world's biodiversity, since it contains two large areas regarded as biodiversity hotspots, the Atlantic Forest in the Coast and the Cerrado (Brazilian savanna) in Central Brazil. A biodiversity hotspot is defined as the area that contains at least 0.5% of the world's 300,000 known plant species as endemics (Myers et al. 2000). The Amazon Forest is another important area in terms of biodiversity; however, its real contribution will only be properly estimated upon further cataloging of its vast numbers of plant, insect and fungi species.

Brazilian biodiversity provides the country with a strategic advantage to develop the rational and sustained exploration of novel natural products of therapeutic value. The almost continental dimension of the country, particularly in the North-South direction, but also in the East-West, comprises a wide range of climates, soil types and altitudes, establishing a unique set of selective pressures for the adaptation of plant life to these various environments. Chemical diversity is also a result of these pressures, arising from the need of plant communities to adapt to the particular abiotic stresses, fauna and microbes that co-exist with them. The largest phytogeographical areas within the Brazilian territory can be divided in: the Amazonian Forest, Atlantic Forest, Araucaria Forest, Cerrado, Caatinga, Pantanal (wetlands), Pampa (grasslands) and the Coastal vegetation. They cover different climates, soils and altitudes, creating a unique set of environmental variables for the specialization of plants and their chemical defenses. Next, a brief general description of these types of vegetation is given, based on Ferri (1980).

The Amazonian Forest is characterized by its diversity of species, many of which unknown to science. Several elements of vegetation have been recognized in the Amazonian Forest. Besides the areas of proper forest, savanna-like regions, grasslands and palm-tree rich areas also occur in the Amazonian region. The predominant element, however, are the forests. The forest areas can be divided into Dry Land Forest ("Mata de Terra Firme"), Transiently

Flooded Forest ("Mata de Várzeas") and Permanently Flooded Forest ("Mata de Igapó"). The Dry Land Forest is distant from rivers and located in higher parts of the landscape. This Forest is permanently green, with little and non-synchronized change of foliage. Density of trees is rather high and leaves are mostly broad. Large trees characterize this forest, such as *Bertholletia excelsa* (castanheira – Brazil nut) and *Castilloa ullei* (caucho). The tree canopy is often continuous, forming a dense obstacle to light, retaining up to 90% of the incident irradiance. Hence, the forest floor is dark, poorly ventilated, and humid. The forest is rich in members of Leguminosae, Moraceae, and Sapotaceae. Palm tree species, such as *Astrocaryum*, *Attalea*, and *Geonoma*, are also frequent. *Triplaris* and *Tococa*, plants that associate with ants, are also well represented. In the understorey Araceae, Generaceae, Piperaceae, Bromeliaceae, and ferns predominate. Transiently Flooded Forest has variable composition, containing elements of the other two types of Forest of the Amazonian region. Species of Moraceae, Sapotaceae and Leguminosae grow into large trees, forming a large part of the canopy. Important species in this forest are *Calycophyllum spruceanum*, *Coumarona odorata* (the coumarin producing Tonka bear tree), *Hevea brasiliensis* (rubber tree), *Dinizia excelsa*, *Andira*, *Parkia*, *Hymenaea*. The Permanently Flooded Forest components include relatively smaller trees (up to 20 m) often producing large support roots and stem perforations (lenticels) for aeration. Typical trees include *Triplaris surinamensis*, *Bombax aquaticum*, and *Macrolobium acaciaefolium*. Many epiphytes are represented in this area, including Piperaceae, Araceae, Orchidaceae, Bromeliaceae, Cactaceae, Gesneraceae, ferns, mosses, and liverworts.

The Atlantic Forest is characterized by a high biodiversity comparable to that of the Amazonian Forest. The proximity to the Atlantic Ocean provides a stable source of humidity, allowing high vegetation density. Trees are dominant, which may reach 30 m in height; there are numerous liana and epiphytes. The shaded and poorly ventilated understorey contains many herbaceous species. A floristic survey of the southern limit of the Atlantic Forest area has been carried out, revealing many species of economic interest, many of which with medicinal properties. Among the upper storey tree species, the "canelas" (*Ocotea* sp.) and other Lauraceae, are valued for their wood and aromatic essences; the "ipes" (*Tabebuia* sp, Bignoniaceae, pau d'arco), appreciated for their wood and phytomedicinal uses; *Euterpe edulis*, the "palmito", from which palm hearts are extracted; the *Cecropia* and *Inga*, two genera that establish associations with ants. Epiphyte species include members of Araceae, Gesneraceae, Piperaceae, Moraceae, and Orchidaceae and Bromeliaceae of high ornamental value. Lower storey trees include *Psychotria umbellata*, which produces psychollatine (formerly known as umbellatine), a novel monoterpene indole alkaloid with morphine-like analgesic activity (Paranhos et al. 2005). At the forest border, members of Leguminosae can be found, many of which often accumulate toxic proteins, amino acids and alkaloids. Herbaceous species include members of families known for accumulating bioactive metabolites (e.g., alkaloids, terpenes, phenolics, peptides, proteins), such as Apocynaceae, Solanaceae, Erythroxylaceae, Lamiaceae, Asteraceae, Leguminosae, and Euphorbiaceae. The herbaceous layer is also rich in Pterydophytes and Bryophytes, some of which have antimicrobial compounds.

The Araucaria Forest is characterized by a relatively regular distribution of rain fall throughout the year in contrast to most other parts of the country, particularly the Western-Central and Northeast regions. Wide differences between seasons with distinct temperatures and photoperiods are other features that are peculiar for Brazilian standards. The occurrence of frost is frequent and snowfall can occasionally happen. Soil conditions also vary, with predominance of basalt-derived purple soils, very rich in nutrients, and stretches of granitic soils much less fertile. As the name suggests, the Araucaria Forest is rich in the Brazilian Pine, *Araucaria angustifolia*. Other notable components of the forest include: *Actinostemum concolor*, *Sorocea ilicifolia*, *Urera baccifera*, *Phytolacca dioica*, *Cedrela fissilis*, *Myrocarpus frondosus*, *Enterolobium timbouva* and species of *Sloanea*, *Ocotea* and *Nectandra*. Another important gymnosperm in this forest is *Podocarpus lambertii*. Taxol has been reported in needles of *Podocarpus gracilior*, the first non-taxaceae to yield this chemical (Stahlhut et al. 1999). Among the important tree species are also included *Ilex paraguariensis* ("mate"), *Symplocus uniflora*, *Ocotea pulchella*, *Rapanea venosa*, *Feijoa sellowiana*, *Eugenia opaca* and associated epiphytes, such as *Polypodium squamulosum* (Pteridophytae), and *Tillandsia tenuifolia* (Bromeliaceae).

The Cerrado is the dominat type of vegetation in Central Brazil. Taking in account its full range, the Cerrado vegetation covers approximately 20% of the Brazilian territory. In most of its range, Central Brazil is characterized by a dry season lasting up to 5 months with rain distributed in the remaining months reaching a total precipitation of approximately 1400 to 1500 mm. In their southern range, Cerrado areas may have slightly lower precipitation and temperatures. The landscape is plain and the soils often acidic and poor in many mineral nutrients. Water availability for the vegetation is generally adequate throughout the year. Several Cerrado plant species have very deep roots, reaching 10 to 18m. In spite of the water availability, several species display scleromorphic characteristics, such as thick barks, shiny and waxy leaves. The soil has relatively high concentration of aluminum, which is toxic to many plants, and, in conjuction with the limitation of some essential minerals, may be responsible for the scleromorphism of various shrubs and small trees. The Cerrado vegetation may be divided in two main components: permanent species (trees and most shrubs) and the transient species (grasses and weeds). An important feature of several species in this type of vegetation is the formation of underground storage organs such as xylopodia (subterranean lignaceous organs with both stem and root characteristics) and bulbs. The loss of leaves and even stems is frequent during the dry season. Examples of species in the Cerrado are: *Andira humilis*, *Eragrostis*, *Paspalum*, *Tristachya*, *Vernonia* (ephemerous species), *Butia leiospatha*, *Strychnos pseudochina*, *Tocoyena formosa*, *Campomanesia*, *Byrsonima*, *Cassia*, *Dalbergia*, *Jacaranda semiserrata*, *Erythroxylum tortuosum* (woody species).

The Caatinga region of the Brazilian Northeast is characterized by high temperatures, relatively low average humidity, and annual rainfall around 250 to 500 mm. The dry season lasts 7 months or more. The winter is the rainy season, in which temperatures are not as high. The soils are of different origins and, as a rule, are chemically fertile, well drained and oxygenated. Water bodies are rarely permanent, drying completely during the summer. Among the dominant species of the Caatinga, one can mention: *Amburana cearensis* (Leguminosae), *Aspidosperma pyrifolium* (Apocynaceae), *Caesalpinia pyramidalis, Caesalpinia microphylla* (Leguminosae), *Cassia uniflora* (Leguminosae), *Croton campestris, Euphorbia phosphorea* (Euphorbiaceae), *Maytenus rigida* (Celastraceae), *Mimosa caesalpinifolia* (Leguminosae), *Nicotiana glauca* (Solanaceae), *Sapium cicatricosum* (Euphorbiaceae), and *Tabebuia caraiba* (Bignoniaceae).

The Pantanal vegetation is located in the West of Brazil, mostly in the states of Mato Grosso and Mato Grosso do Sul. It is a rather complex assembly of environments, showing several forms, including Cerrado-like, grassland, forest, gallery forest (following the course of rivers). During the months of October to March, the rivers of the Paraguai basin raise in level and cause the flooding of extensive areas of soil. During the dry season, from April to September, the water retracts and soils become drained. This transient flooding profile creates peculiar conditions that strongly influence the type of vegetation that is capable of being established. Because of differences in altitude between areas within the Pantanal range, some spots are permanently drained and others remain flooded throughout the year. During the flooding season, soils get fertilized with clay and organic matter from plant decomposition. Examples of plant species from the Pantanal include: trees - *Ficus*, *Pseudobombax*, *Tabebuia*, *Piptadenia*, *Copernicia australis*, *Attalea princeps*, *Inga*, *Cecropia*, *Genipa americana*; several grasses; aquatic plants – *Eichhornia crassipes*, *Pontederia*, *Sagittaria*, *Pistia stratiotes*; marsh plants – *Typha*, various Cyperaceae.

The Pampa corresponds to extensive areas of grasslands in southern Brazil. Besides a large number of species of Poaceae, other species such as members of Leguminosae are frequent. Species of *Baccharis* are also common. Scattered in the grasslands, there are patches of forest known as "capões", often associated with small streams. Gallery forests are also found in the Pampas. In patches of forest several Myrtaceae, *Ficus*, *Smilax*, *Nectandra, Cedrela, Cabralea* and *Schinus* are frequent.

The Coastal areas may be divided in three subtypes: rocky, sandy and muddy shores. In the rocky shores, the lack of soil and the presence of essentially nude rocks constantly battered by the sea, allow the establishment of some macroalgae specifically adapted to these conditions. Sandy shores are colonized by a number of species with tolerance to salt and high irradiance, including *Iresine*, *Remirea*, *Sporobulus*, *Panicum racemosum*. Moving progressively further into the continent other species can be found such as *Salicornia*, *Sesuvium*, *Statice*, *Hydrocotyle*, *Ipomea*, *Canavalia, Sophora*, *Cereus peruvianum* and *Anacardium occidentale* (cashew tree). Muddy shores are often generated by the encounter of fresh and salt water, causing the deposition of colloids and the formation of "mangues". The soils are flooded, hypoxic, salty and unstable. Under these very harsh conditions, only a few highly adapted species can survive. The most frequent species of the "mangues" are *Avicennia tomentosa*, *Rhizophora mangle* and *Laguncularia racemosa*. *Rhizophora* forms a large number of support roots to compensate for the unstable consistency of the soil. *Avicennia* develops a large number of respiratory roots called pneumatophores, which grow upwards (with negative gravitropism), contain perforations for air exchange and are rich in parenchyma tissue containing various air spaces to store oxygen.

3. Brazilian Plant Species as Sources of Chemotherapeutic Agents

3.1. Strategies for Selecting Bioactive Plant Species

New bioactive plant products can be found using random, ethnobotanical, ecological or chemotaxonomical approaches. Random search is often extremely time-consuming and laborious, yielding success rates of approximately 1 new product per 10,000 screened plants. In spite of those disadvantages, random search has proven very valuable in recruiting new chemical entities for the therapeutic arsenal. In a joint effort between the National Cancer

Institute (NCI-USA) and the United States Department of Agriculture (USDA) between 1960 and 1981, more than 114,000 extracts of some 35,000 plants were harvested and tested against a range of animal tumor systems, mostly cell cultures (Cragg et al. 1995). Among the important drugs that have been discovered using this method are taxol, derivatives of camptothecin and homoharringtonine.

The ethnobotanical approach takes avantage of the medical knowledge of traditional societies to select plants for testing bioactivities. Of special interest are those societies and local communities that have been relatively isolated for considerable periods of time turning to the surrounding natural resources as remedies for their ailments. The success rates of this approach are substantially higher than those of random screening. Moreover, the continued use of crude preparations could be roughly compared to small-scale clinical trials, improving the chances of obtaining adequate substances for human use. Tests carried out at the NCI for antitumoral activity using a group of plants chosen on the basis of traditional medicinal use found positive activities in the order of 2 to 5 times higher than random screening (Lewis and Elvin-Lewis 1995). The pharmacopeas of ancient countries (such as China and India) are also powerful resources for obtaining information on traditional use of plants for treating a wide spectrum of diseases and clinical conditions.

Another means of finding leads to plant bioactivities is ecological observation. Low levels of predation or infection in areas infested with herbivores or plant pathogens may indicate the accumulation of toxic defense compounds. Zoopharmacognosy is based on the selection of plant species regularly ingested by animals, mostly primates, for reducing pain, microbial or worm infestations (Berry et al. 1995). This represents a variation of ecological observation that relies on information obtained in ethological investigations.

Phytochemical analysis shows the ocurrence of certain groups of secondary metabolites in botanical families, subfamilies or even lower taxonomical levels. This has been the basis of chemotaxonomy, a method that has been assisting taxonomists in the resolution of certain taxa along with more powerful techniques such as molecular taxonomy, based on the comparison of DNA sequences (Wink 2003). The knowledge that certain taxa are more likely to express biochemical pathways leading to specific groups of secondary metabolites can be successfully used to select plant species for screening bioactivities.

The effectiveness of the above described strategies to find bioactive plant species can be increased by combining two or more of them. For example, a combined approach of ethnobotanical and chemotaxonomical leads has unveiled a number of bioactive species from the genus *Psychotria* in southern Brazil, allowing the identification of new alkaloids with significant pharmacological properties (Elisabetsky et al. 1997, Both et al. 2002, Gregianini et al. 2003, Pasquali et al., 2006).

3.2. Brazilian Plant Species as Sources of Antitubercular and Antimalarial Agents

Approximately 25% of the drugs prescribed worldwide come from plants, whereas 11% of the 252 drugs considered as basic and essential by the World Health Organization derive exclusively from plants (Rates, 2001). In this context, natural products and their derivatives have traditionally been the most common source of drugs and they might represent major basis of innovative therapeutic agents for infectious diseases. In the particular case of malaria and tuberculosis, the identification of plant-derived products could present clear advantages. Firstly, plant extracts and/or compounds might be valuable for contouring the problems of emerging resistance observed with the currently available therapy. In addition, plant-supported therapies are expected to present low costs and less collateral effects. Finally, the R and D model of drug development has not been a valuable tool for identifying new therapeutic agents for both neglected diseases (Olliaro and Taylor, 2003; 2004; Basso et al., 2005).

In the case of malaria, traditional medicines have been employed for thousands of years. Indeed, as mentioned beforehand, most current therapy for malaria is based on the chemical structure of quinine, the active ingredient found in the barks of *Cinchona* ssp., which was first imported into Europe from Peru for antimalarial use in the XVII century by the Jesuits. For almost two centuries the bark was employed for medicine as a powder, extract, or infusion.The dependence on raw material for its extraction and the opportunities presented by its structural elucidation led to the development of fully synthetic and low cost 4-aminoquinoline antimalarials - notably chloroquine and amodiaquine (O'Neill et al. 1998), which have been largely used for treating malaria in the last 50 years. Unfortunately, the use of quinine-derived drugs is clearly associated with the rising of resistant strains of *Plasmodium*. This fact is deeply related to the two-fold increase in the worldwide mortality taxes from malaria in the last 20 years (Olliaro and Taylor, 2004; Wilcox and Bodeker, 2004; Wright, 2005).

Another class of plant-derived products that have been effectively used as antimalarial agents is represented by artemisinin and related compounds. Artemisinin is a sesquiterpene lactone obtained from the leaves and flowers of the Chinese herb "qinghao" (*Artemisia annua*) that was traditionally used for treating fevers (Balint, 2001; Haynes, 2001). Several semi-synthetic derivatives of artemisinin have been used over the past two decades for malaria therapy. These derivatives include artemether (an oil-soluble methyl ester), arteether and artesunate (the water-soluble hemisuccinate salt of dihydroartemisinin), which are all metabolized to dihydroartemisinin - the major active agent in the body (Wright, 2005). All of them are effective against gamethocytes, the sexual stages of *Plasmodium* that infect mosquitoes. It has been demonstrated that multi-drug resistant strains of *Plasmodium falciparum* are susceptible to artemisinin and related compounds (Wilcox and Bodeker, 2000; Balint, 2001). It is worth mentioning that oral preparations of artemisinin and derivatives are rapidly absorbed and they present a very short half-life, what could be responsible for the occurrence of recrudescence of parasites. Because of that, the World Health Organization highly recommends the use of artemisinin (or its derivatives) in combination with quinine-derived drugs for treating both uncomplicated and severe malaria (Balint, 2001; Olliaro and Taylor, 2004).

During the last 20 years, the search for new plant-based antimalarial agents has been widely stimulated, especially by the World Health Organization (Taylor and Berridge, 2006). From these efforts, a series of studies conducted in Africa, Asia and South America indicated that a large array of plants might present antimalarial activity to various extents. Through the 90's, a multidisciplinary research group in Brazil investigated the possible effects of up to 300 plant extracts against malaria (Krettli et al., 2001). The plants assessed in the studies were chosen on the basis of folk medicine of Amazon regions and several kinds of extracts were tested. The evaluation of the extracts was accomplished by means of *in vitro* (*Plasmodium falciparum* blood stage cultures) and *in vivo* (mice infected with *Plasmodium berghei*) approaches. Surprisingly, among all analyzed plants, only some few presented antimalarial activities, such as *Esenbeckia febrifuga*, *Boerhavia hirsuta*, *Acanthospermum australe*, *Tachia guianenses* and *Vernonia brasiliensis* (Tagboto and Towson, 2001; Krettli et al., 2001). In most cases, the plant extracts displayed promising effects, but the percentage of parasitaemia reduction rarely achieved the levels observed for the positive control drug chloroquine (Krettli et al., 2001). Exceptions for these results are the plants belonging to the genus Bidens, especially represented by *Bidens pilosa* (Asteraceae). Of high interest, a study conducted by Brandão et al. (1997) revealed that either ethanolic or methanolic extracts of at least seven Bidens species produced marked reductions of parasitaemia according to assessment in *Plasmodium falciparum* cultures *in vitro*. Furthermore, ethanolic extracts of *Bidens pilosa* obtained from roots, leaves and the whole plant significantly reduced parasitaemia *in vivo* when tested in *Plasmodium berghei*-infected mice (Brandão et al., 1997; Krettli et al., 2001). A recent work of the same group evaluated the possible antimalarial activity of ethanolic extracts of three Cinchona-like plants used to treat fever and malaria in Brazil: *Deianira erubescens*, *Strychnos pseudoquina* and *Remijia ferruginea* (Andrade-Neto et al. 2003). From the three plant species, only *Remijia ferruginea* (an alkaloid-producing species but devoid of quinine) presented activity in mice infected with *Plasmodium berghei*, when administered at the dose of 1000 mg/kg. Other tests carried out with *Bidens pilosa* extracts were also very encouraging (Oliveira et al. 2004). The extracts were active in mice infected with *P. berghei*: doses of up to 500 mg/kg given by the oral route reduced both malaria parasitaemia and mouse mortality. *In vitro* tests against three *P. falciparum* isolates, two chloroquine-resistant and one mefloquine-resistant, showed activity of plants grown in a number of conditions, but higher activities were observed in wild plants. Analysis using thin layer chromatography showed the presence of flavonoids and polyacetylene compounds (Krettli et al., 2001; Oliveira et al., 2004). *In vivo* experimental evidence strongly suggests that alkaloids are the main responsible for the antimalarial activity reported to *Bidens pilosa*, although whole extracts of the plant, containing both polyacetylene and alkaloid compounds, presented most effectiveness in preventing mouse mortality following *Plasmodium berghei* infection (Oliveira et al., 2004). According to the authors, because *Bidens pilosa* is proven to be active against *P. falciparum* drug-resistant parasites *in vitro*, and in rodent malaria *in vivo*, it might represent a good candidate for pre-clinical tests as a phytotherapeutic agent or for chemical isolation of the active compound(s) with the aim of finding new antimalarial drugs.

Studies with plants traditionally used for malaria treatment from various parts of the world (Vietnam, South Africa, and São Tomé and Príncipe) also showed inhibitory activities against chloroquine-sensitive or resistant strains of *Plasmodium falciparum* (Tona et al., 2001; Nundkumar and Ojewole, 2002; Tran et al., 2003). Particularly promising results were obtained with extracts of the following species: *Cassia occidentalis* (root bark), *Morinda morindoides* (leaves), *Phyllanthus niruri* (whole plant), *Coscinium fenestra* (whole plant), *Psidium guajava* (bark), *Vangueria infausta* (leaf), *Struchium spargano-phorum* (leaf), *Cinchona succirubra* (bark), *Tithonia diversifolia* (shoots), *Cedrela odorata* (bark), and *Pycnanthus angolensis* (bark). Once again, related taxa or even some of the same species are available in the Brazilian ecosystems.

It is worth citing that until now more than 1000 plant species have been assessed for their potential as antimalarials throughout the world. In spite of that, for the great part of the studies (approximately 70 %), the obtained data is insufficient or not conclusive to prove effectiveness of the plants (Wilcox and Bodeker, 2004). Another fact that ought to be mentioned is the lack of convincing studies on the active compounds responsible for the antimalarial effects of the extracts. Therefore, additional pre-clinical studies are necessary for evaluating the efficacy and to determine the active principles of the extracts. Attempts should also be made to define the seasonal and/or regional variations, the stability of the active extracts, as well as the possible side effects (safety studies). The alkaloid cryptolepine obtained from the West Africa shrub *Cryptolepis sanguinolenta* can be taken as an illustrative example. Although cryptolepine displayed a potent antiplasmodial effect *in vitro*, it failed to reduce parasitaemia when given by intraperitoneal route to *Plasmodium berguei*-infected mice, exhibiting marked cytotoxic effects. On the other hand, the cryptolepine analogue, 2,7-dibromocryptolepine, presented a superior potency *in vitro* and it was able to markedly reduce the parasitaemia *in vivo*, with no apparent toxicity to mice. Therefore, 2,7-dibromocryptolepine might represent a better candidate for the development of an antimalarial drug, in comparison to the original molecule (Wright et al., 2001; Wright, 2005).

The same concerns made in relation to malaria are also true for tuberculosis. This infectious disease kills about 2 million people per year and it becomes progressively more serious due to the spread of *Mycobacterium tuberculosis* strains resistant to most anti-TB drugs available in the market (Okunade et al., 2004; Basso et al., 2005). Therefore, the search for new antimycobacterial agents (including those derived from plants) clearly represents an urgent need. A screening for growth inhibition of *Mycobacterium tuberculosis* using the microplate Alamar blue assay was carried out with 44 plant species previously chosen on the basis of traditional use against tuberculosis; of these, six species (*Chelidonium majus*, *Pinus brutia*, *Salvia aethiopsis*, *Stachys sylvatica*, *Ulmus glabra*, *Urtica dioica*) showed activity, the last three in the water fraction (Tosun et al., 2004). Some of these taxa or closely related ones are represented in the Brazilian ecosystems. Another study carried out with 22 Mexican plants traditionally used for treating respiratory diseases showed effective inhibition of multi-drug resistant forms of *M. tuberculosis* (Jimenez-Arellanes et al., 2003). Extracts were tested against monoresistant variants of *M. tuberculosis* H37Rv (isoniazid, rifampin, streptomycin, ethambutol resistant) and the hexane extract of *Lantana hispida* showed the best activity. Remarkably, *L. hispida* hexane extract was also active against a group of MDR-TB clinical isolates, whereas it did not inhibit the growth of non-tuberculous mycobacteria. The hexane extract of *L. hispida* was fractionated by column chromatography and one of its fractions (FVI) inhibited the growth of all the MDR-TB clinical isolates at concentrations up to 25 μg/ml. This study further supports the fact that selecting plants by ethnobotanical criteria enhances the probability of finding active species. The genus *Lantana* is well represented in the Brazilian ecosystems and could be further screened for active compounds. Another study conducted by a group of Brazilian researchers assessed the actions of extracts from *Tabernaemontana catharinensis* against *Mycobacterium tuberculosis*, *Mycobacterium avis* and *Mycobacterium kansasii*, being the most pronounced activities observed with the alkaloid-enriched fraction (Pereira et al., 2005). A marked animycobacterial activity was also associated with the glycolipids, sesquiterpenoids and triterpenoids derived from the medicinal Mexican plants *Rumex hymenosepalus* (Polygonaceae), *Larrea divaricata* (Zygophyllaceae), *Phoradendron robinsonii* (Loranthaceae) and *Amphipteryngium adstringens* (Julianiaceae) (Rivero-Cruz et al., 2005). More recent data revealed an *in-vitro* antimycobacterial activity for the compound T-cadinol obtained from the barks of the Peruvian plant *Swartzia polyphylla* DC (Rojas et al., 2006). Additional results showed that hexane and acetone extracts from *Flourensia cernua* DC, a Mexican plant used in traditional medicine for the treatment of respiratory diseases, not only inhibited the growth, but also killed both sensitive and multi-drug-resistant strains of *Mycobacterium tuberculosis* (Molina-Salinas et al., 2006). Considering these pieces of evidence, it is clear that South American biodiversity, including the Brazilian plant resources, represent a good potential source for discovering new effective anti-tuberculosis agents. Figure 8 shows examples of some Brazil-occurring plant species that might represent potential options for the development of antimalarial and antituberculosis chemotherapeutics.

Bidens pilosa

Psidium guajava

Phyllanthus niruri

Swartzia polyphylla

Urtica dioica

Figure 8. Examples of a few plants as potential natural sources of antimalarial (top panel) and antituberculosis agents (bottom panel). The images shown here were from: www.floracyberia.net (*B. pilosa*), www.ruhr-uni-bochum.de (*P. niruri*), www.algonet.se (*P. guajava*), www.biologie.uni-hamburg.de (*S. polyphylla*), de.wikipedia.org (*U. dioica*).

4. Conclusion

Major advances in our understanding of *M. tuberculosis* and malaria parasite biology have been made. Exciting new opportunities for target-based antibacterial and antimalarial drug discovery have been made possible by completion of *M. tuberculosis* and *P. falciparum* genomes. However, the unraveling of more validated biological targets will not necessarily translate into the identification of new chemical entities that are effective against drug resistant pathogens in the long term. To this end, screening of natural compounds may provide new chemical entities active against these pathogens. The search for new drugs against tuberculosis and malaria can be investigated by screening bioative plant products against validated molecular targets. Using a combined strategy of ethnobotanical and chemotaxonomical leads, plant extracts can be tested directly on known immobilized enzymatic targets which represent essential components of the etiological agents of these diseases or involve critical steps in their progression. The advantages of this method are significant: a) the basic mechanism of action is essentially established once a positive binding is obtained, b) the specificity of the bioactive components in the extracts is practically defined, c) a relatively high throughput of samples can be achieved, d) questions of considerable relevance for drug development, such as solubility, can be addressed early in the screening process, since the immobilized enzyme system resembles the cellular environment, e) the selection of plants for testing extracts on the basis of pharmacological and taxonomical leads may increase the chances of positive hits.

In principle, there seems to be a number of strategies through which the unique molecular diversity of natural products can be leveraged in the design of combinatorial libraries. Several types of libraries from natural product-like scaffolds have been reported, such as libraries of natural-product derivatives, natural-product-like compounds prepared by total synthesis, and libraries derived from natural-products (Boldi, 2004). The target-orientated or focused-library approach seeks to elaborate structural modifications onto an existing bioactive natural product scaffold in a parallel, systematic fashion, in order to improve its inherent biological activity or drug-like properties (Kissau et al., 2003). The diversity-orientated approach seeks to leverage the privileged structural motifs of natural product scaffolds to synthesize combinatorial libraries capable of binding a wide range of biological targets (Burke and Schreiber, 2004). It has also been shown that it is possible to synthesize diversity-oriented natural product-based libraries by chemical recombination of complex fragments obtained by chemical degradation of diverse bioactive natural products (Niggemann et al., 2002). In any case, these strategies rely on finding natural chemical compounds that may represent new chemical entities with biological activity.

By no means, the list of molecular targets discussed in our contribution intends to be comprehensive. There are a number of targets described in the literature and new targets are continuously being described. Some of these targets are truly surprising since they are specific to a particular pathogen even though a homologous protein is present in the host. For instance, a diarylquinoline (R207910) has recently been shown to inhibit both drug-sensitive and drug-resistant *M. tuberculosis in vitro*, to have bactericidal activity in a murine TB model, and to have good safety and tolerability in humans (Andries et al., 2005). A puzzling feature of R207910, which represents a member of a new chemical class of antimycobacterial agents, is its exceptional specificity for its drug target: mycobacterial ATP synthase. ATP synthase is a ubiquitous enzyme found in most living organisms, including humans. A new drug target has thus been unveiled for the development of new chemotherapeutic agents and points to an ongoing need for validation of new targets to allow availability of a better arsenal of drugs to kill human pathogens. A new technology, which is called epistatic mini array profiles (E-MAP), holds tremendous promise to assign new function to known genes, to uncover the roles of previously uncharacterized proteins, and to define how biochemical pathways and proteins interact with one another (Schuldiner et al., 2005). The E-MAP technique allows analysis of epistatic interactions – how the activity of one gene affects that of another. Schuldiner et al. (2005) developed a technique called decreased abundance by mRNA perturbation (DamP) that allows to manipulate the half-life of mRNAs, thereby allowing determination of effects of "dialing down" pairs of genes in comparison to "dialing down" a single gene, including genes essential for survival that would otherwise kill the cell.The E-MAP approach could be used as a tool for the development of effective combinations of antibacterial drugs that target proteins produced by interacting genes because, in case the two proteins interact epistatically, inhibiting both proteins should produce a much greater effect than the sum of the two. The E-MAP should also give pharmaceutical researchers clues to the magnitude of possible side effects of drugs under development.

The emergence of drug resistance highlights the ongoing and cyclical need for the development of new therapeutic agents to combat diseases that are all too often neglected because they afflict mainly people living in third-world countries. The assembly of data presented in this chapter strongly indicates that the use of plant based therapies, even alone or in combination with other medicines, could represent an attractive option for malaria and tuberculosis treatment. Undoubtedly, the consistent development of new therapeutic choices on the basis of plant extracts and/or compounds might only be achieved by a significant collaboration involving chemists (natural product characterization), botanists (ethnobotanic and chemotaxonomic studies), biochemists/molecular biologists (target identification) and pharmacologists (pre-clinical studies).

5. References

Acton, N; Roth, RJ. On the conversion of dihydroartemisinic acid into artemisinin. *J Org Chem,* 1992 57, 3610-3614.

Andrade-Neto, VF; Brandão, MGL; Stehmann, JR; Oliveira. LA; Krettli. AU; Antimalarial activity of Cinchona-like plants used to treat fever and malaria in Brazil. *J Ethnopharmacol,* 2003 87, 253-256.

Andries, K; Verhasselt, P; Guillemont, J; Göhlmann, HWH; Neefs, JM; Winkler, H; Gestel, JV; Timmerman, P; Zhu, M; Lee, E; Williams, P; Chaffoy, D; Huitric, E; Hoffner, S; Cambau, E; Truffot-Pernot, C; Lounis, N; Jarlier, V. A diarylquinoline drug active on the ATP synthase of *Mycobacterium tuberculosis*. *Science*, 2005 307, 223-227.

Anyshetti, S; Pulimi, M; Pennathur, G. Potential drug targets in *Mycobacterium tuberculosis* through metabolic pathway analysis. *Comput Biol Chem,* 2005 29, 368-378.

Arav-Boger, R; Shapiro, TA. Molecular mechanisms of resistance in antimalarial chemotherapy: the unmet challenge. *Ann Rev Pharmacol Toxicol,* 2005 45, 565-585.

Asahi, H; Kanazawa, T; Kajihara, Y; Takahashi, K; Takahashi, T. Hypoxanthine: a low molecular weight factor essential for growth of erythrocytic *Plasmodium falciparum* in a serum-free medium. *Parasitology*, 1996 113, 19-23.

Asselineau, C; Asselineau, J; Lanéelle, G; Lanéelle, M-A. The biosynthesis of mycolic acids by Mycobacteria: current and alternative hypotheses. *Prog Lipid Res,* 2002 41, 501-523.

Azevedo, WFJr; Canduri, F; Oliveira, JS; Basso, LA; Palma, MS; Pereira, JH; Santos, DS. Molecular model of shikimate kinase from *Mycobacterium tuberculosis. Biochem Biophys Res Commun*, 2002 5, 142-148.

Bahl, A; Brunk, B; Coppel; RL; Crabtree, J; Diskin, SJ; Fraunholz, MJ; Grant, GR; Gupta, D; Huestis, RL; Kissinger, JC; Labo, P; Li, L; McWeeney, SK; Milgram, AJ; Roos, DS; Schug, J; Stoeckert, CJJr. PlasmoDB: the *Plasmodium* genome resource. An integrated database providing tools for accessing, analyzing and mapping expression and sequence data (both finished and unfinished). *Nucleic Acids Res,* 2002 30, 87–90.

Balint, GA. Artemisinin and its derivatives: an important new class of antimalarial agents. *Pharmacol Ther,* 2001 90, 261-265.

Banerjee, A; Dubnau, E; Quémard, A; Balasubramanian, V; Um, KS; Wilson, T; Collins, D; de Lisle, G; Jacobs, WRJr. inhA, a Gene encoding a target for isoniazid and ethionamide in *Mycobacterium tuberculosis. Science*, 1994 263, 227-230.

Bannister, LH; Hopkins, JM; Fowler, RE; Krishna, S; Mitchel, GH. A brief illustrated guide to the ultrastructure of *Plasmodium falciparum* asexual blood stages. *Parasitol Today,* 2000 16, 427-433.

Barry, CEIII; Lee RR; Mdluli, K; Sampson, AE; Schroeder, BG; Slayden, RA; Yuan, Y. Mycolic acids: structure, biosynthesis and physiological functions. *Prog Lipid Res,* 1998, 37, 143-179.

Basso, LA; Blanchard, JS. Resistance to antitubercular drugs. *Adv Exp Med Biol,* 1998 456, 115-144.

Basso, LA; Santos, DS. Drugs that inhibit mycolic acid biosytheis in *Mycobacterium tuberculosis* – an updqte. *Medicinal Chemistry Reviews – Online,* 2005 2, 393-413.

Basso, LA; Silva, LHP; Fett-Neto, AG; Azevedo, WFJr; Moreira, IS; Palma, MS; Calixto, JB; Filho, SA; Santos, RR; Soares, MBP; Santos, DS. The use of biodiversity as source of new chemical entities against defined targets for treatment of malaria, tuberculosis, and T-cell mediated diseases – a review. *Mem Inst Oswaldo Cruz,* 2005 100, 475-506.

Basso, LA; Zheng, R; Blanchard, JS. Kinetics of inactivation of WT and C243S mutant of *Mycobacterium tuberculosis* enoyl reductase by activated isoniazid. *J Am Chem Soc,* 1996 118, 11301-11302.

Bentley, R. The shikimate pathway – a metabolic tree with many branches. *Crit Rev Biochem Mol Biol,* 1990 25, 307-384.

Berman, PA; Human, L; Freese, JA. Xanthine oxidase inhibits growth of *Plasmodium falciparum* in human erythrocytes *in vitro. J Clin Invest,* 1991 88, 1848-1855.

Berry, JP; McFerren, MA; Rodriguez, E Zoopharmacognosy: a "biorational" strategy for phytochemical prospecting. In Gustine DL and Flores H, *Phytochemicals and Health,* ASPP, Rockville, 1995 p. 165-178.

Biagini, GA; O'Neill, PM; Bray, PG; Ward, SA. Current drug development portfolio for antimalairal therapies. *Curr Opin Pharmacol,* 2005 5, 473-478.

Bissati, KE; Zufferey, R; Witola, WH; Carter, NS; Ullman, B; Mamoun, CB. The plasma membrane permease PfNT1 is essential for purine salvage in the human malaria parasite *Plasmodium falciparum*. *Proc Natl Acad Sci* USA, 2006 103, 9286-9291.

Bloch, K. Fatty acid synthases from *Mycobacterium phlei*. *Methods Enzymol,* 1975; 35, 84-90.

Boldi, AM. Libraries from natural product-like scaffolds. *Curr Opin Chem Biol,* 2004 8, 281-286.

Boshoff, HIM; Reed, MB; Barry, CEIII; Mizrahi, V. DnaE2 polymerase contributes to *in vivo* survival and the emergence of drug resistance in *Mycobacterium tuberculosis*. *Cell*, 2003 113, 183-193.

Both, FL; Kerber, VA; Henriques, AT; Elisabetsky, E Analgesic properties of umbellatine from Psychotria umbellata. *Pharm Biol,* 2002 40, 336-341.

Brandão, MGL; Krettli, AU; Soares, LSR; Nery, CGC; Marinuzzi, HC. Antimalarial activity of extracts and fractions from Bidens pilosa and other Bidens species (Asteraceae) correlated with the presence of acetylene and flavonoid compounds. *J Ethnopharmacol,* 1997 57, 131-138.

Brennan PJ. Structure, function, and biogenesis of the cell wall of *Mycobacterium tuberculosis*. *Tuberculosis*, 2003 83, 91-97.

Breman, JG; Alilio, MS; Mills, A. Conquering the intolerable burden of malaria: what's new, what's needed: a summary. *Am J Trop Med Hyg,* 2004 71, 1-15.

Brindley, DN; Matsumura, S; Block, K. *Mycobacterium phlei* fatty acid synthase – a bacterial multienzyme complex. *Nature*, 1969 224, 666-669.

Brohm, D; Metzger, S; Bhargava, A; Muller, O; Lieb, F; Waldmann, H. Natural products are biologically validated starting points in structural space for compound library development: solid-phase synthesis of dysidiolide-derived phosphatase inhibitors Angew *Chem Int Ed Engl*, 2002 41, 307-11.

Burke, MD; Schreiber SL. A planning strategy for diversity oriented synthesis. *Angew Chem Int Ed,* 2004 43, 46-58.

Butler, MS. The role of natural product chemistry in drug discovery. *J Nat Prod,* 2004 67, 2141-2153.

CDC (Centers for Disease Control and Prevention). Emergence of Mycobacterium tuberculosis with extensive resistance to second-line drugs – worldwide, 2000-2004. *MMWR Morb Mortal Wkly Rep,* 2006 55, 301-305.

Clardy, J; Walsh, C. Lessons from natural molecules. *Nature*, 2004 432: 829-837.

Coggins, JR; Abell, C; Evans, LB; Frederickson, M; Robinson, DA; Roszak, AW; Lapthorn, AP. Experiences with the shikimate-pathway enzymes as targets for rational drug design. *Biochem Soc Trans*, 2003 31, 548-551.

Cohen-Gonsaud, M; Ducasse, S; Hoh, F; Zerbib, D; Labesse, G; Quémard, A. Crystal structure of MabA from *Mycobacterium tuberculosis*, a reductase involved in long-chain fatty acid biosynthesis, *J Mol Biol,* 2002 320, 249-261.

Cole, ST; Brosch, R; Parkhill, J; Garnier, T; Churcher, C; Harris, D; Gordon, SV; Eiglmeier, K; Gas, S; Barry, CE; Tekaia, F; Badcock, K; Basham, D; Brown, D; Chillingworth, T; Connor, R; Davies, R; Devlin, K; Feltwell, T; Gentles, S; Hamlin, N; Holroyd, S; Hornsby, T; Jagels, K; Krogh, A; McLean, J; Moule, S; Murphy, L; Oliver, K; Osborne, J; Quail, MA; Rajandream, M-A; Rogers, J; Rutter, S; Seeger, K; Skelton, J; Squares, R; Squares, S; Sulston, JE; Taylor, K; Whitehead, S; Barrell, BG. Deciphering the biology of *Mycobacterium tuberculosis* from the complete genome sequence. *Nature*, 1998 393, 537-544.

Corbett, EL; Watt, CJ; Walker, N; Maher, D; Williams, BG; Raviglione, MC; Dye, C. The growing burden of tuberculosis: global trends and interactions with the HIV epidemic. *Arch Intern Med*, 2003 163, 1009-1021.

Cragg, GM; Boyd, MR; Grever, MR; Schepartz SA Pharmaceutical prospecting and the potential for pharmaceutical crops – natural product drug discovery at the United States National Cancer Institute. *AnnMissouri Bot Gard,* 1995 82, 47-53.

Cragg, GM; Newman, DJ. Medicinals for the millenia – the historical record. *Annals of the New York Academy of Sciences,* 2001 953, 3-25.

de Voos, JJ; Rutter, K; Schroder, BG; Su, H; Zhu, Y; Barry, CEIII. The salicylate-derived mycobactin siderophores of *Mycobacterium tuberculosis* are essential for growth in macrophages. *Proc Natl Acad Sci* USA, 2000 97, 1252-1257.

DeBarber, A; Mdluli, K; Bosman, M; Bekker, L-G; Barry CEIII. Ethionamide activation and sensitivity in multidrug-resistant Mycobacterium tuberculosis. *Proc Natl Acad Sci* USA, 2000 97, 9677-9682.

Dias, MV; Borges, JC; Ely, F; Pereira, JH; Canduri, F; Ramos, CH; Frazzon, J; Palma, MS; Basso, LA; Santos, DS; Azevedo, WFJr. Structure of chorismate synthase from *Mycobacterium tuberculosis*. *J Struct Biol*, 2006 154, 130-142.

Dobson, CM. Chemical space and biology. Nature, 2004 432, 824-828.

Ducasse-Cabanot, S; Cohen-Gonsaud, M; Marrakchi, H; Nguyen, M; Zerbib, D; Bernadou, J; Daffé, M; Labesse, G; Quémard, A. In vitro inhibition of the *Mycobacterium tuberculosis* β-ketoacyl-acyl carrier protein reductase MabA by isoniazid. *Antimicrob Agents Chemother,* 2004 48, 242-249.

Dye, C. Global epidemiology of tuberculosis. *Lancet*, 2006 367, 938-940.

Dye, C; Scheele, S; Dolin, P; Pathania, V; Raviglione, MC. Global burden of tuberculosis: estimated incidence, prevalence, and mortality by country. *JAMA*, 1999 282, 677-686.

Edwards, G; Biagini, GA. Resisting resistance: dealing with the irrepressible problem of malaria. *Br J Clin Pharmacol,* 2006 61, 690-693.

Elisabetsky, E; Amadorm TA; Lealm M; Nunesm DS Merging ethnopharmacology with chemotaxonomy: an approach to unveil bioactive natural products. The case of Psychotria alkaloids as potential analgesics. *Ciência e Cultura,* 1997 49, 378-385.

Enserink, M. Infectious diseases: source of new hope against malaria is in short supply. *Science*, 307 33.

Espinal, MA. The global situation of MDR-TB. *Tuberculosis*, 2003 83, 44-51.

Evangelista de Duffard, AM; Duffard, R. Behavioral toxicology, risk assessment, and chlorinated hydrocarbons. *Environ Health Perspect,* 1996 104, 353-360.

Feher, M; Schmidt, JM. Property distributions: differences between drugs, natural products, and molecules from combinatorial chemistry. *J Chem Inf Comput Sci,* 2003 43, 218-227.

Fernandes, CL; Breda, A; Santos, DS; Basso, LA; Souza, ON. A structural model for chorismate synthase from *Mycobacterium tuberculosis* in complex with coenzyme and substrate. *Comput Biol Med,* 2006 Mar 31; [Epud ahead of print] PMID: 16584721.

Ferri MG. Vegetação Brasileira (Brazilian Vegetation), Ed. Itatiaia e Ed. da USP; 1980 157 p.

Fonseca, IO; Magalhães, ML; Oliveira, JS; Silva, RG; Mendes, MA; Palma, MS; Santos, DS; Basso, LA. Functional shikimate dehydrogenase from *Mycobacterium tuberculosis* H37Rv: purification and characterization. *Protein Expr Purif,* 2006 46, 429-437.

Fowler, RE; Fookes, RE; Lavin, F; Bannister, LH; Mitchel, GH. Microtubules in Plasmodium falciparum merozoites and their importance in invasion of erythrocytes. *Parasitology*, 1998 117, 425-4233.

Gardner, MJ; Hall, N; Fung, E; White, O; Berriman, M; Hyman, RW; Carlton, JM; Pain, A; Nelson, KE; Bowman, S; Paulsen, IT; James, K; Eisen, JA; Rutherford, K; Salzberg, SL; Craig, A; Kyes, S; Chan, MS; Nene, V; Shallom, SJ; Suh, B; Peterson, J; Angiuoli, S; Pertea, M; Allen, J; Selengut, J; Haft, D; Mather, MW; Vaidya, AB; Martin, DMA; Fairlamb, AH; Fraunholz, MJ; Roos, DS; Ralph, SA; McFadden, GI; Cummings, LM; Subramanian, GM; Mungall, C; Venter, JC; Carucci, DJ; Hoffman, SL; Newbold, C; Davis, RW; Fraser, CM; Barrell B. Genome sequence of the human malaria parasite *Plasmodium falciparum*. *Nature*, 2002 419, 498–511.

Glickman, MS; Jacobs, WRJr. Microbial pathogenesis of *Mycobacterium tuberculosis*: dawn of a discipline. *Cell*, 2001 104, 477-485.

Gomez, JE; McKinney, JD. *M. tuberculosis* persistence, latency, and drug tolerance. *Tuberculosis*, 2004 84, 29-44.

Gornick, P. Apicoplast fatty acid biosynthesis as a target for medical intervention in apicomplexsan parasites. *Int J Parasitol*, 2003 33, 885-896.

Gregianini, TS; Silveira, VC; Porto, DD; Kerber, VA; Henriques, AT; Fett-Neto, AG. The alkaloid brachycerine is induced by ultraviolet radiation and is a singlet oxygen quencher. *Photochem Photobiol*, 2003 78, 470-474.

Gunasekera, SP; Mickel, SJ; Daeffler, R; Niederer, D; Wright, AE; Linley, P; Pitts, T. Synthetic analogues of the microtubule-stabilizing agent (+)-discodermolide: preparation and biological activity. *J Nat Prod*, 2004 67, 749-756.

Hasan, S; Daugelat, S; Rao, PSS; Schreiber, M. Prioritizing genomic drug targets in pathogens: application to Mycobacterium tuberculosis. *PloS Comput Biol*, 2006 2(6), e61.

Haynes, R. Artemisinin and derivatives: the future for malaria treatment? *Curr Opin Infect Dis*, 2001 14, 719-726.

Herrmann, KM; Weaver, LM. The shikimate pathway. *Annu Rev Plant Physiol Mol Biol*, 1999 50, 473-503.

Hingley-Wilson, SM; Sambandamurthy, VK; Jacobs WRJr. Survival perspectives from the world's most successful pathogen, *Mycobacterium tuberculosis*. *Nat Immunol*, 2003 4 949-955.

Jambou, R; Legrand, E; Niang, M; Khim, N; Lim, P; Volney, B. Ekala, MT; Bouchier, C; Esterre, P; Fandeur, T; Mercereau-Pijalon, O. Resistance of *Plasmodium falciparum* field isolates to in-vitro artemether and point mutations of the SERCA-type PfATPase6. *Lancet*, 2005 366, 1960-1963.

Jimenez-Arellanes, A; Meckes, M; Ramirez, R; Torres, J; Luna-Herrera, J; Activity against multidrug-resistant Mycobacterium tuberculosis in Mexican plants. *Phytother Res*, 2003 17, 903-908.

Johnsson, K; King, DS; Schultz, PG. Studies on the mechanism of action of isoniazid and ethionamide in the chemotherapy of tuberculosis. J Am Chem Soc, 1995 117, 5009-5010.

Johnsson, K; Schultz, PG. Mechanistic studies of the oxidation of isoniazid by the catalase peroxidase from *Mycobacterium tuberculosis*. *J AmChem Soc*, 1994 116, 7425-7426.

Kanokmedhakul, S; Kanokmedhakul, K; Kantikeaw, I; Phonkerd, N. 2-Substituted furans from the roots of Polyalthia evecta. *J Nat Prod*, 2006 69, 68-72.

Kapoor, M; Mukhi, PL; Surolia, N; Suguna, K; Surolia, A. Kinetic and structural analysis of the increased affinity of enoyl-ACP (acyl-carrier protein) reductase for triclosan in the presence of NAD^+. *Biochem J,* 2004 381, 725-733.

Kapoor, M; Reddy, CC; Krishnasastry, MV; Surolia, N; Surolia, A. Slow-tight-binding inhibition of enoyl-acyl carrier protein reductase from *Plasmodium falciparum* by triclosan. *Biochem J,* 2004 381, 719-724.

Kappe, SHI; Kaiser, K; Matuschewski K. The *Plasmodium* sporozoite journey: a rite of passage. *Trends Parasitol,* 2003, 19 135-143.

Kicska, GA; Tyler, PC; Evans, GB; Furneaux, RH; Kim, K; Schramm, VL. Transition state analogue inhibitors of purine nucleoside phosphorylase from *Plasmodium falciparum. J Biol Chem,* 2002b 277, 3219-3225.

Kissau, L; Stahl, P; Mazitschek, R; Giannis, A; Waldmann, H. Development of natural product-derived receptor tyrosine kinase inhibitors based on conservation of protein domain fold. *J Med Chem,* 2003, 46 2917-2931.

Koehn, FE; Carter, GT. The evolving role of natural products in drug discovery. Nat Rev Drug Discov, 2005 4: 206-220.

Kolatukudy, PE; Fernandes, ND; Azad, AK; Fitzmaurice, AM; Sirakova, TD. Biochemistry and molecular genetics of cell-wall lipid biosynthesis in mycobacteria. *Mol Microbiol,* 1997 24, 263-270.

Kremer, L; Dover, LG; Morbidoni, HR; Vilchèze, C; Maughan, WN; Baulard, A; Tu, S-C; Honoré, N; Deretic, V; Sacchettini, JC; Locht, C; Jacobs, WRJr; Besra, GS. Inhibition of InhA activity, but not KasA activity, induces formation of a KasA-containing complex in mycobacteria. *J Biol Chem,* 2003 278, 20547-20554.

Krettli, A; Andrade-Neto, VF; Brandão, MGL; Ferrari, WMS. The search for new antimalarial drugs from plants used to treat fever and malaria or plants randomly selected: a review. *Mem Inst Oswaldo Cruz,* 2001 96, 1033-1042.

Kushner, S; Dalalian, H; Sanjurjo, JL; Bach, FL; Safir, SR; Smith, VKJr; Williams, JH. Experimental chemotherapy of tuberculosis. The synthesis of pyrazinamides and related compounds. *J Am Chem Soc,* 1952 74, 3617-3621.

Lack, G; Homberger-Zizzari, E; Folkers, G; Scapozza, L; Perozzo, R. Recombinant expression and biochemical characterization of the unique elongating β-ketoacyl-acyl carrier protein synthase involved in fatty acid biosynthesis of *Plasmodium falciparum* using natural and artificial substrates. *J Biol Chem,* 2006 281, 9538-9546.

Larsen, MH; Vilchèze, C; Kremer, L; Besra, GS; Parsons, L; Salfinger, M; Heifets, L; Hazbon, MH; Alland, D; Sacchettini, JC; Jacobs, WRJr. Overexpression of inhA, but not kasA, confers resistance to isoniazid and ethionamide in *Mycobacterium smegmatis*, M. bovis BCG and M. tuberculosis. *Mol Microbiol,* 2002 46, 453-466.

Lee, ML; Schneider, G. Scaffold architecture and pharmacophoric properties of natural products and trade drugs: application in the design of natural product-based combinatorial libraries. *J Comb Chem,* 2001 3, 284-289.

Lei, B; Wei, C-J; Tu, S-C. Action mechanism of antitubercular isoniazid. *J Biol Chem,* 2000 275, 2520-2526.

Lewis, WH; Elvin-Lewis MP Medicinal plants as sources of new therapeutics. *Ann Missouri Bot Gard,* 1995 82, 16-24.

Lipinski, CA; Lombardo, F; Dominy, BW; Feeney, PJ. Experimental and computational approaches to estimate solubility and permeability in drug discovery and development settings. *Adv Drug Del Rev,* 1997 23, 3-25.

Lopez, AD; Mathers, CD; Ezzati, M; Jamison, DT; Murray, CJL. Global and regional burden of disease and risk factors, 2001: systematic analysis of population health data. *Lancet,* 2006 367, 1747-1757.

Magalhães, MLB; Pereira, CP; Basso, LA; Santos, DS. Cloning and expression of functional shikimate dehydrogenase (EC 1.1.1.25) from *Mycobacterium tuberculosis* H37Rv. *Protein Expr Purif,* 2002 26, 59-64.

Marrakchi, H; Ducasse, S; Labesse, G; Montrozier, H; Margeat, E; Emorine, L; Charpentier, X; Daffé, M; Quémard, A. MabA (FabG1), a *Mycobacterium tuberculosis* protein involved in the long-chain fatty acid elongation system FAS-II. *Microbiology*, 2002 148, 951-960.

Martin, YC; Critchlow, RE. Beyond mere diversity: tailoring combinatorial libraries for drug discovery. *J Com Chem,* 1999 1, 231-50.

McLeod, R; Muench, SP; Rafferty, JB; Kyle, DE; Mui, EJ; Kirisits, MJ; Mack, DG; Roberts, CW; Samuel, BU; Lyons, RE; Dorris, M; Milhous, WK; Rice, DW. Triclosan inhibits the growth of *Plasmodium falciparum* and *Toxoplasma gondii* by inhibition of Apicomplexan Fab I. *Int J Parasitol,* 2001 31, 109-113.

Mdluli, K; Slayden, RA; Zhu, Y; Ramaswamy, S; Pan, X; Mead, D; Crane, DD; Musser, JM; Barry, CEIII. Inhibition of Mycobacterium tuberculosis β ketoacyl ACP synthase by isoniazid. *Science*, 1998 280, 1607-1610.

Middlebrook, G. Sterilization of tubercle bacilli by isonicotinic acid hydrazide and the incidence of variants resistant to the drug *in vitro*. *Am Rev Tuberc,* 1952 65, 765-767.

Mitchison, DA The action of antituberculosis drugs in short-course chemotherapy *Tubercle*, 1985 66, 219-225.

Molina-Salinas, GM; Ramos-Guerra, MC; Vargas-Villarreal, J; Mata-Cardenas, BD; Becerril-Montes, P; Said-Fernandez S. Bactericidal activity of organic extracts from Flourensia cernua DC against strains of Mycobacterium tuberculosis. *Arch Med Res,* 2006 37, 45-49.

Morens, DM; Folkers, GK; Fauci, AS. The challenge of emerging and re-emerging infectious diseases. *Nature*, 2004 430, 242-248.

Morrison, JF; Walsh, CT. The behaviour and significance of slow-binding enzyme inhibitors. *Adv Enzymol Relat Areas Mol Biol,* 1988, 61, 201-301.

Muller, S; Coombs, GH; Walter RD. Targeting polyamines of parasitic protozoa in chemotherapy. *Trends Parasitol,* 2001 17, 242-249.

Myers, N; Mittermeler, RA; Mittermeler, CG; Fonseca, GAB; Kent, J Biodiversity hotspots for conservation priorities. *Nature*, 2000 403, 853-858.

Newman, DJ; Cragg, GM; Snader, KM. The influence of natural products upon drug discovery. *Nat Prod Rep,* 2000 17, 215-234.

Newman, DJ; Cragg, GM; Snader, KM. Natural products as a source of new drugs over the period 1981-2002. *J Nat Prod,* 2003 66, 1002-1037.

Niggemann, J; Michaelis, K; Frank, K; Zander, N; Höfle, G. Natural product-derived building blocks for combinatorial synthesis. Part 1. Fragmentation of natural products from myxobacteria. *J Chem Soc Perkin Trans,* 2002 1, 2490-2503.

Nundkumar, N; Ojewole JA. Studies on the antiplasmodial properties of some South African medicinal plants used as antimalarial remedies in Zulu folk medicine. *Methods Find Exp Clin Pharmacol,* 2002 24, 397-401.

Nunn P. WHO Global TB Programme, Tuberculosis Research and Surveillance Unit of the WHO Global TB Programme; Press Release WHO/74 (www.who.int/gtb/-press/who74.htm), 1997.

O'Neill, P; Bray, P; Hawley, S; Ward, S; Park, B. 4-Aminoquinolines – past, present, and future: a chemical perspective. *Pharmacol Ther,* 1998 77, 29-58.

Oksman-Caldentey, KM; Inzé D. Plant cell factories in the post-genomic era: new ways to produce designer secondary metabolites. *Trends Plant Sci,* 2004 9, 433-440.

Okunade, AL; Elvin-Lewis, MPF; Lewis, WH. Natural antimycobacterial metabolites: current status. *Phytochemistry,* 2004 65: 1017-1032.

Oliveira, FQ; Andrade-Neto, V; Krettli, AU; Brandão, MGL. New evidences of antimalarial activity of *Bidens pilosa* roots extract correlated with polyacetylene and flavonoids. *J Ethopharmacol,* 2004 93, 39-42.

Oliveira, JS; Mendes, MA; Palma, MS; Basso, LA; Santos, DS. One-step purification of 5-enolpyruvylshikimate-3-phosphate synthase enzyme from *Mycobacterium tuberculosis. Protein Expr Purif,* 2003 28, 287-292.

Oliveira, JS; Pereira, JH; Canduri, F; Rodrigues, NC; Souza, ON; Azevedo, WFJr; Basso, LA; Santos, DS. Crystallographic and pré-steady-state kinetics studies on biding of NADH to wild-type and isoniazid-resistat enoyl-ACP (CoA) reductase enzymes from Mycobacterium tuberculosis. *J Mol Biol,* 2006 359, 646-666.

Oliveira, JS; Pinto, CA; Basso, LA; Santos DS. Cloning and overexpression in soluble form of functional shikimate kinase and 5-enolpyruvylshikimate 3-phosphate synthase enzymes from *Mycobacterium tuberculosis. Protein Expr Purif,* 2001 22, 430-435.

Oliveira, JS; Sousa, EHS; Basso, LA; Palaci, M; Dietze, R; Santos, DS; Moreira IS. An inorganic iron complex that inhibits wild-type and an isoniazid-resistant mutant 2-trans-enoyl-ACP (CoA) reductase from *Mycobacterium tuberculosis. Chem Commun,* 2004 3, 312-313.

Oliveira, JS; Sousa, EHS; Souza, ON; Moreira, IS; Santos, DS; Basso, LA. Slow-onset inhibition by an inorganic complex of 2-*trans*-enoyl-ACP (CoA) reductase from *Mycobacterium tuberculosis. Curr Pharm* Des, 2006 2 (invited manuscript Ref. No.: P-138313 – in press).

Olliaro, PL; Taylor, WRJ. Antimalarial compounds: from bench to bedside. *J Exp Biol,* 2003 206, 3753-3759.

Olliaro, PL; Taylor, WRJ. Developing artemisinin based drug combinations for the treatment of drug resistant falciparum malaria: a review. *J Postgrad Med,* 2004 50, 40-44.

Pablos-Méndez, A; Gowda, DK; Frieden, TR. Controlling multidrug-resistant tuberculosis and access to expensive drugs: a rational framework. *Bull. World Health Organ,* 2002 80, 489-495.

Paranhos, JT; Fragoso, V; Henriques, AT; Ferreira, AG; Fett-Neto, AG Regeneration of Psychotria umbellata and production of the analgesic indole alkaloid umbellatine. *Tree Physiol,* 2005 25, 251-255.

Parish, T; Stoker, NG. The common aromatic amino acid biosynthesis pathway is essential in *Mycobacterium tuberculosis. Microbiology*, 2002 148, 3069-3077.

Pasquali, G; Porto, DD; Fett-Neto, AG. Metabolic engineering of cell cultures versus whole-plant complexity in production of bioactive monoterpene indole alkaloids: recent progress related to old dilemma. *J Biosci Bioeng,* 2006 101, 287-296.

Pereira, CG; Leal, PF; Sato, DN; Meireles MA. Antioxidant and antimycobacterial activities of Tabernaemontana catharinensis extracts obtained by supercritical CO2 + cosolvent. J Med Food, 2005 8, 533-538.

Pereira, JH; Canduri, F; Oliveira, JS; Silveira, NJF; Basso, LA; Palma, MS; Azevedo, WFJr; Santos, DS. Structural bioinformatics study of EPSP synthase from *Mycobacterium tuberculosis*. *Biochem Biophys Res Commun,* 2003 312,608-614.

Pereira, JH; Oliveira, JS; Canduri, F; Dias, MVB; Palma, MS; Basso, LA; Santos, DS; Azevedo, WFJr. Structure of shikimate kinase from *Mycobacterium tuberculosis* reveals the binding of shikimic acid. *Acta Crystallogr D Biol Crystallogr,* 2004 6, 2310-2319.

Perozzo, R; Kuo, M; Sidhu, ABS; Valiyaveettil, JT; Bittman, R; Jacobs, WRJr; Fidock, DA; Sacchettini, JC. Structural elucidation of the specificity of the antibacterial agent triclosan for malarial enoyl acyl carrier protein reductase. *J Biol Chem,* 2002 277, 13106-13114.

Pillai, S; Rajagopal, C; Kapoor, M; Kumar, G; Gupta, A; Surolia, N. Functional characterization of β-ketoacyl-ACP reductase (FabG) from *Plasmodium falciparum. Biochem Biophys Res Commun,* 2003 303, 387-392.

Prigge, ST; He, X; Gerena, L; Waters, NC; Reynolds, KA. The initiating steps of a type II fatty synthase in *Plasmodium falciparum* are catalyzed by pfACP, pfMCAT, and pfKASIII. *Biochemistry,* 2003 42,1160-1169.

Projan, SJ. Why is big Pharma getting out of antibacterial drug discovery? Curr Opin Microbiol, 2003a 6, 427-430.

Projan, SJ. Infectious diseases in the 21st century: increasing threats, fewer new treatments and a premium on prevention. Curr Opin Pharmacol, 2003b 3, 457-458

Quémard, A; Sacchettini, JC; Dessen, A; Vilchèze, C; Bittman, R; Jacobs, WRJr; Blanchard, JS. Enzymatic characterization of the target for isoniazid in *Mycobacterium tuberculosis. Biochemistry*, 1995 34, 8235-8241.

Ralph, SA; van Dooren, GG; Waller, RF; Crawford, MJ; Fraunholz, MJ; Foth, BJ; Tonkin, CJ; Roos, DS; McFadden, GI. Metabolic maps and functions of the *Plasmodium falciparum* apicoplast. *Nat Rev Microbiol,* 2004 2, 203-216.

Raman, K; Rajagopalan, P; Chandra, N. Flux balance analysis of mycolic acid pathway: targets for anti-tubercular drugs. *PloS Comput Biol,* 2005 1(15), e46.

Ramaswamy, S; Musser, JM. Molecular genetic basis of antimicrobial agent resistance in *Mycobacterium tuberculosis*: 1998 update. *Tuber Lung Dis,* 1998 79, 3-29.

Raskin, I; Ribnicky, DM; Komamytsky, S; Ilic, N; Poulsy, A; Borisjuk, N; Brinker, A; Moreno, DA; Ripoll, C; Yakoby, N; O'Neal, JM; Cornwell, T; Pastor, I; Fridlender, B. Plants and human health in the twenty-first century. *Trends Biotechnol,* 2002 20, 522-531.

Rates, SMK. Plants as source of drugs. *Toxicon*, 2001 39, 603-613.

Rattan, A; Kalia, A; Ahmad, N. Multidrug-resistant *Mycobacterium tuberculosis*: molecular perspectives. *Emerg Infect Dis,* 1998 4, 195-209.

Rawat, R; Whitty, A; Tonge, PJ. The isonizid-NAD adduct is a slow, tight-binding inhibitor of InhA, the *Mycobacterium tuberculosis* enoyl reductase: adduct affinity and drug resistance. *Proc Natl Acad Sci* USA, 2003 100, 13881-13886.

Reyes, P; Rathod, PK; Sanchez, DJ; Mrema, JE; Rieckmann, KH; Heidrich, HG. Enzymes of purine and pyrimidine metabolism from the human malaria parasite, *Plasmodium falciparum*. *Mol Biochem Parasitol,* 1982 5, 275-290.

Ridley, RG. Medical need, scientific opportunity and the drive for antimalarial drugs. *Nature*, 2002 415, 686-693.

Rivero-Cruz, I; Acevedo, L; Guerrero, JÁ; Martinezm S; Byem R; Pereda-Miranda, R; Franzblau, S; Timmermann, BN; Mata, R. Antimycobacterial agents from selected Mexican medicinal plants. *J Pharm Pharmacol,* 2005 57, 1117-1126.

Rizzi, C; Frazzon, J; Ely, F; Weber, PG; Fonseca, IO; Gallas, M; Oliveira, JS; Mendes, MA; Souza, BM; Palma, MS; Santos, DS; Basso, LA. DAHP synthase from *Mycobacterium tuberculosis* H37Rv: cloning, expression, and purification of functional enzyme. *Protein Expr Purif,* 2005 40, 23-30.

Ro, D-K; Paradise, EM; Ouellet, M; Fisher, KJ; Newman, KL; Ndungu, JM; Ho, KA; Eachus, RA; Ham, TS; Kirby, J; Chang, MCY; Withers, ST; Shiba, Y; Sarpong, R; Keasling, JD. Production of the antimalarial drug precursor artemisinic acid in engineered yeast. *Nature*, 2006 440, 940-943.

Rojas, R; Bustamante, B; Ventosilla, P; Fernadez, I; Caviedes, L; Gilman, RH; Lock, O; Hammond, GB. Larvicidal, antimycobacterial and antifungal compounds from the bark of the Peruvian plant Swartzia polyphylla DC. *Chem Pharm Bull*, 2006 54, 278-279.

Rosental, PJ. Antimalarial drug discovery: old and new approaches. *J Exp Biol*, 2003 206, 3735-3744.

Rozwarski, DA; Grant, GA; Barton, DHR; Jacobs, WRJr; Sacchettini, JC. Modification of the NADH of the isoniazid target (InhA) from *Mycobacterium tuberculosis*. *Science*, 1998 279, 98-102.

Sassetti, CM; Rubin, EJ. Genetic requirements for mycobacterial survival during infection. *Proc Natl Acad Sci* USA, 2003 100, 12989-12994.

Schatz, A; Bugie, E; Waksman, SA Streptomycin, a substrate exhibiting antibiotic activity against gram-positive and gram-negative bacteria. *Proc Soc Exp Biol Med,* 1944 55, 66-69.

Schroeder, EK; Basso, LA; Santos, DS; Souza, ON. Molecular dynamics simulation studies of the wild-type, I21V and I16T mutants of isoniazid resistant Mycobacterium tuberculosis enoyl reductase (InhA) in complex with NADH: towards the understanding of NADH-InhA different affinities. *Biophys J,* 2005 89, 876-884.

Schroeder, EK; de Souza, ON; Santos, DS; Blanchard, JS; Basso, LA. Drugs that inhibit mycolic acid biosynthesis in *Mycobacterium tuberculosis*. *Curr Pharm Biotechnol,* 2002 3, 197-225.

Schuldiner, M; Collins, SR; Thompson, NJ; Denic, V; Bhamidipati, A; Punna, T; Ihmels, J; Andrews, B; Boone, C; Greenblatt, JF; Weissman, JS; Krogan, NJ. Exploration of the function and organization of the yeast early secretory pathway through an epistatic miniarray profile. *Cell*, 2005 123, 507-519.

Seeber, F. Biosynthetic pathways of plastid-derived organelles as potential drug targets against parasitic apicomplexa. Curr Drug Targets Immune *Endocr Metabol Disord,* 2003 3, 99-109.

Service, RF. Surviving the blockbuster syndrome. *Science*, 2004 303, 1796-1799.

Sharma, V. Therapeutic drugs for targeting chloroquine resistance in malaria. *Mini-Reviews in Medicinal Chemistry*, 2005 5, 337-351.

Sherman, IW. Biochemistry of *Plasmodium* (malarial parasites). *Microbiol Rev* 1979; 43: 453-95.

Shi. W; Ting, LM; Kicska, GA; Lewandowicz, A; Tyler, PC; Evans, GB; Furneaux, RH; Kim, K; Almo, SC; Schramm, VL. *Plasmodium falciparum* purine nucleoside phosphorylase. Crystal structures, immucillin inhibitors and dual catalytic function. *J Biol Chem,* 2004 279, 18103-18106.

Slayden, RA; Barry, CEIII. The role of KasA and KasB in the biosynthesis of meromycolic acids and isoniazid resistance in *Mycobacterium tuberculosis*. *Tuberculosis*, 2002 82, 149-160.

Snow, RW; Guerra, CA; Noor, AM; Myint, HY; Hay SI. The global distribution of clinical episodes of *Plasmodium falciparum* malaria. *Nature*, 2005 434, 214-217.

Stahlhut, R; Park, G; Petersen, R; Ma, WW; Hylands P The occurrence of the anti-cancer diterpene taxol in Podocarpus gracilior Pilger (Podocarpaceae). *Biochemical Systematics and Ecology*, 1999 27, 613-622.

Steinrücken, HC; Amrhein, N; The herbicide glyphosate is a potent inhibitor of 5-enol pyruvyl shikimic acid-3-phosphate synthase. *Biochem Biophys Res Comm,* 1980 94, 1207-1212.

Stramentinoli, G. Pharmacologic aspects of S-adenosylmethionine. Pharmacokinetics and pharmacodynamics. *Am J Med,* 1987 83, 35-42.

Surolia, N; Surolia, A. Triclosan offers protection against blood stages of malaria by inhibiting enoyl-ACP reductase of *Plasmodium falciparum*. *Nat Med,* 2001 7, 167-173.

Tagboto, S; Townson, S. Antiparasitic properties of medicinal plants and other naturally occurring products. *Adv Parasitol,* 2000 50, 200-278.

Takayama K; Wang C; Besra GS. Pathway to synthesis and processing of mycolic acids in *Mycobacterium tuberculosis*. *Clin Microbiol Rev.* 2005 18, 81-101.

Tasdemir, D; Lack, G; Brun, R; Rüedi, P; Scapozza, L; Perozzo, R. Inhibition of *Plasmodium falciparum* fatty acid biosynthesis: evaluation of FabG, FabZ, and FabI as drug targets for flavonoids. *J Med Chem,* 2006 49, 3345-3353.

Taylor, S; Berridge V. Medicinal plants and malaria: an historical case study of research at the London School of Hygiene and Tropical Medicine in the twentieth century. *Trans Royal Soc Trop Med Hyg,* 2006 100, 707-714.

Teague, SJ; Davis, AM; Leeson, PD; Oprea, T. The design of leadlike combinatorial libraries. *Angew Chem Int Ed Engl,* 1999 38, 3743-3748.

Ting, LM; Shi, W; Lewandowicz, A; Singh, V; Mwakingwe, A; Birck, MR; Ringia, EAT; Bench, G; Madrid, DC; Tyler, PC; Evans, GB; Furneaux, RH; Schramm, VL; Kim K. Targeting a novel *Plasmodium falciparum* purine recycling pathway with specific immucillins. *J Biol Chem,* 2005 289, 9547-9554.

Tona, L; Mesia, K; Ngimbi, NP; Chrimwami, B; Okond'ahoka; Cimanga, K; de Bruyne, T; Apers, S; Hermans, N; Totte, J; Pieters, L; Vlietinck, AJ. In-vivo antimalarial activity of Cassia occidentalis, Morinda morindoides and Phyllanthus niruri. *Ann Trop Med Parasitol,* 2001 95, 47-57.

Tosun, F; Kizilaym CA; Sener, B; Vural M, Palittapongampim P. Antimycobacterial screening of some Turkish plants. *J Ethnopharmacol,* 2004 95, 273-275.

Towie, N. Malaria breakthrough raises spectre of drug resistance. Nature, 2006 440, 852-853.

Tran, QL; Tezuka, Y; Ueda, JY; Nguyen, NT; Maruyama, Y; Begum, K; Kim, HS; Wataya, Y; Tran, QK; Kadota, S. In vitro antiplasmodial activity of antimalarial medicinal plants used in Vietnamese traditional medicine. *J Ethnopharmacol,* 2003 86, 249-252.

Uhlemann, AC; Cameron, A; Eckstein-Ludwig, U; Fischbarg, J; Iserovich, P; Zuniga, FA; East, M; Lee, A; Brady, L; Haynes, RK; Krishna, S. A single amino acid residue can determine the sensitivity of SERCAs to artemisinin.*Nature Struct Mol Biol,* 2005 12, 628-629.

UNAIDS. AIDS epidemic update: December 2004. Geneva, 2004.

Vieth, M; Siegel, MG; Higgs, RE; Watson, IA; Robertson, DH; Savin, KA; Durst, GL; Hipskind, PA. Characteristic physical properties and structural fragments of marketed oral drugs. *J Med Chem,* 2004 47, 224-232.

Vieth, M; Sutherland, JJ. Dependence of molecular properties on proteomic family for marketed oral drugs. *J Med Chem,* 2006 49, 3451-3453.

Vilchèze, C; Morbidoni ,HR; Weisbrod, TR; Iwamoto, H; Kuo, M; Sacchettini, JC; Jacobs, WRJr. Inactivation of the inhA-encoded fatty acid synthase II (FASII) enoyl-acyl carrier protein reductase induces accumulation of the FASI end products and cell lysis of *Mycobacterium smegmatis*. *J Bacteriol,* 2000 182, 4059-4067.

Waters, AP. Guilty until proven otherwise. Science, 2003 301, 1487-1488.

Webby, CJ; Baker, HM; Lott, JS; Baker, EN; Parker EJ. The structure of 3-deoxy-d-arabino-heptulosonate 7-phosphate synthase from *Mycobacterium tuberculosis* reveals a common catalytic scaffold and ancestry for type I and type II enzymes. *J Mol Biol,* 2005 354, 927-939.

Wickramasinghe, SR; Inglis, KA; Urch, JE; Müller, S; van Aalten, DMF. Kinetic, inhibition and structural studies on 3-oxoacyl-ACP reductase from *Plasmodium falciparum*, a key enzyme in fatty acid biosynthesis. *Biochem J,* 2006 393, 447-457.

Wilcox, ML; Bodeker, G. Plant-based malaria control: research initiative on traditional antimalarial methods. *Parasitol Today,* 2000 16, 220-221.

Wink, M Evolution of secondary metabolites from an ecological and molecular phylogenetic perspective. *Phytochemistry*, 2003 64, 3-19.

World Health Organization. Anti-tuberculosis drug resistance in the world: Third Global Report. 2004.

World Health Organization. Global tuberculosis control: surveillance, planning, financing. 2006.

Wright, CW. Traditional antimalarials and the development of novel antimalarial drugs. *J Ethonopharmacol,* 2005 100, 67-71.

Wright, CW; Addae-Kyreme, J; Breen, AG; Brown, JE; Cox, MF; Croft, SL; Gokcek, Y; Kendrik, H; Phillips, RM; Pollet, PL. Synthesis and evaluation of cryptolepine analogues for their potential as new antimalarial agents. *J Med Chem,* 2001 44, 3187-3194.

Wunberg, T; Hendrix, M; Hillisch, A; Lobell, M; Meier, H; Schmeck, C; Wild, H; Hinzen, B. Improving the hit-to-lead process: data-driven assessment of drug-like and leak-like screening hits. *Drug Discovery Today,* 2006 11, 175-180.

Zabinski, RF; Blanchard, JS. The requirement for manganese and oxygen in the isoniazid-dependent inactivation of *Mycobacterium tuberculosis* enoyl reductase. *J Am Chem Soc,* 1997 119, 2331-2332.

Zhang, Y; Post-Martens, K; Denkin, S. New drug candidates and therapeutic targets for tuberculosis therapy. *Drug Discovery Today*, 2006 11. 21-27.

Zhang, YM; Lu, YJ; Rock, CO. The reductase steps of the type II fatty acid synthase as antimicrobial targets. *Lipids*, 2004 39, 1055-1060.

In: Focus on Biodiversity Research
Editor: Jan Schwartz, pp. 179-198
ISBN: 1-60021-372-3

Chapter 8

SPECIES TURNOVER ALONG GRADIENTS OF NET PRIMARY PRODUCTIVITY

Gary Luck*
Institute for Land, Water and Society
Charles Sturt University, Albury, New South Wales, Australia
Fiona McCallum
Department of Primary Industries, Wodonga, Victoria, Australia

ABSTRACT

Understanding the factors that drive spatial patterns in biodiversity is fundamental to developing effective conservation strategies. The relationships between net primary productivity (NPP) and species richness are well documented in the literature, yet the influence of spatial variation in productivity on species turnover has received relatively little attention. The aims of this study were to: i) compare patterns in species turnover among different taxonomic groups occupying spatial gradients that encompassed low variation in NPP and high variation in productivity; and ii) contrast these patterns with those produced by null models. We examined species turnover for birds, mammals and butterflies along replicate transects of approximately 1000 km split into 10 sampling sites. Six transects encompassed spatial gradients with little variation in NPP and six transects encompassed gradients with high variation in NPP. We calculated three representative indices of turnover and plotted these against distance between sampling sites to determine the decline in similarity among species assemblages with distance. Computer simulations were used to generate random (null) species distributions along transects based on simple 'random distribution models'. These models assumed no differences between species and a consistent decline in species assemblage similarity with distance. Species turnover was greatest along transects with high variation in NPP, especially for birds. Significant turnover for mammals and butterflies was also recorded along many low variation transects. Birds and butterflies exhibited the lowest turnover

* Corresponding author. Ph: +61 2 6051 9945; Fax: +61 2 6051 9897; Email: galuck@csu.edu.au

overall – mammals had high turnover in almost all transects. Patterns in turnover varied depending on the index used. The index emphasising continuity in species composition always recorded the greatest turnover compared to indices representing differences in species identities or richness. A null model assuming a linear decline in species assemblage similarity with distance was able to match observed patterns in turnover in many cases. It was most successful in predicting turnover along high variation transects and for mammals. In many cases, species turnover was predicted from the simple assumption of consistent decline in the similarity between sites with increasing distance irrespective of the gradient in environmental variability. However, important differences occurred between taxonomic groups and gradients of low vs high variation. These differences suggest that protecting a few species rich sites in regions of low variability may capture a substantial proportion of regional diversity for highly mobile groups with large range sizes, although this wouldn't be the case for less vagile and more geographically restricted species. Regions with high variation in NPP require a greater number of reserves owing to strong and consistent turnover irrespective of taxonomic group.

Keywords: Beta-diversity, species turnover, null models, environmental variability, Australia, net primary productivity, birds, butterflies, mammals.

INTRODUCTION

Understanding the factors that underlie patterns in the distribution of species has been a considerable challenge in ecology for decades (MacArthur, 1965; Pianka, 1966; Janzen, 1970; Brown, 1981; Currie, 1991; Gaston, 2000). Latitudinal (see Rohde, 1999; Gaston, 2000; Willig *et al.*, 2003; Hillebrand, 2004 for reviews) and altitudinal (e.g. Lees *et al.*, 1999; Brown, 2001; Sanders, 2002; McCain, 2004) gradients in species richness have received substantial attention and numerous theories have been generated to explain these patterns (Rahbek and Graves, 2001; Willig *et al.*, 2003). The relationship between spatial gradients and species turnover (≈ beta-diversity) has also been explored by many researchers, but has received relatively little attention compared to that afforded to spatial gradients in species richness (e.g. McCoy and Connor, 1980; Harrison *et al.*, 1992; Blackburn and Gaston, 1996; Clarke and Lidgard, 2000; Lennon *et al.*, 2001; Garcillán and Ezcurra, 2003; Koleff *et al.*, 2003b). Variation in species assemblages between individual sites within the same region is recognised as an important spatial pattern in biodiversity (Whittaker, 1960; Wilson and Shmida, 1984; Blackburn and Gaston, 1996; Koleff *et al.*, 2003b).

There is limited knowledge of the mechanisms that may underlie patterns in species turnover. Variation in deterministic processes (e.g. speciation, competition, dispersal, and species' adaptations to environmental variability) between species may play a role, but this has recently been challenged by neutral theory (Hubbell, 1979, 2001; see McGill, 2003; Nee and Stone, 2003; Volkov *et al.*, 2003; Chave, 2004 for extension and discussion). Moreover, empirical patterns of turnover have rarely been compared to a null expectation (but see Koleff and Gaston, 2001; Condit *et al.*, 2002). Understanding the process of species turnover would greatly enhance our knowledge of the factors controlling biodiversity and has direct relevance to conservation planning strategies.

"Beta-diversity" (β) and "species turnover" are often used interchangeably in the literature to refer to the same thing, but this approach is misleading (Vellend, 2001). Beta-diversity is the difference in species composition between study sites. Each site has its own diversity (alpha, α) and β and α-diversity combined contribute to the regional diversity (gamma, γ). Measures of beta-diversity used in this way may not necessarily be spatially explicit, whereas the concept of species turnover implies variation in species composition along a spatial gradient (referring to turnover in space as opposed to turnover in time). Therefore, measures of association between sites must be placed in a spatial context (e.g. related to geographic distance between sites).

High species turnover along a spatial gradient will lead to an increasing decay in the similarity between sites with distance. A decrease in similarity with increasing distance is a well-known geographical phenomenon that is also of interest to biogeographers (Nekola and White, 1999). It is likely that variation in the environment and species characteristics will influence the degree of turnover, but it is not clear if consistent patterns should be expected under certain circumstances or if these patterns differ markedly from those generated at random using a few basic assumptions.

It is well recognised in the literature that alpha diversity varies with net primary productivity (NPP; Waide *et al.*, 1999; Gaston, 2000; Mittelbach *et al.*, 2001; Whittaker and Heegaard, 2003). The relationship is frequently positive, whereby diversity peaks at the highest values of NPP, but hump-shaped relationships are also common where diversity may peak at mid to high values, but exhibit a decline phase at very high values of NPP. Moreover, relationships appear to be scale and taxon specific (Whittaker and Heegaard, 2003). Despite its prominence in studies of alpha diversity, the influence of variation in NPP on species turnover along spatial gradients remains largely unexplored. In this study, we examine patterns of species turnover among various taxonomic groups occupying environments with different gradients in NPP. We also compare empirical patterns with those derived at random assuming a constant decay in similarity between sites with distance and ignoring differences between species. Our primary interest is not in absolute turnover, but turnover relative to the productivity gradient and taxonomic group.

We make a number of predictions about the degree of turnover likely to occur given variation in productivity gradients. For spatial gradients with low variation in NPP we predict:

i. low turnover in species composition;
ii. no difference in the degree of turnover among the different taxonomic groups; and
iii. no difference between actual patterns in turnover and those generated by our null models that assume no differences between species and decay in similarity is a random process influenced solely by the distance between sites.

For spatial gradients with high variation in NPP we predict:

iv. high turnover in species composition;
v. higher turnover for less mobile groups (i.e. turnover for butterflies > mammals > birds); and

vi. significant differences between actual patterns in turnover and those generated by the null models.

METHODS

Species Data

Species richness and turnover with distance were assessed for birds, mammals and butterflies within continental Australia. Distribution data were obtained for birds from Barrett *et al.* (2003), for mammals from Strahan (1995), Menkhorst and Knight (2001), McKenzie, N.L., Burbidge, A.A. and Baynes, A. (unpublished) and state-based wildlife atlases, and for butterflies from Braby (2000). These data were collected at a spatial resolution of 1° grid cells (= the sampling grain) defined by latitude and longitude.

Transects

Transects consisted of 10 contiguous grid cells (= the sampling extent, ≈ 1000 km), running west-east, that spanned different regions with contrasting variation in NPP (see below). Each transect was confined to a single latitudinal band, but transects were not contiguous latitudinally or longitudinally. NPP data (presented in tC ha^{-1}) were obtained from http://audit.ea.gov.au/ANRA/atlas home.cfm.

The transects were selected using stratified random sampling. Stratification was based on the level of variation in productivity across a region and whether it was in northern or southern Australia (to account for broad climactic differences). Four transect types were chosen classified as: low variation south (LVS), low variation north (LVN), high variation south (HVS) and high variation north (HVN). There were three replicates of each transect type (resulting in 12 transects).

These transects sampled spatial gradients in NPP rather than gradients defined by topography or latitude because we were primarily interested in species turnover in relation to spatial variation in productivity independent of altitudinal or latitudinal gradients. The spatial extent selected for the analysis is broadly relevant to the taxonomic groups studied and, for high variation transects, encompassed the necessary variation in NPP for meaningful comparisons.

Changes in habitat type will influence NPP and species turnover with turnover likely to be high between sites in different habitats. Our LV transects spanned minimal change in broad habitat type (encompassing deserts and grasslands), whereas HVS transects spanned deserts, grasslands, temperate woodlands and temperate forests, and HVN transects spanned deserts, grasslands and subtropical forests.

Analysis of Species Turnover

We calculated three characteristic measures of turnover for each transect based on pair-wise comparisons of each cell. This resulted in matrices of turnover measures for the actual and randomly generated data (see below). Different equations may emphasise the different components of changes in species assemblages across sites so we considered it instructive to include representative measures. The following classifications and descriptions were obtained from Koleff *et al.* (2003a) who reviewed a number of turnover measures.

Measure β_1 (Routledge, 1977) is influenced by differences in species identities between sites more than differences in species richness. It can be expressed as:

$$\beta_1 = \frac{(a+b+c)^2}{(a+b+c)^2 - 2bc} - 1$$

where, when comparing paired sites, a equals the total number of species common to both sites, b equals the total number of species occurring in the second site, but not the first, and c equals the number occurring in the first, but not the second.

β_2 (Wilson and Shmida, 1984) is a measure of the continuity in species composition between sites, but is not influenced by the relative magnitude of species gains and losses. It can be expressed as:

$$\beta_2 = \frac{b+c}{2a+b+c}$$

Note that β_2 is equivalent to 1 - the Sørensen coefficient.

β_3 (Lennon *et al.*, 2001) is simply a measure of the difference in species richness between sites. It can be expressed as:

$$\beta_3 = \frac{2|b-c|}{2a+b+c}$$

The turnover matrices incorporate the spatial structure in the data (i.e. provide values for cells that differ in their spatial relationship) and are a representation of species turnover. We also calculated the Euclidean distance between each pair of cells in a transect to derive a distance matrix (this was the same for each transect). Hence, a given transect had the same distance matrix as all other transects, but a different turnover matrix.

Following the methods of Whittaker (1960, 1972) and Vellend (2001), we plotted log-transformed values of the turnover indices against Euclidean distance for each transect. This provides a graphical representation of the decline in similarity with distance. These authors also fit straight lines to their data using linear regression (and used a randomisation procedure to test for significant differences between regression slopes – see Vellend (2001) for details), but our data were not suitable for regression analysis in most cases. Therefore, we simply present the data in scatterplots.

We tested our data using two randomization procedures (which is appropriate for non-independent data). First, we determined if there was a significant relationship between species turnover and distance using Mantel's test (Mantel, 1967). Second, we determined if there was a significant difference in the pattern of species turnover among taxonomic groups, transects and actual vs random data by comparing differences observed in nature with a distribution of randomly generated differences. Distance and turnover matrices were randomised 1000 times with a measure of association (Spearman rank correlation) recorded at each iteration, for each taxonomic group in each transect. Actual differences between transects were then compared to the randomly generated distribution of differences to test for significance. The alpha level was kept at 0.05 because we believe this is the more conservative approach in this instance (i.e. a smaller alpha level would lead to fewer significant differences between observed and randomly generated patterns, providing support for our null models).

Random Species Distribution Models

Computer simulations were used to generate random species distributions based on pre-determined turnover 'models'. We simulated species distribution for each taxonomic group across four transects; one in each variability type. Simulated distributions were generated from actual data from the first replicate in transect types LVS, LVN, HVS and HVN. For example, to simulate the distribution of birds across a LVS transect, we used the observed number of bird species in the first replicate of transect type LVS as the species pool from which to generate the random distribution. A species was chosen at random, with replacement, from the species pool and placed within the transect following certain rules (see below). The total number of random selections was based on the total number of individuals (birds) actually occurring in the transect. This was repeated for each taxonomic group in each transect type.

The placement of species within a transect (i.e. within one of the ten grid cells) was determined by a turnover model. We used two models: exponential and linear. In the exponential model, the first individual was placed in the transect at random. The second individual was also placed at random unless it was the same species as the first. If it was the same, it was placed in a neighbouring grid cell with an assigned probability for each cell that declined exponentially with distance from the cell the species already occupied. For example, if the species already occurred in the first cell in the transect, there was a higher probability that it would be placed in the second cell compared to the third compared to the fourth and so on. This probability declined exponentially with distance. This approach attempted to simulate the decline in similarity between cells with distance. The same approach was followed for the linear model, except that the probability of species occurrence in a cell declined linearly with distance from any cell it already occupied.

The purpose of these models is to facilitate comparison of patterns in turnover observed in nature with those derived at random. If actual patterns deviate significantly from a random expectation this provides a stronger justification for invoking potential mechanistic explanations for those patterns. These are not entirely null models in that they assume a consistent decline in similarity between sites with distance. Therefore, the question being asked is whether actual patterns in turnover are simply a result of sites being further apart and the probability that species assemblage similarity will decline with distance, or are driven by other factors. For example, turnover along a spatial gradient that encompasses multiple habitat types and contains species with varying vagility and range sizes should be significantly different to the turnover generated by our random models if these factors are important in defining the level of turnover.

RESULTS

Variation in NPP

Low variation (LV) transects are characterised by a small variation in the range of NPP values occurring along the spatial gradients, evidenced by coefficients of variation ranging from 12.8-32.2%. In contrast, high variation (HV) transects had coefficients of variation ranging from 67.4-87.5% (Table 1). For LV transects, the values of NPP were generally comparable across the entire spatial gradient, fluctuating within a tight range, whereas for HV transects there was a substantial and consistent increase in NPP with distance (Figure 1). Hence, the productivity gradient was much steeper for HV transects, as expected, and consequently a more substantial decay in species assemblage similarity with distance would be expected.

Table 1. Summary statistics (SD = standard deviation) and coefficient of variation (CV) for NPP values across all 12 transects in each variability type

Transect type	Replicate	Range	Mean	SD	CV
LVS	1	0.17 0.34	0.23	0.05	22.61
	2	0.25-0.38	0.31	0.04	12.76
	3	0.21 0.45	0.29	0.09	32.16
LVN	1	0.58-0.99	0.79	0.13	16.11
	2	0.38-0.97	0.54	0.17	31.59
	3	0.18-0.32	0.24	0.05	22.36
HVS	1	0.85-6.05	2.56	1.72	67.38
	2	0.65-7.90	3.25	2.38	73.09
	3	0.83-7.07	2.89	2.29	79.08
HVN	1	0.10-3.35	1.47	1.16	79.28
	2	0.37-4.00	1.34	1.13	84.38
	3	0.22-3.61	1.18	1.03	87.49

LVS = low variation south;
LVN = low variation north;
HVS = high variation south;
HVN = high variation north

Species Turnover

Species Richness

There were distinct patterns in species richness with distance, especially for HV transects where species richness increased markedly from west to east mimicking the trends in NPP

(Appendix 1). Patterns were less obvious for LV transects. These results likely influenced the patterns observed in species turnover with the addition of new species with distance across HV transects. However, the strongest patterns in turnover generally did not occur using β_3 the index representing differences in species richness (see below). This suggests true turnover in species assemblages with distance rather than simply adding new species.

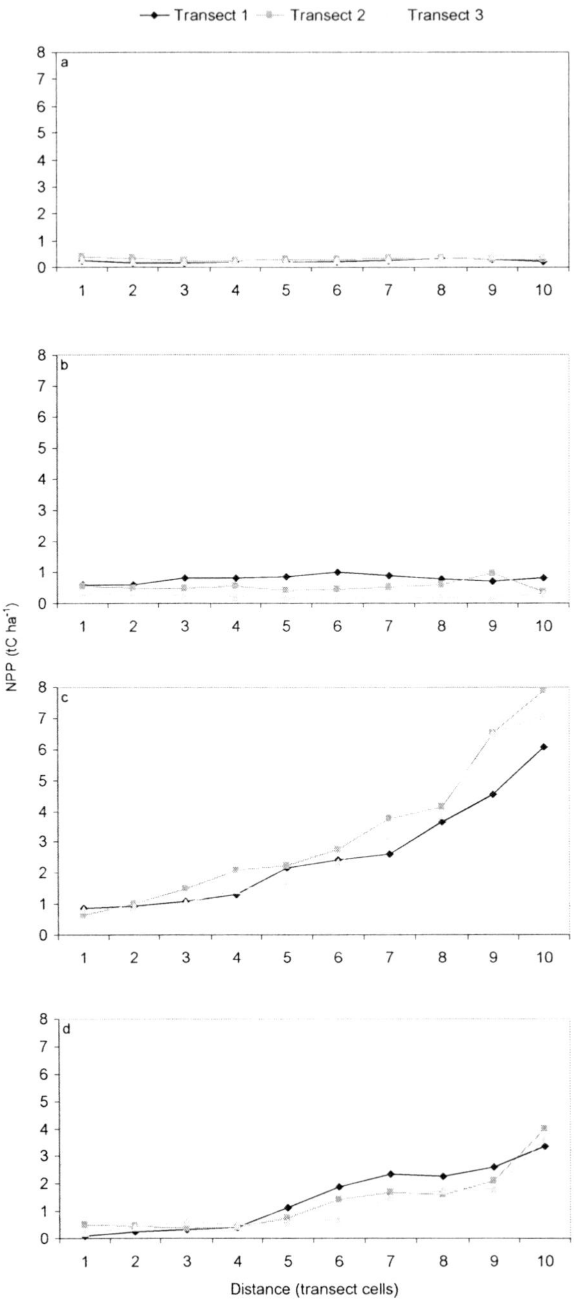

Figure 1. Differences in NPP values with distance along replicate transects of: a) LVS; b) LVN; c) HVS; and d) HVN.

Comparisons among Transects

As expected, species turnover was greatest along HV transects. This was most evident for birds (Figure 2) with turnover for mammals and butterflies still differing significantly from zero across many LV transects (Figures 3 and 4). Nevertheless, the pattern of turnover was most consistent across HV transects with a strong decline in species community similarity with increasing distance. This is emphasised by pair-wise comparisons across all transects using randomisation tests (ignoring differences between indices). When patterns of turnover differed significantly between transects, this was largely driven by differences between LV and HV transects (i.e. birds, 69% of all differences, $n = 226$; mammals, 59%, $n = 135$; and butterflies 51%, $n = 173$).

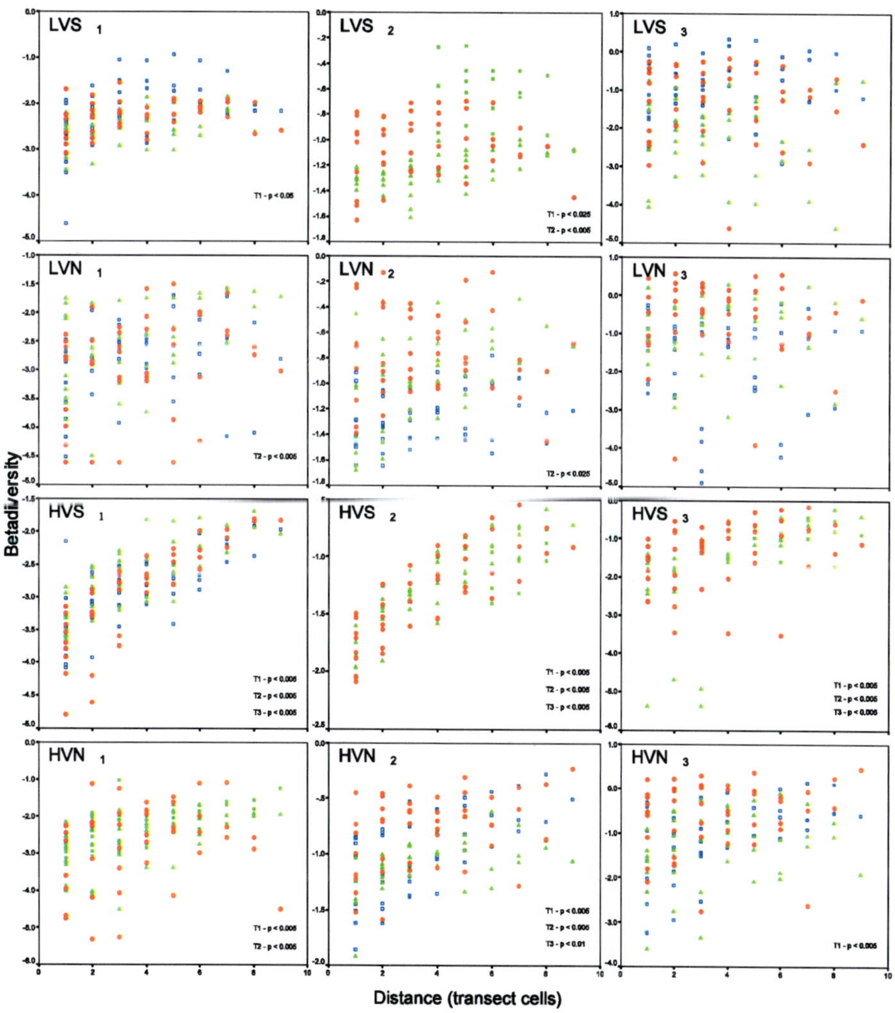

Figure 2. Relationship between species turnover and distance for birds. Graphs show the level of similarity between species assemblages in each transect cell – higher values = lower similarity. Replicate transects are represented by □ = transect 1(T1), ▲= transect 2 (T2) and ● = transect 3 (T3). P values are only provided for significant relationships. Subscript numbers refer to indices β_1 to β_3 respectively.

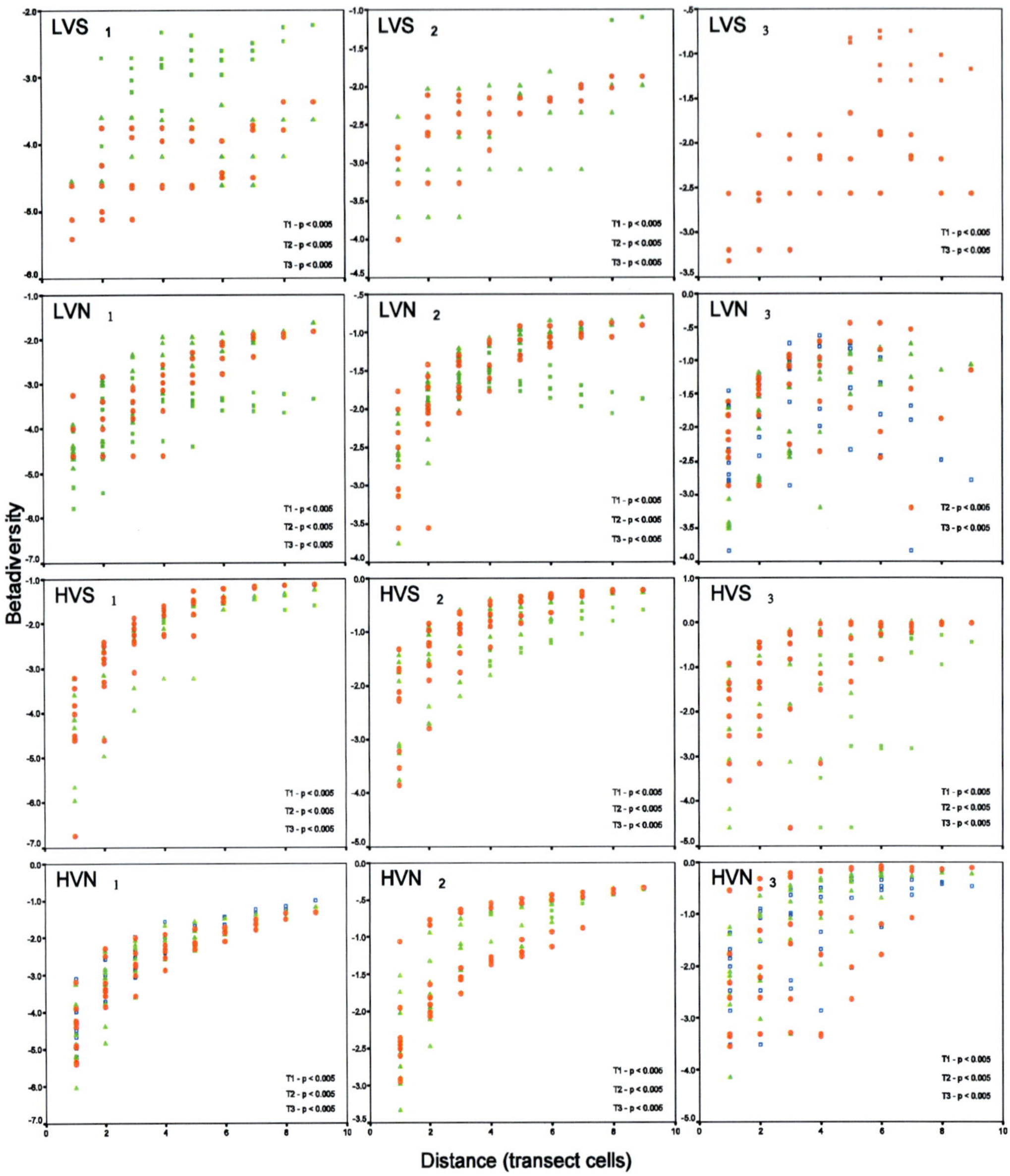

Figure 3. Relationship between species turnover and distance for mammals. Graphs show the level of similarity between species assemblages in each transect cell – higher values = lower similarity. Replicate transects are represented by □ = transect 1(T1), ▲= transect 2 (T2) and ● = transect 3 (T3). Subscript numbers refer to indices β_1 to β_3 respectively. Data for LVS transect 2, using β_3 were dominated by zeros (i.e. little turnover) and are not plotted. P values are only provided for significant relationships.

Within a variation class, there were minor differences in species turnover between north and south transects (i.e. HVN vs HVS: birds 14% of all differences, 38% of all comparisons (n = 81); mammals 5% and 8%, respectively; and butterflies 15% and 32%, respectively: LVN vs LVS: birds 6% and 16%, respectively; mammals 16% and 27%, respectively; and butterflies 8% and 17%, respectively).

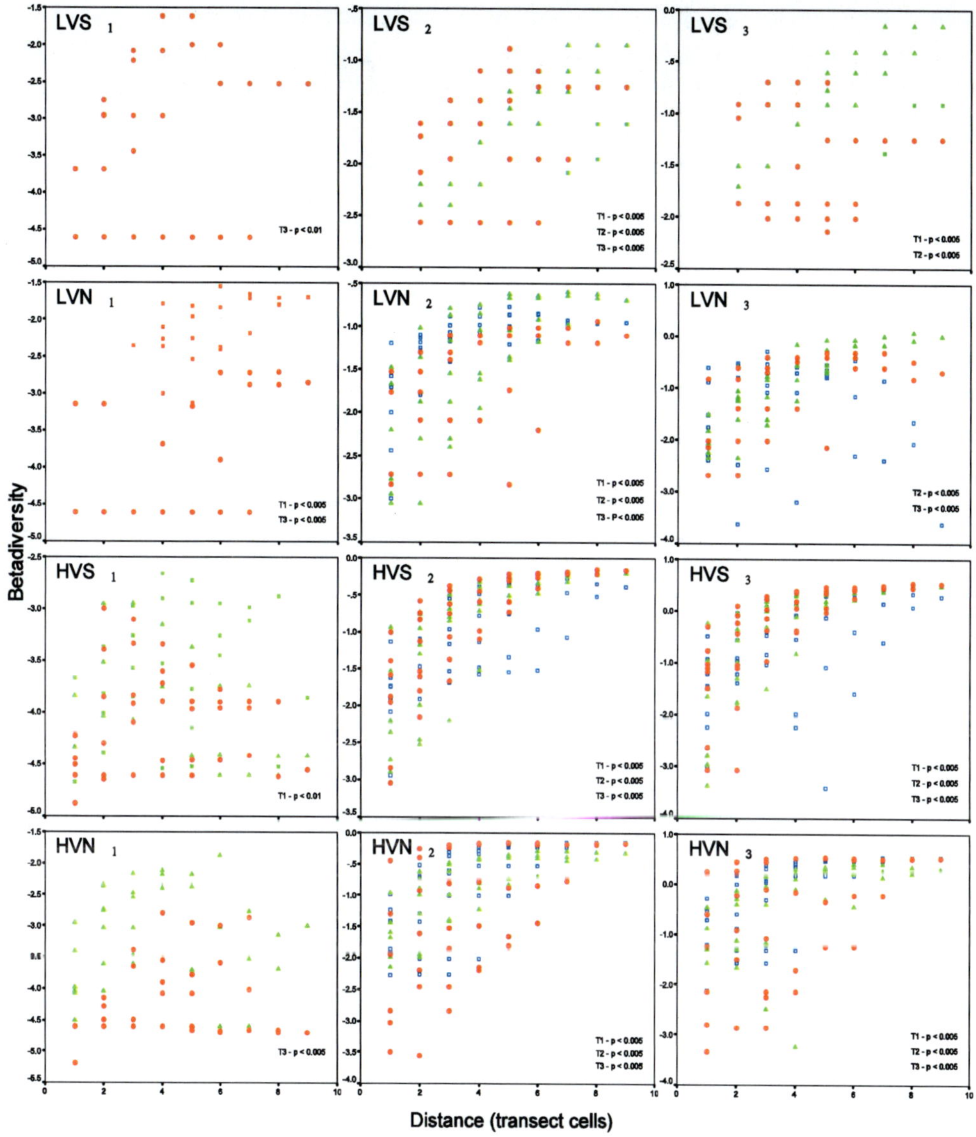

Figure 4. Relationship between species turnover and distance for butterflies. Graphs show the level of similarity between species assemblages in each transect cell – higher values = lower similarity. Replicate transects are represented by □ = transect 1(T1), ▲= transect 2 (T2) and ● = transect 3 (T3). Subscript numbers refer to indices β_1 to β_3 respectively. Data for LVS transects 1 and 2, LVN transect 2, and HVN transect 1, all using β_1, were dominated by zeros (i.e. little turnover) and are not plotted. Note that data for LVS transect 3 and LVN transect 3, both using β_1, also have a high number of zeros, even though they are plotted and a significant relationship was identified. P values are only provided for significant relationships.

Comparisons among Taxonomic Groups

Comparing across taxonomic groups, birds and butterflies had the lowest turnover overall. Indeed, for butterflies, there was no turnover whatsoever across some cells in certain transects (mostly LV – Figure 4). Birds generally had low turnover in LV transects (Figure 2). On the other hand, mammals had high turnover in almost all transects. Pair-wise comparisons between taxonomic groups using randomisation tests showed that spatial gradients in mammal turnover differed significantly from birds in the majority of LV transects (61% of all LV transects) regardless of the index used; and differed significantly from butterfly turnover, but almost exclusively in HV transects (Table 2). The decline in similarity with distance was almost always stronger for mammals. When comparing birds with butterflies, butterfly turnover was strongest, but significant differences were mostly confined to LV transects (59% of all significant differences).

Table 2. Significant differences in turnover among taxonomic groups for all transects and indices (e.g. LVS1 = low variation south, replicate 1)

	Birds vs Mammals	Birds vs Butterflies	Mammals vs Butterflies
LVS1 β_1			
LVS2 β_1			
LVS3 β_1			
LVS1 β_2	M		
LVS2 β_2			
LVS3 β_2	M	BU	
LVS1 β_3	M	BU	
LVS2 β_3		BU	BU
LVS3 β_3	M	BU	
LVN1 β_1	M	BU	
LVN2 β_1	M		M
LVN3 β_1	M		
LVN1 β_2		BU	
LVN2 β_2	M	BU	
LVN3 β_2	M	BU	
LVN1 β_3			
LVN2 β_3	M	BU	
LVN3 β_3	M	BU	
HVS1 β_1			M
HVS2 β_1		B	M
HVS3 β_1		B	M
HVS1 β_2			
HVS2 β_2			
HVS3 β_2			
HVS1 β_3			
HVS2 β_3			
HVS3 β_3		BU	
HVN1 β_1			M
HVN2 β_1	M	B	M
HVN3 β_1	M		M
HVN1 β_2			
HVN2 β_2			
HVN3 β_2	M		
HVN1 β_3		BU	
HVN2 β_3		BU	
HVN3 β_3	M	BU	

M = mammal turnover was greater (i.e. a larger correlation coefficient) in pair-wise comparisons; B = bird turnover was greater; and BU = butterfly turnover was greater. Non-significant differences are not indicated.

Comparisons among Indices

Important differences in turnover patterns were observed depending on the index used. For example in butterflies, turnover was weak or non-existent in many cases when using β_1 (emphasising differences in species identities), whereas for the same transects, turnover was significant in every case using β_2 (emphasising continuity in species composition; Figure 4). These relationships were not consistent across taxonomic groups; for birds, β_3 (differences in species richness) recorded the lowest turnover. What was consistent, however, was that β_2 always recorded the highest turnover (i.e. significant turnover was recorded most often and there was never a situation where turnover was recorded as significant by the other two measures and not by β_2).

When examining significant differences in turnover between indices across all transects, β_3 differed from β_1 and β_2 most often for birds (35% and 47% of all comparisons (n = 144), respectively, compared to 30% for β_1 vs β_2) and mammals (28% and 28%, respectively, compared to 13% for β_1 vs β_2). For butterflies, β_1 vs β_2 differed the most (48%), compared to β_3 vs β_1 (43%) and β_3 vs β_2 (13%). Hence, on average, the results obtained using β_3 tended to differ more from β_1 and β_2 than the latter two indices did from each other.

Model Performance

The random models all performed similarly when predicting patterns of turnover across transects and taxonomic groups. However, the exponential model tended to overestimate turnover in many cases, leaving the linear model as the one most closely fitting the observed data. Therefore, we only present the results for the linear model. Moreover, the results are confined to the first replicate transect in each variation class, as these data were used to parameterise the model.

Table 3. Significant differences in turnover between modelled and actual results for each taxonomic group in the first replicate of each transect variability type.

	Birds	Mammals	Butterflies
LVS β_1	H		
LVS β_2	H		
LVS β_3			H
LVN β_1	H		
LVN β_2	H		
LVN β_3			H
HVS β_1			H
HVS β_2			
HVS β_3			
HVN β_1			H
HVN β_2			
HVN β_3			L

H = the random model predicted higher turnover (i.e. a larger correlation coefficient) than actually occurred; L = the model predicted lower turnover than actually occurred. Non-significant differences are not indicated.

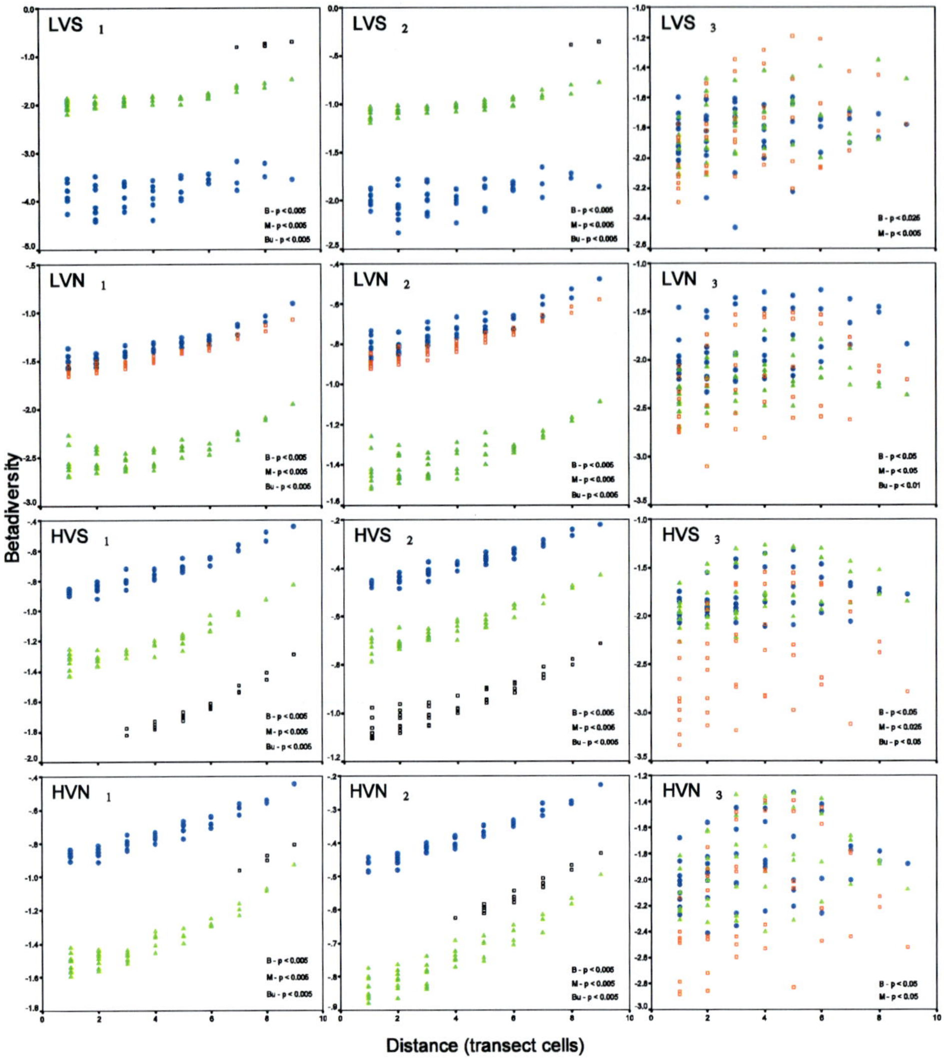

Figure 5. Relationship between species turnover and distance for the linear model. Graphs show the level of similarity between species assemblages in each transect cell – higher values = lower similarity. Different taxa are represented by □ = birds, ▲= mammals and • = butterflies. P values are only provided for significant relationships. Subscript numbers refer to indices β_1 to β_3 respectively.

We determined significant differences between actual and simulated data using randomisation tests with Spearman rank correlation coefficients as the measure of association (see Methods). This analysis identified differences in the pattern of turnover, but was unable to detect differences in the magnitude of turnover (e.g. the same correlation coefficient can be generated between values in a tight range, compared to those in a much broader range). Comparing slopes generated from linear regression can assess magnitude of turnover, but our data were mostly not suitable for such analyses. However, magnitude of turnover can be interpreted visually by comparing data presented in Figures 2 to 5.

In all cases, except two, the linear model predicted a significant rate of species turnover with distance for each taxonomic group in each transect (Figure 5). The two exceptions were for butterflies in the LVS and HVN transects using the β_3 index. The patterns of turnover were obvious and consistent for all transects using the β_1 and β_2 indices, but were less so using β_3.

There was remarkable congruence between the simulated and actual patterns of turnover for mammals (Table 3). Indeed, there were no significant differences between observed and predicted patterns. The model also performed well in predicting patterns of bird turnover, especially for HV transects. It was less successful at predicting butterfly turnover with no consistent patterns for LV or HV transects. When differences were significant, the model almost always predicted higher turnover than that actually observed (Table 3).

Discussion

We begin the discussion by re-visiting our predictions. For spatial gradients with low variation in NPP we found:

i. relatively low turnover compared to high variation transects, but still a higher degree of turnover than we expected;
ii. differences in the degree of turnover among the different taxonomic groups, which was unexpected; and
iii. actual patterns of turnover consistent with the predictions of the null model in most cases, although it predicted higher turnover for birds in 4 out of 6 transects.

For spatial gradients with high variation in NPP we found:

iv. consistently high turnover as expected;
v. turnover differed between taxonomic groups, but mammals had the highest turnover followed by butterflies and then birds (where we predicted butterflies > mammals > birds); and
vi. few differences between actual patterns in turnover and those predicted by the null model, where we expected significant differences to occur.

Using a simple null model, we were able to predict patterns of turnover for certain taxonomic groups occupying particular environments. The model was most successful at predicting turnover for mammals – regardless of the gradient in NPP or turnover index used. It almost always predicted a strong decline in species assemblage similarity with distance (especially with indices β_1 and β_2; Figure 5). Hence, when such a decline actually happened in nature, the modelled and observed patterns matched. Indeed in HV transects, random and observed patterns of turnover only differed significantly in three out of 18 transects. The main conclusion from our modelling is that species turnover is largely predictable, especially for strong gradients in NPP, and that the patterns of turnover can be explained by randomly assigning species to geographic space with a simple assumption of consistent decline in similarity with distance between sites.

Koleff and Gaston (2001) found that mid-domain models, that randomly assign species ranges (Colwell and Hurtt, 1994), were a poor fit to actual patterns of turnover for parrots and woodpeckers across the New World. Condit *et al.* (2002) found that the adequacy of fit of null models to species turnover varied with scale of observation. The test of our model is limited because we were only able to compare patterns in turnover rather than rate or magnitude. However, visual inspection of actual and random patterns of turnover (Figures 2 to 5) suggests some comparability in the magnitude of similarity decay.

It is important to recognise the role that variation in NPP appeared to play in influencing turnover. Strong patterns of turnover were most consistently recorded in HV transects for all taxonomic groups, but especially birds (when comparing to LV transects). Variation in NPP is correlated with environmental heterogeneity in this study, whereby low variation transects encompassed only one to two major habitat types of similar structure (desert and grassland), whereas high variation transects covered three to four major types. Environmental (habitat) heterogeneity has long been recognised as a primary driver in promoting species diversity (MacArthur and MacArthur, 1961; Finch, 1989; Daniels *et al.*, 1992; Kerr and Packer, 1997; Willig *et al.*, 2003; Tews *et al.*, 2004), but its contribution to influencing turnover is becoming more widely acknowledged (e.g. Nekola and White, 1999; Condit *et al.*, 2002; Garcillán and Ezcurra, 2003).

Our study was conducted across a relatively large spatial extent, using a coarse grain size. For such a spatial extent, one would expect factors like vagility and geographic range size to influence turnover patterns more than factors operating at smaller extents (e.g. local competition; Harrison *et al.*, 1992). We expected butterflies to have a higher degree of turnover than mammals owing to differences in body size and perceived levels of site fidelity and habitat specificity (e.g. relationships with specific host plants). However, this expectation seems misguided. Certainly birds appear to be the most mobile of the groups in our study and the patterns in turnover support this. Turnover in mammals may have been greater than expected because many nonvolant mammals in Australia are relatively small and probably don't move great distances.

However, if vagility is driving patterns in turnover we would expect similar results in LV and HV transects, because there is no reason to suspect vagility varies consistently with variation in NPP. This interpretation is limited because vagility can vary greatly within a taxonomic group and further work is required that places species in cross-taxonomic 'vagility groups' (although data for such classifications are lacking).

Small range sizes should correspond with higher turnover (Blackburn and Gaston, 1996; Gaston and Williams, 1996). Range sizes (measured in contiguous grid cells with a mean for all species) in HV transects were smaller than LV transects for all taxa (birds: 5.98 SD± 3.32 and 6.45 SD± 3.56, respectively; mammals 4.28 SD± 2.03 and 6.85 SD± 2.81; and butterflies 3.22 SD± 1.74 and 5.85 SD± 3.09). This may have contributed to the greater turnover recorded in HV transects. However, ranking taxa from largest to smallest average range size equals birds > mammals > butterflies, contrary to what would be expected given the patterns in turnover (the position of mammals and butterflies should be switched). We don't use actual geographic range sizes for each species, but extent of occurrence within a given transect, as we believe this is more relevant when comparing to turnover patterns.

Dispersion of range sizes would also influence the level of 'true' turnover (i.e. species replacement) across a spatial gradient (Harrison *et al.*, 1992). True turnover would most likely occur when ranges are distributed evenly in space. In HV transects especially, our results are influenced by the fact that species ranges are concentrated in the eastern end of the transect (represented by high species richness in these areas).

Direct comparisons between studies of turnover are hindered by differences in methods, spatial grain and extent, and taxa (Koleff *et al.,* 2003a). Nonetheless, some general conclusions can be made. Our results concur with Harrison *et al.* (1992) in that vagility didn't appear to influence turnover as much as range size and environmental variation (i.e. variation in NPP). Habitat specialisation may also be a factor – turnover should be greater among species with more specialised habitat requirements (≈ smaller niche breadth). NPP is strongly correlated with temperature and rainfall (although numerous factors influence NPP – see Eamus, 2003). Hence, species with a restricted niche breadth and narrow tolerance to variation in these factors should show high turnover. In our study, we would expect butterflies to have the greatest environmental restrictions, especially if associated with host plants that also have a limited distribution. The fact that butterfly turnover was weak in LH transects probably reflects the low species richness in these areas (spanning arid and semi-arid landscapes; Appendix 1).

Spatial gradients in species richness influence turnover, but contrary to the results of Lennon *et al.* (2001) and Koleff and Gaston (2002) the most consistent patterns of turnover in our study were recorded across spatial gradients with high (as opposed to low) total species richness (although we measure turnover in different ways and using different spatial configurations). Certainly, it doesn't appear inevitable that low species richness should result in high turnover; especially if the spatial gradient spans an area of low environmental variability (Lennon *et al.* (2001) also raise the issue of individual species association with low species richness areas. If this occurs, low species richness areas would only represent a subset of regional diversity – resulting in low turnover).

With minimal landscape change (in biotic and abiotic factors), the same species should be able to exploit the full extent of a spatial gradient. In our study, high species richness and high environmental variation were correlated making it difficult to tease apart the individual influence of these factors. However, there were substantial differences in species richness between LVS and LVN transects for all taxa (Appendix 1). Given the similar levels of productivity variation across these transects, the difference in species richness did not result in considerable differences in patterns of turnover. This further emphasises the role of environmental variation.

It wasn't our purpose to examine in detail the performance of the turnover indices we used – this has been done elsewhere (Koleff *et al.*, 2003a.). Rather, we were interested in how the representation of patterns of turnover varied with the different indices, and to ensure that the patterns we observed were not just an artefact of the index used. Not surprisingly, the patterns represented by indices β_1 and β_2 (the first focussing on species replacement only and the second combining species replacement and richness gradients) were most similar (although not for butterflies), and both of these generally differed most strongly from the patterns represented by β_3 (focussing on changes in species richness only). We concur with Koleff *et al.* (2003a) that indices that measure species gains and losses are probably the most useful for studies that wish to represent true turnover across space.

The apparent strong relationship between turnover and productivity has important implications for land and conservation management. Since remote sensing or modelling can estimate values such as NPP, it may also be possible to indirectly estimate patterns in turnover for a given taxonomic group along a spatial gradient (i.e. without direct sampling and assuming knowledge of the size of the regional species pool). However, it is important to note that in HV transects, NPP increased consistently across the spatial gradient (undoubtedly contributing to the consistent decline in species assemblage similarity). The null model we employed may not perform so well with greater spatial variability in NPP or heterogeneity (i.e. a range of peaks and troughs). It may be best applied to gradients that span low to high NPP (e.g. desert – grassland – woodland – forest).

The conservation implications of our results are that in landscapes with low variability, protecting a couple of species rich areas may capture a large proportion of regional diversity for highly mobile taxonomic groups with primarily large range sizes (e.g. birds). However, it appears this wouldn't be adequate for butterflies or small mammals. More reserves are required in high variability landscapes regardless of focal taxon, owing to strong and consistent turnover.

ACKNOWLEDGMENTS

This study was funded by a Charles Sturt University Seed Grant. We thank David Newth and Simon McDonald for assistance with computer simulations and data analysis. Julian Reid, David Roshier, Natasha Schedvin, Peter Spooner and David Watson made valuable comments on the chapter.

REFERENCES

Barrett, G., Silcocks, A., Barry, S., Cunningham, R. and Poulter, R. (2003). *The new atlas of Australian birds.* Hawthorn East, Victoria: RAOU.

Blackburn, T.M. and Gaston, K.J. (1996). The distribution of bird species in the New World: patterns in species turnover. *Oikos*, 77, 146-152.

Braby, M.F. (2000). *Butterflies of Australia: their identification, biology and distribution.* Collingwood, Victoria: CSIRO.

Brown, J.H. (1981). Two decades of homage to Santa Rosalia: toward a general theory of diversity. *American Zoologist,* 21, 877-888.

Brown, J.H. (2001). Mammals on mountainsides: elevational patterns of diversity. *Global Ecology and Biogeography,* 10, 101-109.

Chave, J. (2004). Neutral theory and community ecology. *Ecology Letters,* 7, 241-253.

Clarke, A. and Lidgard, S. (2000). Spatial patterns of diversity in the sea: bryozoan species richness in the North Atlantic. *Journal of Animal Ecology,* 69, 799-814.

Colwell, R.K. and Hurtt, G.C. (1994). Nonbiological gradients in species richness and a spurious Rapoport effect. *American Naturalist*, 144, 570-595.

Condit, R., Pitman, N., Leigh, E.G., Chave, J., Terborgh, J., Foster, R.B., Nunez, P., Aguilar, S., Valencia, R., Villa, G., Muller-Landau, H.C., Losos, E. and Hubbell, S.P. (2002). Beta-diversity in tropical forest trees. *Science*, 295, 666-669.

Currie, D.J. (1991). Energy and large-scale patterns of animal- and plant-species richness. *American Naturalist,* 137, 27-49.

Daniels, R.J.R., Joshi, N.V. and Gadgil, M. (1992). On the relationship between bird and woody plant species diversity in the Uttara Kannada district of south India. *Proceedings of the National Academy of Sciences* (USA), 89, 5311-5315.

Eamus, D. (2003). How does ecosystem water balance affect net primary productivity of woody ecosystems. *Functional Plant Biology,* 30, 187-205.

Finch, D.M. (1989). Habitat use and habitat overlap of riparian birds in three elevational zones. *Ecology*, 70, 866-880.

Garcillán, P.P. and Ezcurra, E. (2003). Biogeographic regions and β-diversity of woody dryland legumes in the Baja California peninsula. *Journal of Vegetation Science,* 14, 859-868.

Gaston, K.J. (2000). Global patterns in biodiversity. *Nature*, 405, 220-227.

Gaston, K.J. and Williams, P.H. (1996). Spatial patterns in taxonomic diversity. In K.J. Gaston (Ed.), *Biodiversity: a biology of numbers and difference* (pp. 202-229). Oxford: Blackwell.

Harrison, S., Ross, S.J. and Lawton, J.H. (1992). Beta diversity on geographic gradients in Britain. *Journal of Animal Ecology,* 62, 151-158.

Hillebrand, H. (2004). On the generality of the latitudinal diversity gradient. *American Naturalist,* 163, 192-211.

Hubbell, S.P. (1979). Tree dispersion, abundance and diversity in a tropical dry forest. *Science*, 203, 1299-1309.

Hubbell, S.P. (2001). The unified neutral theory of biodiversity and biogeography. Princeton: Princeton University Press.

Janzen, D.H. (1970). Herbivores and the number of tree species in Central American forests. *American Naturalist,* 104, 501-524.

Kerr, J.T. and Packer, L. (1997). Habitat heterogeneity as a determinant of mammal species richness in high-energy regions. *Nature*, 385, 252-254.

Koleff, P. and Gaston, K.J. (2001). Latitudinal gradients in diversity: real patterns and random models. *Ecography*, 24, 341-351.

Koleff, P. and Gaston, K.J. (2002). The relationships between local and regional species richness and spatial turnover. *Global Ecology and Biogeography,* 11, 363-375.

Koleff, P., Gaston, K.J. and Lennon, J.J. (2003a). Measuring beta diversity for presence-absence data. *Journal of Animal Ecology,* 72, 367-382.

Koleff, P., Lennon, J.J. and Gaston, K.J. (2003b). Are there latitudinal gradients in species turnover? *Global Ecology and Biogeography,* 12, 483-498.

Lees, D.C., Kremen, C. and Andriamampianina, L. (1999). A null model for species richness gradients: bounded range overlap of butterflies and other rainforest endemics in Madagascar. *Biological Journal of the Linnean Society,* 67, 529-584.

Lennon, J.J., Koleff, P., Greenwood, J.J.D. and Gaston, K.J. (2001). The geographical structure of British bird distributions: diversity, spatial turnover and scale. *Journal of Animal Ecology,* 70, 966-979.

MacArthur, R.H. (1965). Patterns of species diversity. *Biological Review,* 40, 510-533.

MacArthur, R.H. and MacArthur, J.W. (1961). On bird species diversity. *Ecology*, 42, 594-598.

Mantel, N. (1967). The detection of disease clustering and a generalized regression approach. *Cancer Research,* 27, 209-220.

McCain, C.M. (2004). The mid-domain effect applied to elevational gradients: species richness of small mammals in Costa Rica. *Journal of Biogeography,* 31, 19-31.

McCoy, E.D. and Connor, E.F. (1980). Latitudinal gradients in the species diversity of North American mammals. *Evolution*, 34, 193-203.

McGill, B.J. (2003). A test of the unified neutral theory of biodiversity. *Nature*, 422, 881-885.

Menkhorst, P.W. and Knight, F. (2001). *A field guide to the mammals of Australia. South Melbourne,* Victoria: Oxford University Press.

Mittelbach, G.G., Steiner, C.F., Scheiner, S.M., Gross, K.L., Reynolds, H.L., Waide, R.B., Willig, M.R., Dodson, S.I. and Gough, L. (2001). What is the observed relationship between species richness and productivity? *Ecology*, 82**,** 2381-2396.

Nee, S. and Stone, G. (2003). The end of the beginning for neutral theory. *Trends in Ecology and Evolution,* 18, 433-434.

Nekola, J.C. and White, P.S. (1999). The distance decay of similarity in biogeography and ecology. *Journal of Biogeography,* 26, 867-878.

Pianka, E.R. (1966). Latitudinal gradients in species diversity: a review of concepts. *American Naturalist,* 100, 33-46.

Rahbek, C. and Graves, G.R. (2001). Multiscale assessment of patterns of avian species richness. *Proceedings of the National Academy of Sciences* (USA), 98, 4534-4539.

Rohde, K. (1999). Latitudinal gradients in species diversity and Rapoport's rule revisited: a review of recent work and what can parasites teach us about the causes of the gradients? *Ecography*, 22, 593-613.

Routledge, R.D. (1977). On Whittaker's components of diversity. Ecology, 58, 1120-1127.

Sanders, N.J. (2002). Elevational gradients in ant species richness: area, geometry, and Rapoport's rule. *Ecography*, 25, 25-32.

Strahan, R. (Ed) (1995). *The Mammals of Australia.* Sydney: Reed New Holland.

Tews, J., Brose, U., Grimm, V., Tielbörger, K., Wichmann, M.C., Schwager, M. and Jeltsch, F. (2004). Animal species diversity driven by habitat heterogeneity/diversity: the importance of keystone structures. *Journal of Biogeography*, 31, 79-92.

Vellend, M. (2001). Do commonly used indices of β-diversity measure species turnover? *Journal of Vegetation Science,* 12, 545-552.

Volkov, I., Banavar, J.R., Hubbell, S.P. and Maritan, A. (2003). Neutral theory and relative species abundance in ecology. *Nature*, 424, 1035-1037.

Waide, R.B., Willig, M.R., Steiner, C.F., Mittelbach, G.G., Gough, L., Dodson, S.I., Juday, G.P. and Parmenter, R. (1999). The relationship between primary productivity and species richness. *Annual Review of Ecology, Evolution and Systematics,* 30**,** 257-300.

Whittaker, R.H. (1960). Vegetation of the Siskiyou mountains, Oregon and California. *Ecological Monographs*, 30, 279-338.

Whittaker, R.H. (1972). Evolution and measurement of species diversity. *Taxon*, 21, 213-251.

Whittaker, R.J. and Heegaard, E. (2003). What is the observed relationship between species richness and productivity? *Ecology*, 84, 3384–3390.

Willig, M.R., Kaufman, D.M. and Stevens, R.D. (2003). Latitudinal gradients of biodiversity: Pattern, process, scale, and synthesis. *Annual Review of Ecology, Evolution and Systematics,* 34, 273-309.

Wilson, M.V. and Shmida, A. (1984). Measuring beta diversity with presence-absence data. *Journal of Ecology*, 72, 1055-1064.

In: Focus on Biodiversity Research
Editor: Jan Schwartz, pp. 199-300

ISBN: 1-60021-372-3

Chapter 9

THE ORIGIN AND FUNCTIONS OF BIODIVERSITY IN INFECTIOUS AND NON-INFECTIOUS DISEASES

Sergey N. Rumyantsev[1,*] and V.K. Gerasimov[2]
[1]Andent Inc., Waukegan, IL, USA
[2]Russian National Library, 20 Sadovaya ul., Saint-Petersburgm 191069, Russia

ABSTRACT

Despite considerable immunological, epidemiological, genetic and pharmacological efforts, various diseases continue to take their toll in death and disability. Although the fear of infectious epidemics has begun to be controlled, contemporary humanity may become the victims of rapidly growing global epidemics of very dangerous non-infectious diseases including atherosclerosis, heart and liver diseases, stroke, diabetes, high blood pressure, obesity, different kinds of cancer, gout, disease of the joints, schizophrenia, senescence, etc. Many questions about the nature of the diseases do not receive the answers they deserve. Moreover many very important questions have not been formulated. Nothing, for example, is known about the factors and mechanisms of epidemic spread of non-infectious diseases or the origin of extraordinary diversity in clinical manifestations and severity of such diseases. This paper is designed to formulate and examine a new paradigm exploring the genesis and role of biodiversity in determination of the spread, pathogenesis, and individual variation in the signs, course and severity of both infectious and non-infectious diseases.

INTRODUCTION

Despite considerable immunological, epidemiological, genetic and pharmacological efforts, various diseases continue to take their toll in death and disability. Although the fear of infectious epidemics has begun to diminish, contemporary humanity may become the victims

* E-mail: rumyan1@yahoo.com

of rapidly growing global epidemics of very dangerous noninfectious diseases including atherosclerosis, heart and liver diseases, stroke, diabetes, high blood pressure, obesity, different kinds of cancer, gout, disease of the joints, schizophrenia, senescence, etc. Many questions about the nature of the diseases do not receive the answers they deserve. Nothing, for example, is known about the factors and mechanisms of epidemic spread of noninfectious diseases or the origin of the extraordinary diversity in clinical manifestations of such diseases. The knowledge of the diseases' causes and origin as well as of the rules of their spread, development and manifestations continue to form the main current functions of medicine and biomedical sciences. Biodiversity is an indispensable characteristic of any disease. It reveals itself at every feature of a disease.

This paper is designed to formulate and examine a new paradigm exploring the genesis and role of biodiversity in determination of the spread, pathogenesis, and individual variation in the severity of diseases. Currently, many biologists focus on various aspects of biodiversity at the species and population level. However, another form of biodiversity is often overlooked— the secret internal diversity of each individual body that is revealed in pathology conditioned by microbial aggression or varied ecological and physiological agents revealing hereditary heterogeneity patterns in cell clones that differ for each type of agent. Unfortunately, we know little about this type of biodiversity and, until now, have done little to explore this area. By studying this phenomenon, we can learn much about species, population and individual features of diseases.

In this article, the key findings concerning ecology, molecular ecology, genetics as well as cell and molecular level of biodiversity in both infectious and non-infectious diseases are summarized and interpreted mainly from the unified evolutionary viewpoint with certitude that this approach can help to decipher the undiscovered secrets of the diseases epidemiology and pathogenesis, as well as clinical course and severity. The present paper reviews the existing knowledge dealing with the host genetics and its implication in the pathogenesis of diseases. The discussion includes genetic association with the clinical disease and the genetic restriction of cells that appear to play an important role in the pathogenesis of diseases. Because of space constraints, only selected works, presumably reviews have been cited.

The Origin and Molecular Constitution of Genetic Immunity

Background

Immunity means the capacity to withstand something, especially a state of increased resistance against the effect of an ecological or physiological agent, e.g. the natural capacity of a living organism to withstand infectious disease. As an all-biological phenomenon, immunity involves an extremely complex set of surveillance agents and systems which geared to the maintenance of a healthy existence.

Three systems of immunogenesis – constitutional (genetic), phagocytic and lymphatic - are known today (Figure 1). All the systems are of innate origin. The organisms that possess one, two or all of the systems, should form them according to the genetic program of individual ontogenesis.

Figure 1. The main principles of immunogenesis: (1) innate, (2) phagocytic, and (3) lymphatic [14].

Constitutional and phagocytic mechanisms for protection are inherent. They develop according to the genetic program of individual ontogeny, but without individual interaction with any particular parasite or parasitic product. Constitutional defense exists in the body before the attack of disease-causing agents will happen. In contrast, the mechanisms of resistance provided by lymphatic system are reactive. They only develop relevant agents of defense in a response to a stimulus, i.e. following the interaction with a particular parasite or with relevant components of that parasite. For these reasons, constitutional resistance represents the most fundamental level of immunity to infection [1-4].

Phagocytic systems of self-defense consist of the basic cleaners of the body--macrophages, natural killer cells and Europhiles--rather than the T cells, which must first be exposed to a pathogen in order to know to attack it. These are the cells usually known to fight bacteria. Recently the anticancer function of phagocytic system has been discovered [5].

Vertebrate animals possess three levels of immunity. Constitutional and phagocytic mechanisms serve as a first line of defense and if they are ineffective, the affected organism can be rescued by the influence of specific immunoglobulin's and other agents that are formed mainly by the innate systems of reactive immunogenesis especially by its lymphoid component. Key mechanisms of innate immunogenesis are proteolytic cascades, leading to localized blood clotting and opsonization, and transient production of antimicrobial peptides including immuneglobulines [6]. Invertebrates have only constitutional, phagocytic and initial mechanisms of reactive resistance. In plants, protozoa, fungi, bacteria and viruses, defense is performed mainly by constitutional and initial factors of reactive self-defense. Constitutional immunity is a general biological principle that provides relevant antimicrobial defense to all types of living beings, including humans (Figure 1) This type of resistance is of primary importance not only in its phylogenetic origins (Figure 1) but also for its extraordinary current significance [4, 13, 14].

One of the implicit dogmas of present-day immunology is that only immunoglobulins and lymphocytes are responsible for the strength of antimicrobial immunity. Recently, however, the key protective function of the lymphatic system has begun to be doubted, because its response to a first microbial attack is more inert and lazy [1, 9] than that of infectious pathogenesis. For instance, the incubation period of avian influenza A (H5N1) may be longer than that of other known human influenzas. Nevertheless it is to short. In 1997, most cases occurred within two to four days after exposure [10]. An infected individual begins to develop a remarkable antibody response following primary infection only by day six. If the infected organism is not dead at the moment, its lymphatic system provides him with many acquired protective agents that are considered to be able to stop the primary infection.

The resistance provided by the lymphatic system develops slowly and is unable to counteract fast-replicating microbes immediately [7]. Its mechanism for protection is reactive and begins to function only after infection develops an optimal quantity of the microbial antigens required for stimulation of reactive immunogenesis. Moreover, the adaptive response of the lymphatic immunogenic system includes the transport of microbial antigens into the lymphoid organs, the recognition and differentiation of antigens by T- and B-lymphocytes, clonal expansion and differentiation of effecter cells, and biosynthesis of specific immunoglobulins and their transport to infected organs.

The main paradigm of twentieth-century immunology was impotent to ask and answer the questions about the origin and functions of interspecies, intra-species, individual and intra-individual featufes that formed the discussed kind of biodiversity observed both in infectious and non-infectious diseases. This was because the role of the lymphatic system of reactive immunogenesis in the determination of this biodiversity seems to be more than problematic. The discovery of constitutional immunity has not been involved in the mainstream of present-day immunology. The nature of such genetic variations remained undiscovered, and the true number of susceptible individuals among humankind remained elusive up to the end of 20th century.

Main Principles

Constitutional immunity is the suite of individual defenses against pathogens that is genetically predetermined and ensures immediate protection without requiring prior stimulation and followed reactive response [4, 8, 11]. They are inherent and present in the body before the attack of disease-causing agents. The manifestations of innate immunity can be observed in species, populations and individuals, as well as in cells and subcellular structures.

Most mechanisms of genetic immunity functions on the cellular and molecular levels of the microbe–victim interaction. The genetic immunity of an organism is provided by the relevant properties of its cells, namely their pathogenetically competent molecular structures. These properties can be revealed by relevant procedures, for instance, on the step of the virus adhesion on the cell envelope, which is the key step of infection by intracellular parasites [4, 8, 12, 13].

Self-reproduction is a major requirement for microbial aggression. Pathogenic microbes are parasites, they survive inside other organisms as a result of biosyntetic processes carried out by victims. This requires the ability to invade the victim, to penetrate its surfaces, to reproduce in the victim, to interfere with its defences and to damage the victim. Microbes penetrate organisms and enter cells exclusively through particular molecular processes controlled by specific chemical (molecular ecological) substances (adhesins, toxins, enzymes, cytolysins, polynucleotides and so all). Infectious disease develops as a result of the damages induced in the affected body molecular structures by molecular ecological agents of pathogenic microorganism. Every host-parasite interaction is specific. It involves mutual molecular recognition of victim and parasite, mediated by chemically complementary molecules. Since the results of microbial chemical aggression may be extraordinary dangerous, every living organism must have immunity against it in order to survive.

To affect a victim every infectious agent uses unique assortment of unique molecular ecological agents which are able to affect unique targets in the body of victim attacked by the agent. For instance, different receptor structures are involved into interaction of human cells with influenza viruses and meningococci. Accordingly, constitutional antimicrobial immunity encompasses several specific and common structures possessed by the attacked organisms and functioned in defence against molecular ecological aggression of harmful micro-organisms. These structures are presented with a variety of molecular barriers (lack of specific cell receptors, modification of specific cell receptor, modification of specific nutrients, lack of a specific nutrient, presence of specific antibiotic and\or poison, nonspecific mechanisms). Its study illustrates not only extraordinary diversity of molecular anatomy of various living forms but also the sense and the origin of this diversity [4, 14].

Lack or Deficit of Cell Receptors

Certain biomolecules expressed on cell surfaces play a key role in the attachment of micro-organisms as the first, i.e. clue step in infection. This role may be related to the presence on the cell of molecules acting as receptors for chemically complementary adhesive molecules of microbes. The presence, quality, quantity, and density of molecular targets play a vital role in the phenomena of constitutional predisposition or constitutional immunity to infection. If adhesion of a parasite onto victim cells could be restricted, infection would be interrupted and disease prevented.

The non-sensitivity of cells, e.g. of animal cells to adhesion of N.meningitidis indicates that some cell surface molecules unique to cells of some humans could act as receptors. On the other hand the attachment of Neisseria meningitidis to cell surfaces involves lectin-like sites on the microbial surface which bind to specific ganglioside radicals on affected cells. This concept was illustrated by ganglioside-inhibited binding of meningococcus to erythrocytes as well as to epitheliocytes and leukocytes [15]. According to the data, it is becoming increasingly clear that certain peculiarities of the molecular constitution of cells can contribute to whether or not a child or adult will develop meningococcal infection. The revealing of age and individual distinctions is the first step to decoding of their molecular basis.

The phenomenon of newborns resistance to meningococcal infection may be associated with some age constitutional peculiarities of the organism, namely with impotence of cellular receptor structures, responsible for the agent adhesion that is the key stage of the infectious process. The spectrum of disease incidence and severity in N.meningitidis meningitis also can be attributed to heterogeneity of human population in susceptibility of cells to adhesion of the infectious agent. The conjoint effects of the victims ontogenesis and followed heterogeneity in the ability of their cells to adhesion of N.meningitidis can help explain some intriguing epidemiological and clinical observations. However extended investigations at the level of individuals are necessary in order to draw comprehensive conclusions. Whether an individual infected with N.meningitidis will develop disease or not, will depend on a variety of risk factors. Certainly, adhesive ability of victim's cells is one of them. The second, third and other factors could be found on transmembrane and intracellular stages of pathogenesis.

Choleraic toxin is able to interact only with ganglioside macromolecules that contain the subunits of ceramide, lactose, galactosamine, galactose, and the radical of sialic acid joined with the molecule of lactose. But the same toxin does not interact with gangliosides of other types; for instance, those that do not contain galactose or have an additional radical of sialic acid in the end position [16]. Receptors for botulinic and tetanic toxins as well as for influenza virus possess different ganglioside compositions (Figure 2).

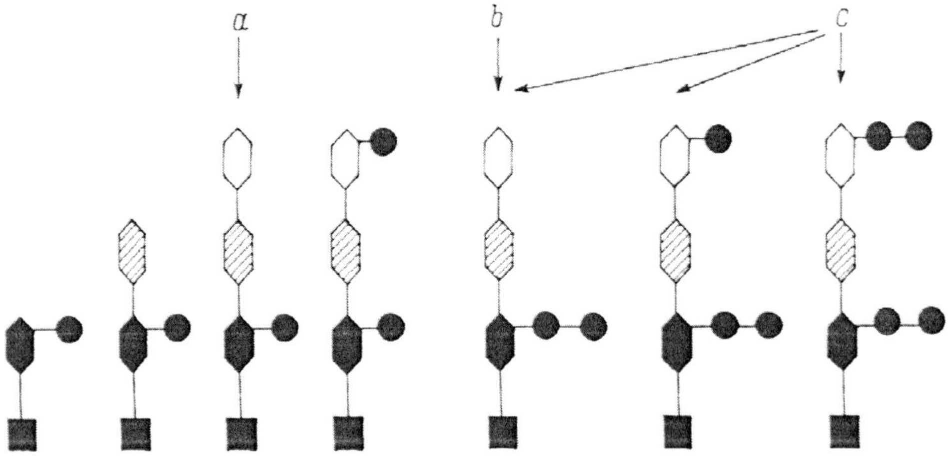

Figure 2. Dependence of organism' vulnerability by microbial agents on the presence of gangliosides (1-7) in cell membranes. Choleraic (a), botulinic (b) and titanic (c) toxins [190].

The presence or absence of specific cell surface receptors can influence the victim range and tissue tropism of viruses. Receptors for poliovirus, for example, are known to confer susceptibility on human and some primate cells while cells of any other species lack such receptors and are immune to the infection. A strong correlation was found between receptor expression on cells and their constitutional susceptibility or immunity to encephalomyocarditis (EMC) virus infection. Cell lines which did not express detectable receptors were not susceptible to infection. In contrast, cell lines with demonstrable receptors were capable of being infected [17].

Rabbit diarrheagenic Escherichia coli-1 (E.coli RDEC-1) is only able to induce diarrhea in rabbits. The adherence of RDEC1 to intestinal brush borders is a species-specific event, occurring with the rabbit but not with guinea pig, rat, or human brush borders. Futhermore, after the onset of the diarrhea phase, RDEC-1 was found to have colonized the ileal and cecal regions of these rabbits heavily with minimal colonization found in the jejunum. This species and tissue specificity of in vivo colonization and infectivity clearly correlates with the specificity of in vitro adherence. The receptors for RDEC-1 are selectively localized on the brush border surface of rabbit ileal epithelial cells and are not present on the basolateral membranes of epithelial cells or on any of the other membranes of the underlying tissue. This remarkable degree of species, tissue, and cell specificity possessed by RDEC-1 indicates that the rabbit intestinal mucosa contains receptors specific for RDEC-1, which are not present in other tested mammalian species [18].

In contrast, the non-adhesive strain of E.coli, HS, failed to colonize rabbit ileum and the extent of cecal colonization by HS was 10,000-fold less than that seen with RDEC-1. The inability of RDEC-1 to colonise the ileum and caecum of rats and guinea -pigs demonstrated the lack of RDEC-1 receptors on the mucosal surface in these animals, which is likely to be a factor of their constitutional immunity to the diarrheagenic effect of RDEC-1: the failure of RDEC-1 to colonise their intestines, due to a lack of receptors in these animals would lead to rapid clearance of nonadherent RDEC-1 [18].

Modification of Cell Receptors

Bacteria have evolved various means of self-defence against viral attackers at almost all stages of the bacteriophage life cycle. The most effective defence is to prevent any productive contact between the virus and the bacteria. This can be done by mutation of bacteriophage receptors on surface of the bacteria or by secretion of a barrier which prevents the adhesion of the bacteriophage, such as a capsule or slime layer.

Sendai virus has surface glycoprotein spikes which can adsorb on specific receptors of cells attacked by the virus. The receptors on erythrocyte membranes contain sialic acid. Sendai virus cannot adsorb to erythrocytes which have been treated with sialidase. Gangliosides can serve as Sendai virus receptor and a multiplicity of receptors was needed for virus binding to ensure that the virus particle is bound to enough:. no fewer than 3,000 such receptor molecules are required for the attachment of a single virion. The binding of the virus to the ganglioside is specific, tight and not readily reversible [19].

Gangliosides appear to serve as viral receptors which can interact specifically with the glycoprotein spike on the viral surface. Ganglioside GM1, which lacks a terminal sialic acid radical does not act in a virus adhesion, whereas mono-, di-, and trigangliosides with terminal sialic radicals do. The terminal sialic acid radical is required for receptor activity [19].

The biomolecular target of a microbial toxin plays a key role in the evolution of spotted disease in sugar cane; the target biomolecule exists in nature in two forms, differing in amino acid structure. One variant is incapable of interacting with the toxin because the plant possessing the modified membrane protein is not susceptible to this disease and is constitutionally resistant to it. This particular structure involves a change in only four amino acid positions out of the 110 amino acids of the given membrane protein [20].

Modification of Specific Nutrient

Similarly, small changes in the amino acid content of hemoglobin gives rise to immunity to Plasmodia and creates constitutional immunity to malaria [21]. African green monkey (Cercopithecus torquatus) kidney cell lines (Vero, CV-1) have approximately 10,000 to 20,000 surface diphtheria toxin-binding sites per cell and are exceptionally sensitive to this toxin. The increased resistance to diphtheria toxin in a CHO cell line is attributed to a mutation giving rise to receptors with reduced affinity for toxin [17].

The erythrocytes of sheep and oxen, wich have a high content of sphingomyelin and a low lecithin content, are lysed by parasitic leptospirae more rapidly than the erythrocytes of other animal species. The lysis of rabbit and rat erythrocytes, in which the amount of lecithin is greater than that of sphingomyelin, appeared after a longer incubation than was required for lysis of sheep erythrocytes. The enzymatic degradation differed not only between the different types of phospholipids, but also, in some cases, between phospholipids of the same type present in the erythrocytes of different species. Thus, lysolecithin in rat erythrocytes, in contrast to that in rabbit and sheep erythrocytes, was only degraded very slowly by the leptospirae. The phosphatidyl ethanolamine and phosphatidyl serine in sheep erythrocytes appeared to be resistant to degradation by leptospirae but, when present in rat erythrocytes, these phospholipids were degraded [22].

Absence of Specific Nutrient

Resarch on brucellosis, which is mainly a disease of cattle (caused by Brucella abortus), sheep and goats (Brucella melitensis) and pigs (Brucella suis) provides a demonstrative example of identification of a nutritional factor of constitutional immunity. Infection takes several very mosaic forms. In the non-pregnant female the brucellae cause a debilitating disease but do not reproduce abnormally in any particular tissue. In pregnant animals they reproduce extensively in the foetal side of the placenta and the chorion, leading to abortion. In males there is a localisation in the genitalia which causes infertility. The stimulant of the reproduction was identified as erythritol, a foetal product concentrated in the those structures. Erythritol, rather than glucose, is a preferred nutrient for B.abortus, B. melitensis and B.suis. B.abortus infection in new-born calves was enhanced by erythritol. The absence of this component of molecular constitution of some species and their tissues is responsible for the restriction of brucellae reproduction. This is a major factor of constitutional immunity to brucellosis [23].

Presence of Specific Antibiotic

At the culminant stage of the bacteriophage life cycle the virus injects its genome polynucleotide into the bacterial body. Once bacteriophage DNA has been injected into bacterial cell, the victim's DNA restriction enzymes ensure the survival of the bacterium. Bacteria use DNA restriction and modification enzymes to maintain own genetic stability when foreign genome polynucleotides are introduced into their bodies. The enzymes within the victim may destroy the polynucleotide, thus, DNA restriction enzymes provide the antiviral immunity of bacteria.

The bacteriophages possess antirestriction mechanisms, and are able defend themselves by inhibition of restriction enzymes by DNA modifications or by their own proteins. The anti-restriction mechanisms play very important roles in the immunity of viruses. Bacteriophages

T3 and T7 protect their DNA from restriction by producing antirestriction proteins. Although of differnt molecular constitution, the two phage proteins act using the same mechanism. The most striking feature of bacteriophage anti-restriction is the wide variety of specific mechanisms possessed by different viruses [24]. Heritable changes in the molecular organisation of bacteria, for example, the appearance of metabolites which can neutralise one or another antibiotic, inhibit the effective action of certain antibiotic agents.

Genetic immunity to influenza virus infection is associated with the existence of Mx gene which controls biosynthesis of Mx protein. The mouse Mx1 protein acts at an early stage of the viral infection by inhibiting primary transcription of the virus genome. The human MxA protein performs a subsequent act in the cytoplasm of the cells infected by influenza virus [25].

Constitutional immunity of Western red cedar wood (Thuja plicata) to insect attack is attributed to the presence of monoterpenes, of which metyl thujate is toxic to larvae of the black carpet beetle, the furniture beetle, and the case-making beetle. The diterpene gossypol decreases weight gain in larvae of the tobacco budworm and pink bollworm. The terpenoids inhibit or inactivate acetylcholinesterase and cause acetylcholine to accumulate at the cholinergic site, producing continuous stimulation of insect cholinergic nerve fibers throughout the central and peripheral nervous system, followed by paralysis and death of the animal. Acetylcholinesterase in resistant animals has evolved to become inhibited more slowly than that in susceptible animals. The release of HCN from living cells of plants plays an important role in plant defence against the damage that can be caused by grazing animals. Cyanogenesis plays an important role in plant defence mechanisms against herbivores, in particular against molluscs [26].

Thus, the pathogenic effects of a parasitic invader molecules on the molecular targets of a victim is provided by the genetically determined molecular structure of the victim's target, thereby defining constitutional susceptibility or constitutional immunity to infection or invasion. As a result of this property of molecular constitution, viability and propagation of invaders can be deflected or restricted.

Specificity of Constitutional Immunity

It would be metabolically impractical for any living being to develop a new chemical against every species of parasite which threatens, but living organisms needs to do so very frequently. For this reason, constitutional immunity is mainly very specific, as is acquired immunity, which is induced by the lymphoid system of reactive resistance. The specific constitutional barriers include only chemical mechanisms.

The idea of specificity operating between pathogenic micro-organisms and the target of these micro-organisms is encompassed by species specificity, population and individual specificity, tissue specificity, and the specificity of cell clones and molecular targets. Specificity of constitutional immunity is determined by specificity of the micro-organism-victim molecular ecology. Specificity is not surprising, as the molecular ecological agents and modes of interaction and application of the various microorganism-victim systems differ significantly.

The specific immunity of contemporary cultivated potato stocks to late blight is derived from hybridization of wild resistant of potato with cultivated varieties. Immune plants are almost immune to some species of the fungus, while susceptibility to other species is unaffected. Specific resistance is controlled by at least nine dominant genes [27]. The insecticide (e.g. dieldrin) resistance mechanism seems to be the same in all species of test insects used. Dieldrin is ineffective against resistant insects because their receptors have become incompatible to this insecticide [28].

Non-specific Constitutional Mechanisms

Constitutional immunity may also be non-specific. The non-specific innate host defenses include both physical and chemical barriers. Non-specific constitutive defenses include both physical and chemical barriers: physical barriers may be mechanical (e.g. ciliated respiratory epithelium, mucous layer), energetic (e.g. elevated body temperature) or electrostatic mechanisms; non-specific chemical mechanisms include acid pH, lack of nonspecific nitrients and the effects of non-specific enzymes, antibiotics, xenobiotics and poisons.

The fitness disadvantage of resistance in on insecticide-free environment is behavioral: resistant insects are less active and less responsive to stimuli. Insects which were homozygotes for resistance spent less time searching for hosts, and were less responsive to predator movement than heterozygotes and susceptible insects. The effecfs of fitness were always recessive and never modified by the genetic background. In addition, female homozygotes were less responsive to oviposition stimuli and male homozygotes had limited mating success. Insecticide resistance gives the insect an enormous advantage in the presence of insecticide, regardless of the other susceptibilities in the insect [28].

Opsonins, i.e. factors other than immunoglobulin G and C3b fraction of complement, can enchance phagocytosis. In neonates, opsonins stimulated phagocyte function and the inflammatory response which determines the innate immunity to infection. Interferon, the viricidal protein, provides antiviral immunity of many cells which synthesized that constitutional factor. However, the sensitivity of various viruses to interferon is not uniform. In the skin, secretion from the sebaceous glands contains fatty acids that may kill some microbial species [29].

Genetic of diversity in the incidence of infections. The effectiveness of constitutional immunity (which might be rephrased as genetically determined resistance) is very high, and ordains the future of many, if not all, species and populations. Genetic immunity play significant role in the incidence and severity of infectious diseases. An epidemic of tuberculosis in a herd of cattle descended from two bulls was studied. Investigations revealed that 62% of the descendants of one bull were infected with tuberculosis whereas the descendants of the other bull developed only a 4% incidence of tuberculosis [30]. Webster was the first to develop the Mendelian ratios for the inheritance of immunity to microbial infections by mating constitutive immune and constitutive susceptible strains of mice [31]. Trough selective breeding, he developed strains of mice which differed in resistance to St. Louis and louping ill flaviviruses. He showed that immunity is inherited as a single, dominant factor .

Genes which controll constitutional immunity to some infections in mice have been identified and mapped to a specific chromosomal location. Thus, constitutional immunity of mice to Rickettsia tsutsugamushi, the etiologic agent of scrub typhus, is controlled by a single dominant gene (Ric) located on chromosome 5. The inherent immunity of mice to Mycobacterium intracellulare is regulated by a gene located on chromosome I. This gene is linked to Bcg gene of the chromosome 1. Resistance to group B arbovirus infection is inherited in mice as a dominant, autosomal allele. The BRVR, PRI and C3H/RV inbred strains are homozygous for this form of resistance. No other inbred strains tested are able to resist infection with group B arboviruses. Levels of complement were the same in resistant and susceptible mice. Wild mice collected from widely separated geographical areas were resistant to the infection. Thus, the gene controlling resistance to group B arbovirus infection may well be a widely distributed natural gene which endows selective advantage in the presence of a lethal virus infection [32].

Inherent immunity is determined by a single dominant gene in cases of myxoviruses and herpes viruses in mice, as well as mousepox virus. Inherent susceptibility of mice to polioma virus and mouse hepatitis virus (MHV) is also determined by a single dominant gene. MHV-2 infection in mice provides an example of susceptibility inherited as a dominant infactorial trait. Instances in which immunity either immunity or susceptibility is determined by a single dominant gene have generally offered the best prospects for elucidating the microbial V.S. host interactions involved: when two or more genes are involved, genetic evaluation is more difficult [33]. Polygenic traits have been more difficult to evaluate.

Constitutional immunity to Salmonella typhimurium, the agent of murine typhoid, is controlled by at least three distinct genes (Ity, Lps and xid). The first of these genes appears to be the major determinant of immunity while the other two are mutant alleles that render mice susceptible [34].

Genes that control constitutional immunity to some infections in humans and animals have been identified and mapped to a specific chromosomal location [33, 35, 36]. The CCR5 gene 32-base pair deletion provides strong constitutional immunity of human homozygotes to HIV infection [37]; in the heterozygous state it may provide relative immunity, thus delaying the progression of HIV infection to AIDS in affected individuals [38].

The immunity of leafhoppers to maize streak virus is controlled by a single sex-linked gene which regulates the resistance of gut epithelial cells to the viral infection. [39]. Genes which controllconstitutional immunity to bacterial spot disease in the pepper plant have been identified and mapped to a specific chromosomal location. Peppers carrying the gene Bs1 are resistant to infection with Xanthomonas capestris, which causes this condition. Predisposition and host ability to restrict parasitic invasion, development and survival are the important immunological factors in the epidemiology of helminths that may be under genetic control [40].

Evolution of Constitutional Immunity

Pathogenic effects of microbes and microbial molecules on molecular targets in the victim body are provided by the genetically determined molecular structure of the targets, thereby defining constitutional susceptibility or constitutional immunity to infection. As a result of this property of molecular constitution, the viability and propagation of intruders can be either allowed or aborted. This kind of immunity is provided by constitutional incongruence between acting molecules of infectious agents and their molecular targets in the attacked organism. The success or failure of microbial aggression depends on the precise molecular congruence of infectious agents with the molecular make-up of the attacked organism, especially that of its receptors. Structural modifications of relevant molecular attributes of a victim are able to reduce or even eliminate the success of microbial intrusion [1, 4, 14]. The incongruence arises as a result of mutations and is propagated in a population, thanks to natural selection, as an evolutionary adaptation to life-threatening environmental agents.

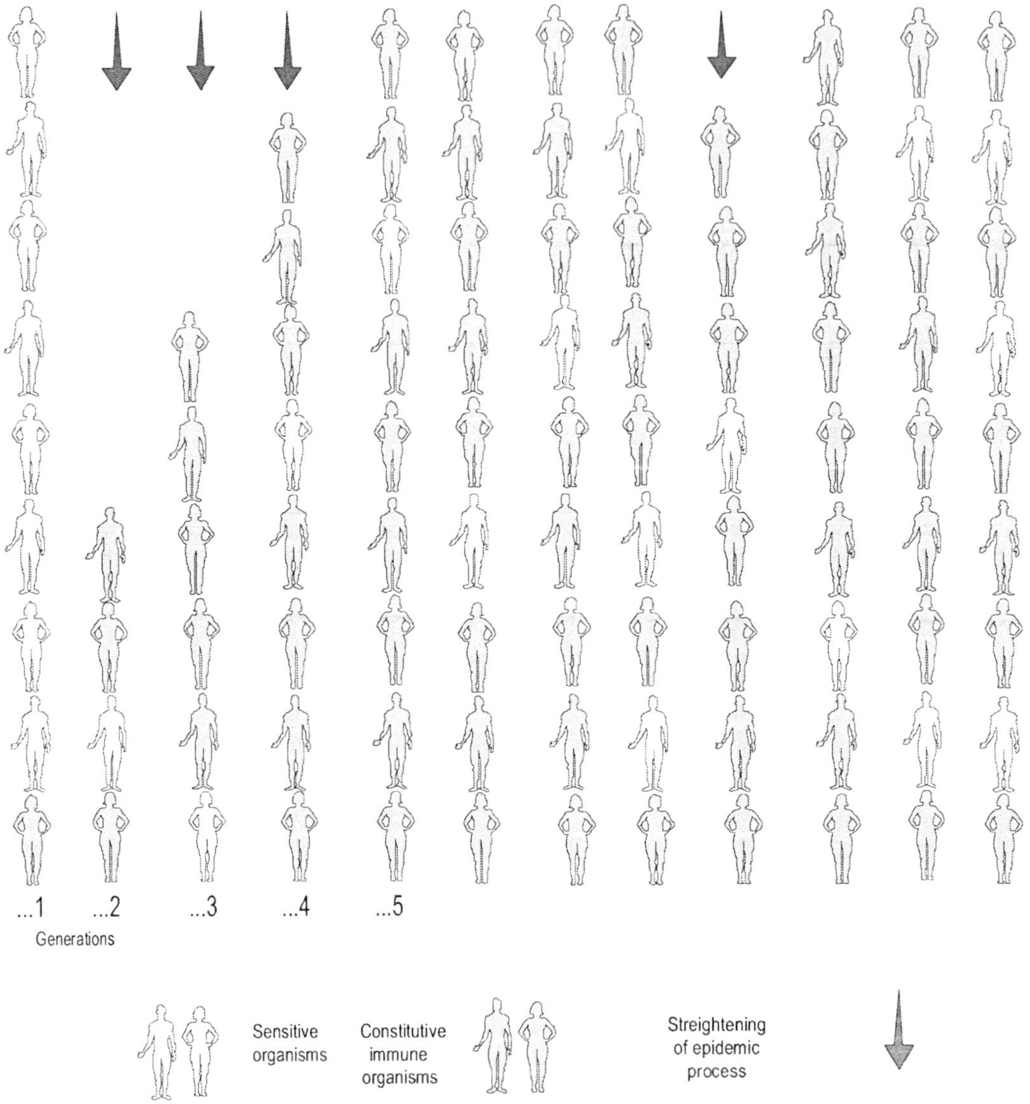

Figure 3. Evolution of an epidemic process: transformation of a population genetic immunity of in the fight against infectious agent [42].

Certain stereochemical modifications, as well as chemical modular compositional variations, can arise as a result of mutations. These may diminish, even completely exclude, the possibility of intermolecular interaction between microbes and their victims, thus lessening the success of microbial aggression. The organisms possessing this modified molecular constitution are incapable of interacting with the microbe; they are not susceptible to this disease but are constitutionally resistant to it. These organisms are not infected by the corresponding microbe, but, in contrast, continue to develop normally and produce healthy progeny while other individuals of the same species which do not have this molecular structure become ill and die. On repeated exposure to a given pathogen, the progeny of resistant mutants will finally predominate in a population; an individual protective mutation becomes a property of a group, then of a population, and finally of the species (Figure 3). The selective principle of the appearance and spread of antimicrobial resistance lies at the base of natural constitutional immunity, which is characterised by specific molecular constitutional factors that exclude the possibility of interaction between parasite and victim.

The possibility and intensity of evolving constitutional immunity is determined by ecological prerequisites for regular interaction of potential victims with populations of infectious agents [3, 14]. According to their different ecological features, living beings have been differentially subjected to interaction with different infectious agents. Populations of domestic poultry do not regularly contact the sources of influenza virus H5N1 in its artificial environment and have not experienced intensive natural selection for genetic immunity to the agent (Figure 4, a).

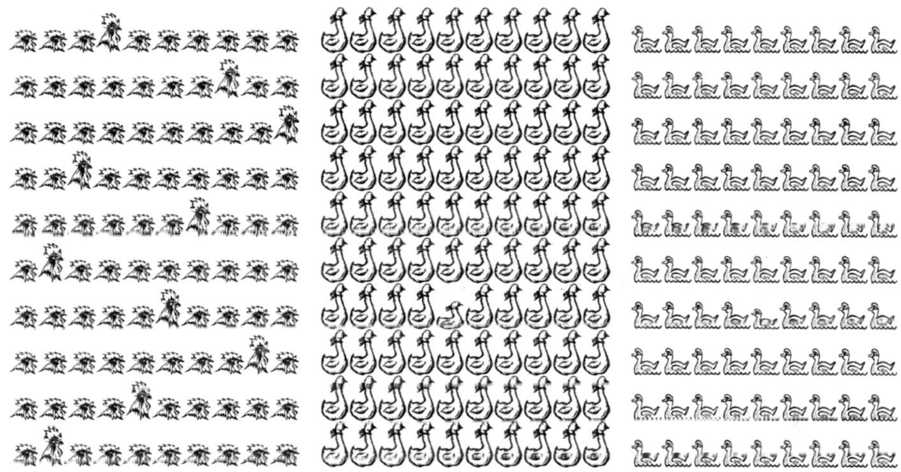

A B C

Figure 4 . Avian influenza virus H5N1 is highly pathogenic for domestic poultry (A; mortality up to 90%), but nearly nonpathogenic for wild ducks and geese (B, C; mortality less than 1%) [148]. [] – unaffected animals; [] - affected animals.

In contrast, the presence of genetic immunity to influenza in most ducks, geese and other aquatic birds [Rumyantsev, 2006] shows that they had regular interaction with the sources of relevant infectious agents in their natural environments, especially in the hydrosphere and had been subjected to natural selection for genetic immunity to influenza (Figure 4 b, c). Genetic immunity to influenza infection in aquatic birds can be considered to be the result of their evolutionary adaptation to inevitable coexistence with this virus over a great many generations [{Webster RG, 1992 562 /id}].

The evolution of genetic immunity to most of the existent infections [3, 14], could have been induced by humans' continual, inevitable and intensive carnivorous contact with various potentially infected animals. It is known, for example, that diseased animals are caught by hunters more easily than healthy ones. The adaptation could have been performed by means of natural selection among both victim species, leading to formation of its genetic immunity, and populations of the parasites, leading to formation of new pathogenic variants being able to break down the victims' immunity [3, 4, 14].

The selection has continued over millions of years, and species living today presumably possess genes for natural defense against relevant infectious agents. As a consequence, constitutional immunity currently seems to be a leading factor in the immunological structure of the population that determines the main parameters of epidemic processes. Nevertheless, one should take into account that, in fact, all "viral" and "bacterial" cell receptors exist not to interact with microbial pathogens, but to provide important physiological functions of relevant physiologically active molecules. The evolutionary adaptation of harmful microbes to a parasitic way of life could have been performed by the exploitation of pre-existent physiological receptors. The subsequent evolution and spread of genetic immunity to infections could have created a constitutional genetic immunity to the molecular physiological agents, thus providing the spread of relevant forms of pathology [4, 41]. The regularities in genetics, together with the data on the phenomena of evolutionary resonance and conjugated evolution, reveal the possibility of changes in the strength ratio in parasite–victim systems, thus entailing a change in the sensitivity to infectious agents and next intensification of epidemic processes [4, 42]

Observation of Biodiversity in Infectious Diseases

Biodiversity in infectious diseases displays itself firstly on the level of their causes and consequences. There exists a very diverse plethora of infectious agents; each of them able to induce only specific but very diverse affections at selected but various species. Accordingly, the two main sections of this first part of our investigation are devoted to epidemiological observations and clinical observation. Diversity of affected species in their relation to an infectious agent forms especially important level of biodiversity in infectious diseases. Consequently, our research of epidemiological data begins with the discussion of interspecies diversity in the incidence of infections, followed by the sections of inter-population, inter-individual, age and cell levels of diversity in the incidence of infections. The research of epidemiological diversifications is further continued by investigation of subcellular and molecular levels of diversity as well as of the origin of this kind of biodiversity.

The data of foregoing section are deepened by purposeful investigation of clinical manifestations of infectious diseases, including interspecies and inter-individual diversity in the course and severity of infectious diseases. The three final paragraphs this first part of our investigation are devoted to intra-individual diversity in the location and size of infectious affections as well as to its cellular and molecular levels and origin. They are of special novelty and interests, especially for theoretical and applied aspects of both medical and general genetics. The concluding part of infectious section is devoted to the functions of biodiversity in infectious diseases. The contribution of biodiversity to the epidemiology and severity of infectious diseases are interpreted in special paragraphs followed by the investigation of the contribution to infectious pathogenesis and biomolecular evolution.

Epidemiological Observations

Interspecies Diversity in the Incidence of Infections

Most of diseases are induced by infection with microbial parasites (viruses, bacteria, fungi, protozoa) that are specialized to survive as a result of biosynthetic processes carried out by other organisms. To perform their self-reproduction, parasitic microbes cause many disabilities and even deaths within various natural groups of victims. Microbes may destroy forests, crops, domestic and wild animals, including insects, reptiles, birds, fish and other forms of life, although with different levels of intensity. Even microorganisms themselves are the subject to attack by other microbial species. Some species become affected when attacked by microbes. Many other species are preserved by their innate features.

Influenza

Influenza viruses can infect representatives of a variety of species, including humans, dogs, pigs, horses, sea mammals, aquatic birds, poultry, and passerine birds but with extraordinary different levels of intensity. The possible involvement of many other vertebrata and invertebrata animals as well as protozoa has not been investigated. Although many different subtypes of influenza A viruses (H1 to H14 and N1 to N9) are perpetuated in aquatic birds, particularly in migrating waterfowl, all the registered human pandemics were induced by just three of them (H1N1, H2N2, H3N2). Other subtypes of influenza A virus, as well as types B and C, induced neither pandemic nor epidemic conditions among humans, only small outbreaks.

Recently the avian influenza H5N1 strain has been proclaimed a priori as highly pathogenic (deadliest) for human [43, 44]. In reality, since this variant of influenza virus has been identified at 1998 only nearby 100 lethal cases of H5N1 influenza virus infection have been registered around the World population of 6,5 billions. At the same time this kind of influenza has already resulted in rapidly fatal diseases among poultry, with a mortality approaching 90–100% and intentional annihilation of millions of domestic hens throughout the world.

Aquatic birds occupy an unique and very important position in the natural history of influenza viruses [45], harboring influenza viruses of all known HA and NA subtypes. Nevertheless, they mostly reveal no disease signs. The relevant morbidity is very low among them, though they are considered to be natural reservoirs of influenza infection. Like in other genetically immune species [4], among the species and populations of aquatic birds susceptibility to influenza disease involves only minimal number of individuals, whereas the majority of representatives reveal innate immunity (Figure 4 b, c).

Influenza viruses of aquatic avian origin have also initiated the outbreaks of influenza infection in aquatic mammals, such as seals and whales [45]. In 1979 to 1980, approximately 20% of a harbor seal population died of a severe primary viral pneumonia [45]. Influenza virus H7N7 was found in high concentrations in the lungs and brains of the dead seals. However, the other 80% of the population escaped death, thanks to their own innate mechanisms of defense. In this case, innate immunity to influenza involved the majority of individuals too.

In human this virus showed the potential to cause the lightest form of influenza (conjunctivitis), but did not able to spread further to other humans. Both genetically and biologically, the virus H7N7 is closely related to those from different aquatic avian influenza virus strains. It replicated poorly in aquatic avian species and did not produce any disease signs in it. In contrast, the same virus manifested itself as highly dangerous for poultry during an outbreak in Dutch poultry in 2003. To restrict the spread of epidemics some 30 million birds have been slaughtered [XXX].

Anthrax

Anthrax most commonly occurs in cattle, sheep, goats, and other herbivores, which are infected by ingesting spore from the soil or inhaling the spores containing dust. Sheep, mice, guinea pigs, rabbits and monkeys are susceptible to experimental anthrax, while dogs, horses, cats, rats and pigs are not (Figure 5)(Table 1). Anthrax spores are registered in soil samples from throughout the world. [46] Large- anthrax epizootics in herbivores have been reported; during a 1945 outbreak in Iran, 1 million sheep died.[47] Algerian sheep are resistant to anthrax. Serious anthrax illness in humans is uncommon.[48] Naturally occurring cases of anthrax in humans is acquired following contact with anthrax-infected animals or anthrax-contaminated animal products. Humans can become infected when they are exposed to infected animals as well as to tissue from infected animals or by inhaling the spores containing dust. It is not contagious from one person to another.

Table 1. Levels of susceptibility to anthrax infection in a selection of vertebrata species [42]

Species	Grade of ecological interaction with B. anthracis	LD50 (number of spores)
White mice	1	5
Guinea pigs	2	50
Macaca rhesus	3	3,000
Rabbits	3	6,000
Rats Fisher	4	700,000
Black rats	4	1,500,000
Pigs	5	1,000,000,000
Dogs	5	10,000,000,000

An accident inside a biological weapon production factory in Sverdlovsk, U.S.S.R., caused military-grade anthrax to waft out in aerosol plume over a city. The deadly aerosol cloud that formed then proceeded from the military facility over sixty km to the southernmost suburb [49]. In the distance of four km from the factory the cloud could kill some peoples whereas in the distance between four km and sixty km the cloud could not kill human but only sheep [49], which are fare more susceptible to anthrax infections than humans, dogs, pigs and other carnivorous are [42].

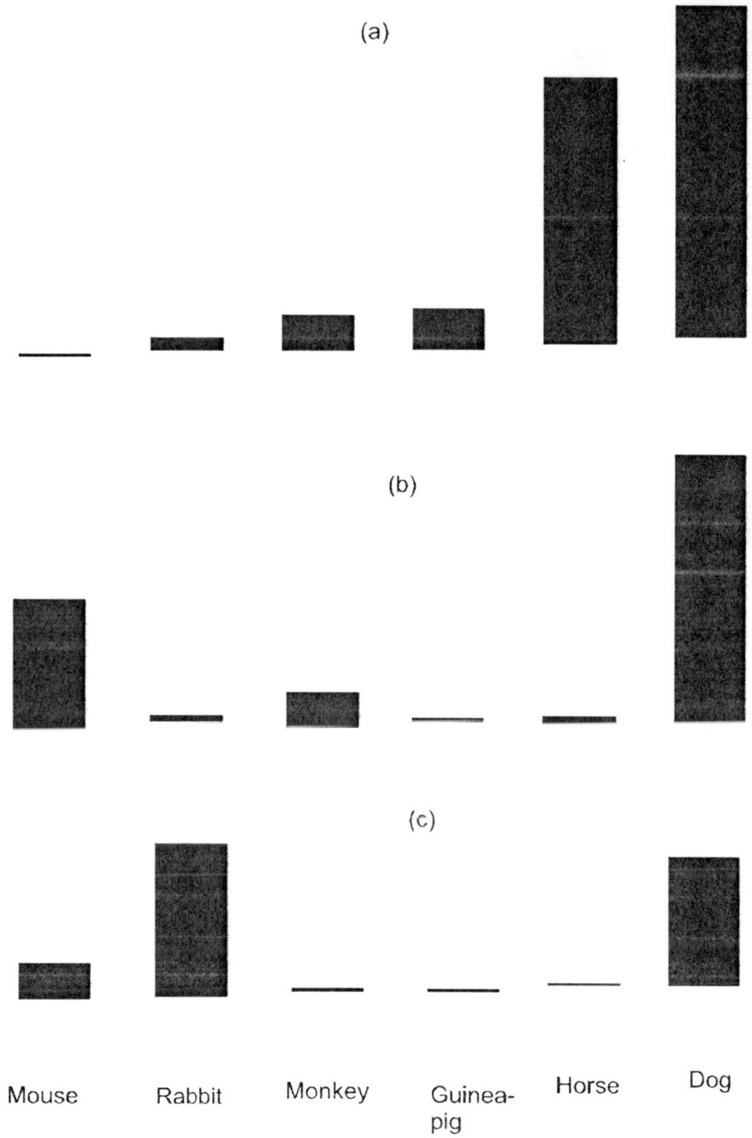

Figure 5. The variation in degree of constitutional immunity in a selection of mammalian species: a) to B. antracis (antrax), b) to Cl. botulinum toxin, c) to Cl. tetani toxin [14].

Botulism

The disease can affect the representatives of many species including fishes, birds and mammalian either in natural environment or in experiments but with different grades of incidence and prevalence. Herbivores (guinea pigs, rabbits and horses) are highly sensitive to botulinic toxins of all types for which tests have been performed. Of the carnivores, dogs and cats are very immune to all types of botulinic toxin. The Rhesus monkeys (Macaca mulatta) are resistant to type D botulism but sensitive to the other types. Mice possess constitutional immunity to type B toxin but are highly sensitive to other types of botulism (Figure 5) (Table 2). Human susceptibility to botulism is analogous to those of the dog [3]).

Table 2. Levels of sensitiviy to botulinus toxins types A, B, C, D, E, and F in a selection of vertebrata species [42]

Animal species	Minimal lethal doses of toxins					
	A	B	C	D	E	F
Herbivorous animals						
Guinea pig	1.0	2.0	1.6	0.7	0.8	1.2
Rabbit	1.0	0.5	o.6	1.5	1.2	1.0
Horse	0.4	-	2.0	-	0,6	-
Omnivorous animals						
Mouse	5.0	2,500	5.0	20.0	1.2	2.0
Macaca rhesus	4.0	36.0	2.0	2,000	4.0	0.5
Dove	200.	100.0	20.0	2,000	25.0	-
Carnivorous animals						
Mink	10.0	>200,000	10.0	-	>250.	-
Dog	500.	150,000	5,000	>800,000	125.0	-
Cat	500.	250,000	4,300	300,000	500,	-
Hen	10.	20.	2,000	100,000	25.	-

Tetanus (Lockjaw)

Donkeys, horses, monkeys and guinea pigs are very susceptible to tetanus, whereas hedgehogs, rabbits, dogs, cats and hens posses constitutional immunity to this toxin. At the middle of the range of sensitivity to tetanus are goats, sheep, rats and mice [3]. The minimum human lethal dose of tetanus toxin estimated on experiments with guinea pigs is 2.5 nanograms per kilogram of body weight or 175 nanogram for A 75-kg human.

Diphtheria

Rats and mice in contrast to monkeys, chickens, pigeons, rabbits and guinea pigs can be affected by diphteria toxin only after the dose is increasing 50,000-times that is deadly for guinea pigs. Both rats and mice have natural immunity to this toxin although there is no apparent evidence of specific immunoglobulins in the body of the rat and mouse [50].

Foot-and-mouth Disease

Horses, dogs, cats, birds and humans in contrast to cows and more than fifty other mammalian species have inherited resistance to foot-and-mouth disease [4]. Between 1956 to 1970 foot-and-mouth disease affected tens of millions of domestic animals in a region of Soviet Union. More than 200,000 people were involved in the care of these sick animals but only 21 light cases of foot-and-mouth disease were recorded in humans. Thus only one in 10,000 of the exposed human population was infected with the virus; for the remainder this virus was harmless and there was no foot-and-mouth virus antibody in these people [51].

Anthrax

Malaria

This disease is transmitted between humans and some animals by mosquitoes The malarial parasite, Plasmodium knowlesi, is highly pathogenic for the Rhesus monkey, but induces only a chronic or much less lethal infection in Macaca fascicularis [52].

Infection with H.pylori

Helicobacter pylori is present only in humans and has adapted to the stomach environment. Around the world, millions of people have suffered from chronic stomach ailments. For generations, medical science taught that these painful and often disabling conditions were caused bynon-infectious causes first of all by emotional stress and dietary factors. It is now firmly established that the diseases of this kind are of infectious origin. Which non-infectious disease will be next? Barry J. Marshall and J.Robin Warren, who challenged prevailing dogmas and first described the association between *Helicobacter pylori* and gastritis have been awarded the Nobel Prize in physiology or medicine 2005 for their discovery of the bacterium *Helicobacter pylori* and its role in gastritis and peptic ulcer disease". Robin Warren observed small curved bacteria at the stomach antrum in about 50% of patients. He made the crucial observation that signs of affection were always present in the gastric mucosa close to where the bacteria were seen. Barry Marshall became interested in Warren's findings and together they initiated a study of biopsies from 100 patients. After several attempts, Marshall succeeded in cultivating *Helicobacter pylori* from several of biopsies. Together they found that the organism was present in almost all patients with gastric inflammation, duodenal ulcer or gastric ulcer. Dr. Marshall took the courageous step of experimenting on himself, deliberately infecting himself with helicobacter to prove it was the cause of these common ailments. He confirmed this infectious agent damages gastric tissue leading to gastric ailments when introduced intragastrically [53]. Based on these results, they proposed that *Helicobacter pylori* is involved in the aetiology of these group of diseases.

Inter-population Diversity in the Incidence of Infections

Natural resistance as well as very high susceptibility to this or that infectious agent is not adequate in all representatives of a species. Some stocks, populations, subpopulations, families and individuals are naturally immune . For instance, the Immunity/susceptibility of potato to the late blight (phytophtorose) varies from one stock to another. All of animal species that have been investigated before have been found resistant to meningococcal infection.

Influenza

The incidence of influenza in humans revealed the tendency to change in time and from population to population. Spanish influenza, the first of known influenza pandemics has demonstrated thisvariations very clear as well as two last of three known human pandemics of influenza followed in 1957 ("Asian influenza" H2N2), and in 1968 ("Hong Kong influenza" H3N2). Both of the pandemics also spread rapidly around the world but caused substantially less morbidity and mortality. Nevertheless, the virus did not lose its virulence: after the virus entered in 1973 a virgin population of Yamamoto Indians in Southern Venezuela the arisen epidemics killed 27% of inhabitants [54]. The current annual influenza A epidemic takes 2000-fold fewer victims per year than in the past. Nevertheless, the virus did not lose its ability to kill humans,: According to a recent estimate of WHO, 0,008% of the current worldwide population is dying of influenza per year [{Stohr, 2002 495 /id}].

Since the Hong Kong influenza pandemic in 1968, i.e., for nearly forty years, the world has anticipated the gravest possible danger of another devastating influenza attack [43, 44]. As we move away in time from the previous pandemics, everything seems to indicate that we're on the edge of the next one. Although the next influenza pandemic is considered to be inevitable, nobody can calculate when and which kind of influenza virus will be able to induce it and how many people will be diseased and killed.

Salmonella Infection

Reported mortality associated with Salmonella infection varies between different ethnic populations. For instance, the incidence of mortality due to S. typhi infection in Indonesia and New Guinea is higher than in other countries in southeast Asia [56, 57], although all S. typhi isolates from different areas of the world can be classified as clones that have spread from a single source [58].

Recent report on an inherent basis for genetic immunity to Salmonella infection in animals contained information on a number of different lines of chickens that have been shown to be either resistant or susceptible to systemic salmonellosis caused by S. enterica serovar Gallinarum in adult birds, or by S. enterica serovar Enteritidis and S. enterica serovar Typhimurium in young chickens. Immune lines show only a moderate pathology and low mortality rates, whereas susceptible lines display extensive pathological changes and higher levels of mortality following Salmonella infection. Genetic immunity to the salmonelloses in chickens was dominant and not linked to sex, MHC, or Slc11a1 (formerly known as Nramp1), which leads to resistance in mice and other species. A novel locus-encoding immunity to salmonellosis has been identified on chicken chromosome 5 and designated SAL1. [59].

Plague

The rodents , Meriones meridianus, which live on the left bank of the lower river Volga are 70,000-fold more resistant to the plague than representatives of the same species which live on the right bank of the same river [4].The Black Death, which spread across Europe between December 1347 and December 1350, was perhaps the most famous disease outbreak in the history of mankind. It reduced various European populations by between 10-75 percent. In 1347, bubonic plague entered and devastated: England. More than 50 percent of the four million people, who then formed the population of England, died. Nevertheless, although in some cities, the plague carried off more than 75 percent of the inhabitants, in London less than 10 percent of the population died from this infection [60]. As dramatically as the plague entered, it disappeared. Not vaccination or antibiotics were involved.

Smallpox

An extensive range in susceptibility to smallpox has been revealed between populations through the effects of variolation. Variolation is the procedure of artificial, deliberate infection of healthy persons with the smallpox virus (as by intracutaneous inoculation with virus containing pus from a person recovering from a mild type of smallpox). It was practiced worldwide up to the end of 19th century as "prophylaxis" against the severe form of smallpox. Most European recipients of the variolation procedure became ill with very light form of smallpox. However, some died of black smallpox, the heaviest form of the disease. In 1837, thirty healthy Native American young women were intentionally infected with the smallpox according to instructions for varioltion. The inoculators apparently hoped to immunize the women with pus from a patient recovering from a mild type of smallpox. All the women died of the heaviest form of the disease [61].

Europeans brought smallpox infectious agents in their bodies and belongings, thus inducing unintentional biological "attacks" against Native Americans. The "attacks" were very effective, even though they were performed without any special bioweapon devices. In contrast, the same infection was not as dangerous for Europeans' population as for Native Americans [61], Within a period of several years, during the latter half of the 17th century, a maximum of 3,000 smallpox deaths (1 percent of the 3O0,000-person population) occurred in London.[62]. The rest of the population (297,000, i.e.. 99 percent escaped death without vaccination or medicines [42] (Figure 6).

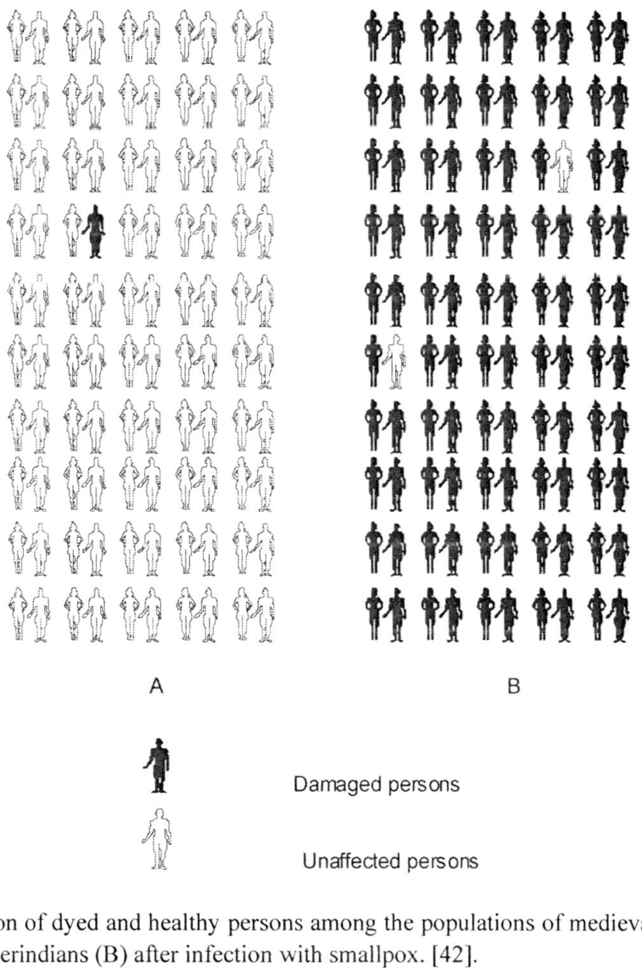

Figure 6. Correlation of dyed and healthy persons among the populations of medieval London (A) and XIX Century's Amerindians (B) after infection with smallpox. [42].

HIV/AIDS

At the end of 2005 up to 40,3 million people, i.e. nearly 0,6 % of current Earth population were living with HIV. During 2005, 3,1 million people (0,048% of the world population) died of HIV-related illnesses, and 4,9 million people (nearly 0,075 %) were newly infected with HIV. Epidemic of HIV infection exists everywhere in the World but with different levels of intensity (Table 3).

Table 3. HIV prevalence by region in 2005 [234]

Region	Adult prevalence rate (%)
Sub-Saharan Africa	7.2
Caribbean	1.6
East. Europe/Centr. Asia	0.9*
North America	0.7
South/South-East Asia	0.7
Latin America	0.6
Oceania	0.5
Western/Central Europe	0.3
North Africa/Middle East	0.2
East Asia	0.1
Global	0,6

* According to UNAIDS, *AIDS Epidemic Update,* December 2005.

Sub-Saharan Africa is the worst-affected region. It has just over 10% of the world's population, but is home to more than 60% of all people and more than three quarters (77%) of all women living with HIV. Adult prevalence in this region consists of 7,2 %. [234]. In contrast, the populations of North Africa, Middle East, Western and Central Europe are characterized by the lowest levels of the HIV-epidemics intensity. The number of registered people living with HIV continues to rise. The steepest increases have occurred in Eastern Europe (Russia), Central Asia, and East Asia but in several countries it began to decline.

The epidemic is not homogeneous within any region with some countries fare more afflicted than others. Thus, in the Caribbean, the second-most affected region in the world (1,6 %), national adult HIV prevalence reveals substantial differences in the extent and intensity of its epidemics. HIV prevalence surpasses 1% in Barbados, Dominican Republic, Jamaica and Suriname, 2% in the Bahamas, Guyana and Trinidad and Tobago, and exceeds 3% in Haiti. In Cuba, on the other hand, prevalence is yet to reach 0.2%.

In Latin America (index of prevalence 0,6 %), the highest HIV prevalence, however, is found in the countries of Belize, Guatemala and Honduras where approximately 1% of adults or more are infected. In contrast, among population of Chile, Ecuador and Bolivia the prevalence of HIV infection is between 0,1-0,5 %. Even at the country level there are wide variations in the incidence of infection between different local populations.

Malaria

In contrast to many native members of the Euro-Asian population, most West Africans and many Afro-Americans are inherently immune to infection by Plasmodium vivax. Within the period of European conquest, much of coastal Central America harbored large native Indian populations. European slaves from Africa brought malaria with them across the Atlantic Ocean. Once malaria was introduced, Indians could no longer live there. They were extraordinary susceptible to malaria. As a result, by the beginning of the twentieth century, these areas were inhabitable only by Africans because they have genetic immunity to malaria. Thus, subsequent infection changed Central American' population demographics [63].'

H. Pylori Infection

Helicobacter pylori colonizes the stomach in about 50% of all humans. In developed countries infection is considerably less common than in developing countries where virtually everyone may be infected. The overall prevalence of H. pylori was 76%, but the prevalence in the remote rural community was 91%, compared with 60% in the urban community. The odds of having H. pylori were six times greater for rural than for urban participants (odds ratio [OR], 6.34; 95% CI, 3.89-10.33). Further, the overall odds of H. pylori infection in males (rural and urban combined) were greater than for females (OR, 1.61; 95% CI, 1.02-2.54). In both communities, the prevalence of infection remained relatively constant after the age of 10. The prevalence of H. pylori in the two Indigenous communities was two to three times higher than that in the non-Indigenous Australian population and higher than that shown in previous studies in Indigenous Australians [64]. Mice lacking the Prptz gene are immune to gastric ulcer induction by Helicobacter pylori [65].

Inter-individual Diversity in the Incidence of Infections

Meningococcal Infection

There is massive epidemiological evidence, that different individuals have different susceptibility to meningococcal infection. In most humans meningococci can not been found even on the surface of nasopharyngeal mucosa. But the same infectious agents can be carried as harmless commensals in the nasopharynx of about 1% of the population. In some individuals, namely in only a small proportion of the carriers, however, the same microbe causes devastating and fatal illness [66].

Influenza

The Spanish influenza H1N1 pandemic (1918–1919) annihilated between 20 to 40 million people [67], i.e., nearly 1-2% of the worldwide population existing at that time [42]. The other 98% escaped death without vaccination or specific medication (Figure 7) but through their own innate mechanisms of self-defense. Thus, the 1918 infecting agent was extremely pathogenic but only for a minority of the human world population, whereas innate immunity to influenza involved, in this case, the majority of individuals. The deadly risk from Spanish flu for most people can be assessed as very low.

Figure 7. The Spanish influenza H1N1 pandemic (1918–1919) has annihilated near 2% of current worldwide population [148].

We meet here extraordinary individual diversity in natural susceptibility of humans to influenza infection. Like any other human infectious agent [3, 4, 42], the influenza virus can't cause illness in most members of the current world population. Some organisms attacked by influenza viruses fall ill, whereas most individuals display natural immunity even to primary infection. Only in a minimal proportion of current human population are the same microbes able to cause fatal illness. That is why any kind of influenza virus cannot spread from a diseased person to any other person, but only to a susceptible one.

Anthrax

Four consequences of exposition to anthrax infection occur in humans. Some healthy people were reported to carry hundreds of anthrax spores in their noses and throats without incurring any disease [68]. For reasons that are unclear for many experts, these incidents of anthrax contamination failed to produce illness. Serious anthrax illness in humans is uncommon [48]. The most common naturally occurring form of anthrax infection has mainly cutaneous (skin) manifestations, with an estimated 2,000 cases reported annually around the World [69]. This form is nasty but secure. In the United States, 224 cases of cutaneous anthrax were reported between 1944 and 1994, i.e. nearly 4.3 cases yearly. The largest reported epidemic occurred in Zimbabwe between 1979 and 1985, when more than 10,000 human cases of anthrax were reported, nearly all of them cutaneous [70] .

In contrast to analogous infection of herbivorous animals, the naturally occurring inhalational anthrax is a rare cause of human disease. Only eighteen cases of inhaled anthrax were reported in the United States from 1900 to 1978, with the majority occurring in special risk groups, including goat hair mill or goatskin workers and wool or tannery workers. Historically, wool sorters at industrial mills were at highest risk of inhalational anthrax. Two of the eighteen cases were laboratory associated [69].

Gastrointestinal, or swallowing, anthrax follows ingestion of insufficiently cooked contaminated meat and produces a more serious intestinal disease which is frequently fatal. This form is uncommonly reported. [69, 71]. However, gastrointestlnal outbreaks have been reported in America and Asia [72]. In contrast to many other infections, a human patient with any form of anthrax cannot infect someone else.

In an April morning of 1979, an accident inside a biological weapons production factory in Sverdlovsk, U.S.S.R., caused military-grade anthrax to waft out in aerosol plume over a city of 1.2 million [73]. The source.of the aerosol was at the military microbiology facility located in the southern area of the city. On the day of the accident, an operator in the facility switched on a ventilation system without switching on the filtration system and thus allowed the military-grade anthrax aerosol to rush out of the work box [74]. There was a northerly wind blowing from the facility to the southern city limit. The aerosol cloud that formed then proceeded from the military facility over sixty km to the southernmost suburb [49].

Most of the fatally affected people lived or worked in a zone up to one km wide and four km long, extending from the source of the aerosol to the southern city limit. Experimental evidence suggests that following an outdoor aerosol release, persons indoors could be exposed to a similar threat as those outdoors.Approximately 7,000 people lived in the area of doubtless deadly concentration, precisely marked by the daytime location of people who contracted the anthrax disease. Inside the zone, the effect resulted in a total of approximately 68 fatalities from anthrax and eleven survivors of a light form of the disease [49]. Thus, the accident demonstrated the doubtless lethal potential of anthrax aerosols on 1 percent of exposed peoples (Figure 2). Thus, near only 1 percent of the population of 7,000 people contracted the anthrax :infection as the result of the accident. The rest of exposed population (approximately 99%) could counteract disease by means of their constitutional features.

HIV/AIDS

HIV - infection can affect and destroy a variety of human cells, tissues and organ. The range of its affection can be individually variable inducing a variety of clinical forms. These forms can either stop spontaneously or continue its successive development. The possibility and rate of clinical disease progression varies widely between individuals and has been shown to be affected by many factors first of all by the degree of victim genetic susceptibility [38]. Four of these forms (the asymptomatic one, the primary HIV infection, clinical latency and AIDS i.e. acquired immunodeficiency syndrome) are considered as basic ones.

The primary HIV infection is often not recognized because most afflicted individuals (80 to 90%) develop a nonspecific acute syndrome characterized by flu-like symptoms of fever, malaise, headache, myalgia, lymphadenopathy or pharyngitis, and sometimes a rash [75]. This form of HIV infection is often misdiagnosed as one of the more common infectious diseases with the same symptoms. Within an average of three weeks after transmission of HIV-1, a broad HIV-1 specific response occurs that displays seroconversion. The developed production of specific antibodies can reduce the number of viral particles in the blood stream, marking the start of the infection's clinical latency stage. The duration of clinical latency can vary individually between two weeks and 20 years.

The most severe syndrome of HIV infection has been named as AIDS. Acute HIV infection progresses over time to clinical latent HIV infection and then to early symptomatic HIV infection and later, to AIDS, which is identified on the basis of certain opportunistic infections that are common in people with AIDS. Fortunately, not all persons infected with HIV have progressed to AIDS.

Salmonella Infection

Like any infectious agent, Salmonella cannot cause illness in all members of an observed species or population. Salmonella can be carried as harmless commensals in some sectors of the population. In some individuals, however, the same microbes cause illness while others display constitutional immunity to primary Salmonella infection. Volunteer challenge studies have established individual variations of minimal infective dose for S. typhi from 100,000 to 1,000,000,000 microbial cells (Table 4) [76]. According to a WHO estimate, annually there are more than 17,000,000 Salmonella infection cases, associated with about 600,000 deaths [77], i.e. 0,01 % of the world human population.

Table 4. Artificial infection of humans with abdominal typhoid [76]

Dose		Effect	
Number of microbes	Number of injurious doses	Number of unaffected persons (%)	Number of injured persons (%)
1,000	0,01	100	0
100,000	1	72	28
10,000,000	100	50	50
10,000,000,000	10,000	5	95

Thus the differences in host immunity or susceptibility to infectious disease as well as the disease severity cannot be attributed solely to the virulence of infectious agents. Variations in response to infection, often associated with constitutional features of a body make-up and polymorphism in the genome, should now be took in account at the first line.

In most current human populations attacked by any infection, individuals can be conveniently divided into at least four categories [78]. Any of the categories may involve either a few individuals or the majority of a population or species under consideration. As in any infection, in the case of Spanish influenza, for instance, the categories could be presented as follows: i) homozygous, totally resistant organisms which contain no susceptible structures and cannot be damaged (up to 80%) ; ii) heterozygous, mildly susceptible organisms in which a few susceptible foci appear and the infection runs a benign or asymptomatic course; iii) heterozygous organisms in which the number of constitutionally susceptible structures is high, and the infectious process develops in a severe form, and iv) homozygous, absolutely defenseless individuals with death occurring within 1 week (near 1-2%).

Age Diversity in the Incidence of Infections

A general characteristic of many infections is the changed resistance to disease that develops in the course of growth and maturation of the possible victim. The changes are particularly dramatic in neonatal period. Most infections, for instance influenza, can be either very severe, middle or light both in neonatal and postnatal period. Meningococcal infection and anthrax present two of the rare exceptions to this common law.

Meningococcal Infection

Under one month children are resistant to meningococcal infection [79, 80] in contrast, for instance, to influenza infection that is not uncommon at the same age [81]. In healthy children under 2 years N.meningitidis could never be found on the surface of nasopharyngeal mucosa. Beginning from the third year of life, the situation begin to change, and in the nasopharynx of young adults meningococci are found far more often. But after the fourth decade the rate of N.meningitidis carriers decrease [82]. About 15 per cent of all cases of childhood meningitis are due to the meningococcus [80], whereas children under one month very seldom are known to fall ill with meningococcal infection. In a large series of 211 cases of neonatal meningitis, N.meningitidis was an uncommon organism. Only two of 211 were due to N.meningitidis [79].

Anthrax

Analogous paradoxical increasing of individual constitutional susceptibility with age has been revealed by anthrax infection. All of 68 citizens have been killed during Sverdlovsk' accident were over twenty-three years old [42].

Tick Born Encephalitis

The incidence of tick born encephalitis (TBE) is clearly higher in the elderly people than in younger age groups. In Germany, Austria, Sweden and Lithuania more than 50% of TBE cases affected the age group over 50 years [83].

Respiratory Syncytial Virus Infection

Infant ferrets are susceptible to respiratory syncytial virus infection in both the upper and lower respiratory tracts. In the nose, viral replication is restricted to the surface respiratory epithelium. In the lungs, viral replication is of a lower order of magnitude and is localized in the alveolar cells. The pattern of viral replication in nasal tissues is independent of the age of animal at infection, whereas the pattern in lung tissues shows a striking age dependence, with viral replication progressively decreasing as a function of age. The virus population is reproduced in cells of respiratory epithelium, and in lungs it is located in alveolar cells. At adults ferrets replication of viruses in lungs do not happen any more while in f the top respiratory ways it is carried out irrespective of age. At humans the given infection amazes alveolar cells of lungs also only in the childhood whereas infection of the top respiratory ways could happen at all age groups [84].

Malaria

The relative incidence of malaria caused by the two dominant species (P. falciparum and P. vivax) varied with age: P. vivax predominated in the early years of life and was not superseded by P. falciparum until after the age of 2 years [85].Where P.falciparum is holoendemic, high parasite rates and densities are observed in children between the ages about six month and four years. About 10% of children are dying directly or indirectly from malaria. But the majority of children survive to school age, and by that time parasite rates and counts have fallen.

Newborn and three week old mice are absolutely defenseless with regard to the virus of yellow fever, but by adulthood they become completely resistant [86]. Newborn mice are very sensitive to herpes virus, but by 4 weeks of age this sensitivity is replaced by constitutional resistance [87]. Similar age related changes are observed in resistance to mengo virus and encephalomyocarditis. Coxsackie virus B type 4 caused a generalized infection, affecting several organs in newborn mice. In contrast, in adult mice only one target organ, i.e. the pancreas is permanently affected [88]. Older snakes of Bothrops atrox species appeared also particularly susceptible to an viral respiratory epizootic that swept a snake farm in Zurich, Switzerland [89]...

Infections generally change in frequency and severity with age and this can not be explained only by the decline in efficiency of the lymphoid system immunogenic response. The increase in susceptibility of the elderly to certain infectious diseases can be explained by the changes in molecular constitution of organs, tissues, cell clones, cells and extracellular structures, that occur with age. There are developmental changes in the molecular constitution of cells which are correlated with developmental differences in their physiological and ecological characteristic, and this may cause the age-related differences in susceptibility to relevant agents.

Cell level of Diversity in the Incidence of Infections

Most infectious agents are specialized to survive as intracellular parasies and their attacks involve at least three sequential steps: (1) Adhesion of microbe to plasma cell membrane; (2) translocation of the microbe into cell plasma; (3) reproduction of the microbe in the cytoplasma of the cell. Only microbes being capable to adhere to cell' outer membrane are capable of penetrating the membrane and introducing themselves into cytoplasma of the cell [8]. Adhesion of intracellular parasites on the victim cell's surface is a starting, i.e. key step in the development of relevant infectious process.

Virus Infection in Microbes

Individual cells can be either sensitive or resistant to the effects of the one or another infectious agent. The phenomenon of cellular resistance to viruses has been studied, for example, in bacteria that have been attacked by specific viruses (bacteriophages). Under these circumstances, some bacteria are resistant, survive and multiply in conditions which kill others of the same strain. Progeny of resistant bacteria are also resistant. Bacteria possess some means of defense against viral attack. The bacterial virus, for its part, has evolved means of defense against bacterial immunogenic agents.

Infection in Plants

Cells of both susceptible and immune maize plants which have been cultured in vitro bear the same relationship to pathogenic organisms as their natural counterparts [90]. Animal cells in culture also exhibit either immunity or susceptibility to microbes and their products. This relationship is quite specific - one and the same cell can be resistant to one strain of microbe and also sensitive to another [17].

Meningococcal Infection

A crucial property of the pathogenicity of meningococi is their ability to adhere to human target cells [91]. The potential of N.meningitidis to cause systemic disease is dependent on its ability to attach to relevant tissues and cells. Most N.meningitidis strains express pili responsible for interaction with and adherence to epithelial and mesenchymal cells, especially to endothelial and blood cells [91-93]. The characteristics of adherence which allow meningococcus to reside on cell surfaces are of interest because these microbes have been associated with very heavy disease states.

Expressive interspecies and individual variations of cell adherence of meningococci serogroups A, B and C have been found by testing the cells of 5517 humans and 120 individuals of 11 animal species (8 mice, 8 guinea pigs, 5 rats, 1 Syrian hamster, 24 rabbits, 17 goats, 17 sheep, 17 donkeys, 5 horses, 3 cows, 14 hens. Most investigations have been performed with individual specimens of erythrocytes. Leucocytes and buccal epithelial cells were also used in some experiments. Erythrocytes have been chosen because clue step of meningococcal infection, i.e. intrusion of infectious agents into mesenchymal cells is performed by specific adhesin (haemagglutinin), which is responsible for virions' adhesive binding to victim's cell receptors [94].

The adhesive interaction of N.meningitidis with cells of all of animals tested did not occur. All the applied concentrations of meningococcal adhesins of the three serogroups failed to adhere to the cells of all of the animals used for the proof. All samples of animal cells were extremely resistant to adhesion of N.meningitidis serogroups A, B and C (index of adhesion < 0,06).

In human, the cell susceptibility to meningococcal adhesion was examined in: 103 new-born children until 7 days age; 20 children at the age of 2-4 weeks; 19 children at the age of 2-3 months; 23 children at the age of 4-12 months; 44 children at the age of 2-3 years; 5286 adults at the age of 18-60 years; 11 persons during pregnancy (17-36 weeks) and 11 persons during childbirth.

Cells of the adult people, persons in pregnancy and in childbirth included, displayed individually variable adhesion of meningococci. The individual ability of 137 surveyed adult men and wumen to adhesion of meningococci was characterized by indices of adhesion from 0,12 up to 2,0, being distinguished more than in 16 times. In a large group (5149 young men) tested for adhesion of N.meningitidis B the indices of adhesion varied from 0,12 up to 16,0 being distinguished more than in 128 times. Meanwhile the adhesin treatment of some human' erythrocytes resulted in agglutination only when the adhesin concentration was more than 128 times that used for erythrocytes of other individuals. Among the pregnant women and women during childbirth various degree of sensitivity also had come to light.

The revealed range of individual distinctions demonstrated impressive diversity of adults in sensitivity to adhesion of meningococci. Some adults displayed sensitivity to all three adhesins tested (A, B and C), whereas others were sensitive to one or two of them. The variations of individual susceptibility to meningococcal adhesion was also observed by testing of human leukocytes and epitheliocytes [15]. Individual ability of adult to adhesion of meningococci was stable over period of 5 years. Cells of the four human blood types (ABO) failed to reveal significant differences. Examination of the blood plasma did not reveal any humoral antiadhesive capacity.

The adhesion of meningococci on the new-born children's cells was in contrast with thouse of adults (Table 5). Compared cells of older individuals, the erythrocytes of new-born children until 7 days age were extremely resistant to adhesion of N.meningitidis serogroups A, B and C (index of adhesion < 0,06). All the concentrations of meningococcal adhesins failed to agglutinate the erythrocytes of all new-borns tested. None of 103 new-born children had an index of adhesion equal to or more than 0,06. Plasma from new-born children as well as from pregnant and other adult persons did not reveal any antiadhesive humoral factors in them. Adhesion of N.meningitidis to human red blood cells appears to be age dependent.

Table 5 .Ontogenesis of children erythrocytes sensitivity to adhesion of *N. meningitidis* groups A,B and C [94]

Menin-gococ-ci	The num-ber of in-dividuals tested	Incidence of indices of adhesion (%)						Range of individu -al distinc-tions
		0,06	0,12	0,25	0,5	1,0	2,0	
1st week								
A	103	0	0	0	0	0	0	-
B	103	0	0	0	0	0	0	-
C	103	0	0	0	0	0	0	-
2nd week								
A	6	83	17	0	0	0	0	2
B	6	0	0	0	0	0	0	-
C	6	0	0	0	0	0	0	-
3rd week								
A	5	80	0	20	0	0	0	4
B	5	40	60	0	0	0	0	2
C	5	80	20	0	0	0	0	2
4th week								
A	9	56	11	22	11	0	0	8
B	9	22	56	0	22	0	0	8
C	9	78	11	11	0	0	0	4
2nd-3rd month								
A	19	32	42	26	0	0	0	4
B	19	47	21	21	11	0	0	8
C	19	74	21	5	0	0	0	4
4tth-12th months								
A	23	4	48	14	17	17	0	16
B	23	13	39	26	13	9	0	16
C	23	26	56	0	4	14	0	16
2nd-3rd years								
A	44	7	32	34	20	7	0	16
B	44	14	50	25	11	0	0	16
C	44	9	43	30	14	4	0	16

The tests of the new-born children's cells by influenza viruses were not in agreement with above observations. The individual cell susceptibility to adhesion of influenza viruses type A and B was examined in 20 new-born children and 29 adults at the age of 20-60 years.

Erythrocytes of all new-born children and adults were sensitive to adheson of influenza A and B viruses but in different extent. Like analogous cell of adult peoples erythrocytes of children had individually variable sensitivity to agglutination by influenza viruses just at the moment of birth. The indices of adhesive ability of the adults and children did not in essence differ from each other. In the two age groups tested , both high and low individual sensitivity to viral adhesion was revealed. Thus, the adhesion of influenza viruses to human red blood cells is age independent.

The minimal susceptibility of some children to adhesion of meningococci and its individual variability appeared only after the 1st week of life when children's erythrocytes did not respond to maximum concentration of adhesin A only in 1 of 6 individual tested. The range of population diversity was minimal (Table 5; range of individual distinctions: 2). Meanwhile all the concentrations of meningococcal adhesins of serogroups B and C failed to adhere on the erythrocytes of all children tested during second week of their life; individual differences were not revealed (Table 5, index of adhesion: <0,06).

During the third week of life erythrocytes of all children were agglutinated by the maximum concentration (16 units) of adhesins A,B, and C: moreover 20% of them responded also to a concentration of 4 units of ahesin A (index of adhesion: 0,25), 60% to 8 units of adhesin B (index of adhesion: 0,12), and 20% to 8 units of adhesin C (Table 5). The range of variability in the third week of life consists of 4 for adhesin A and 2 for adhesins B and C.

There was a reduction in the proportion with low indices of adhesion and an increase in the proportion of persons with higher parameters of adhesiveness observed in the subsequent development. The indices of adhesive ability of the children increased sharply, especially during the first three months of life. Thus the susceptibility of erythrocytes to the adhesion of meningococci serogroups A, B and C at first (up to 3 month) sharply and then slowly increases (Table 5).

During the second to fourth weeks after birth, the interaction with the same adhesins showed a decrease in the lower indices of adhesion and corresponding increase in the frequency of higher levels. In all other age groups both high and low indices of individual sensitivity to viral adhesion were revealed. In the period of 1-3 months of age the cells of most children have reached a level of adhesion equal 0,12- 0,25. None of the samples from the 1-3 months age group showed the indices of adhesion equal 1 and 2. The range of individual variation reached the value of 16 only after three months of life. The erythrocytes of some children were agglutinated by every of the adhesins tested (A, B and C); others were sensitive to one or two.

In the 4-12 months age the range of distinctions consisted of more than 16 times, and 47 % of the samples had the indices of adhesion equal 0,25-1,0. The maximum index of adhesion (2 or more) was found out only in adults (18-60 years). Amongst 209 children examined during their first three years (Table 5), there were no individuals with a high index of adhesion (2,0). Among 137 adults, there were 11 (8 %) samples of cells with high indices of sensitivity. In another adult group of 5149 young men 47 % demonstrated indices of adhesion of 2 or more. Thus during the postnatal period of ontogenesis the cells become either high or low in individual sensitivity to meningococcal adhesion

Thus, of twelve cell species tested, only human cells were able to adhere N.meningitidis, and the cells of various individuals reacted differently. In older age groups the meningococcal adhesion had the place, but among some individuals only. The range of individual diversity increased from absolute absence in one week old children up to maximal values (128 times) in adults. The maximum adhesion of N. meningitidis occurred only with cells of some but not all of individuals tested. The adhesion of meningococci serogroups A, B and C to human red blood cells is individual and age dependent. The number of human cell specimens adhered by the meningococci increased during postnatal ontogenesis and then was comparatively stable.

In view of the above data it seems to be not a surprise that cells from different persons showed impressive differences in their capacity to adhere N.meningitidis. We can therefore hypothesise that resistance of cells from some individuals to adhesion of meningococci may correlate with their individual resistance to N.meningitidis infection. The discussed form of human cells individuality can explain why meningococcal disease develops in only a small proportion of individuals contracted the causative bacteria [94].

The lack of animal cells to adhere meningococci correlates with natural immunity of animals to N.meningitidis infection. Evidence of the weak adhesive interaction of N.meningitidis with the newborn cells correlates with low level of meningococcal infection among newborn infants and children under one month of age. Lack of meningococcal adhesion among animal species in which the disease does not occur indicates this might be a marker of natural immunity to this pathogen.

Thus evidence that immunity of cells to adhesion of meningococci correlate with immunity of hole body to infection is also provided by the aggregate of facts that all organism categories known to be resistant to natural infection consists of cells insensitive to adhesion. In contrast, the organisms known to be susceptible to natural infection contain susceptible cells. At this stage of investigation evidence for the correlation between of the tested organism adhesive potency and its resistance to meningococcal infection are provided by the fact that all species and human age categories considered to be resistant to infection are resistant to adhesion too. The next step should be realised by investigation of persons affected by various forms of meningococcal infection [94]..

From these data one could infer that specific function of meningococcal adhesion receptors do not exist in erythrocyte plasma membrane of resistant animals and children under one month. On the other hand, these receptors may present either in various amounts or have a various activity or a different location or another structure in the erythrocyte plasma membranes of some children over one month and some adults [94].

Influenza

In contrast, the cells of newborns as well as those of adults can adhere influenza viruses types A and B although with individually different intensity. In the line with these results, it is well known that children under one month can fall ill with influenza infection not very seldom. These results could be interpreted as meaning that influenza virus receptors are present at the human cell at the moment of birth whereas the receptors of N.meningitidis adhesion develop later in the course of ontogenesis than those of influenza viruses [94].

Species and individual variations have been revealed also in relation to various cells' (erythrocytes, leucocytes and tracheal epithelium) adhesion by influenza virus types A (H0N1, H1N1, H3N2), B and C. Most experiments have been performed with a quantitative test for virus adhesion on the erythrocytes of a selection of individual representatives of 12 vertebrata species (Leghorn chickens, dogs, guinea pigs, mice, Syrian hamsters, horses, donkeys, pigs, cats, goats, sheep, rabbits) and in the groups of 40 new born children, 100 healthy adult humans of 20-60 years old and 29 aged men and women of 60-80 years old [95]. Testing has been performed at the end of 70th of XX Century. Erythrocytes have been chosen because influenza infection's clue step, i.e. intrusion of virions into mesenchymal cells is performed by specific haemagglutinin, which is responsible for virions' adhesive binding to victim's cell receptors (Webster RG, 1992 562 /id). Erythrocytes represent the most convenient and widely approved model for studying microbial adhesion (Goldhar J., 1995 77 /id). The conditions of the test performance excluded the influence of immunoglobulins and other soluble factors of antiviral activity. As a result, very expressive species and individual differences in cell sensitivity to attachment of influenza viruses were found (Table 6).

Table 6. Species and individual differences in cell sensitivity to adhesion of various influenza viruses [148]

The cells tested			Indices	Intensity of their of interaction with influenza viruses				
	Species	Number of individuals		A H0N1	A H1N1	A H3N2	B	C
I	Chickens*	8	a	128	256	256	256	12
			b	64..256	128..512	128..256	64..256	0..16
			c	4	4	2	4	16
	Humans	100	a	160	128	208	32	4
			b	16..640	16..512	64..640	8..128	2..32
			c	40	32	10	16	16
	Dogs	5	a	128	128	64	128	0
			b	64..128	128	64.. 128	64..128	0..2
			c	2	1	2	2	2
II	Guinea pigs	5	a	64	64	64	64	0
			b	32..64	16..128	32.. 64	64	0
			c	2	8	2	1	1
	Mice	6	a	16	64	96	12	8
			b	8..32	32..128	64..128	2..64	8..32
			c	2	4	2	32	4
III	Syrian hamsters	7	a	2	4	64	2	-
			b	0..2	2..4	32.. 256	2..4	
			c	2	2	8	2	
	Horses	6	a	2	3	128	4	1
			c	0..4	2..8	32..256	0..64	0..2
			c	4	4	8	64	2
	Donkeys	5	a	0	2	64	64	0
			b	0	0..2	8..256	32..64	0..2
			c	1	2	32	2	2
	Pigs	3	a	2	2	256	2	-
			b	2	2	128..256	2	
			c	1	1	2	1	
IV	Cats	12	a	0	2	64	64	2
			b	0..16	0..8	32..128	16..128	2
			c	16	8	4	8	1
	Goats	5	a	0	8	64	64	2
			b	0..4	0..64	32..64	32..64	0..2
			c	4	64	2	2	2
	Sheep	5	a	0	16	64	32	0
			b	0..2	8..64	16..128	16..128	0..2
			c	2	8	8	8	2
V	Rabbits	5	a	4	4	0	2	2
			b	0..32	2..32	0..4	2..16	2
			c	32	16	4	8	1

a – median value; b – extreme variants; c – range of individual distinctions.

The highest levels of cell sensitivity to adhesion of most viruses tested, except virus type C, have been revealed in Leghorn chickens, The lowest indices of cell sensitivity to all tested viruses were found in rabbits. The highest levels of cell immunity to all viruses, except H3N2 and, partially, type B, have been revealed in other 8 species. The greatest diapason of individual differences was observed in humans (in relation to all viruses, up to 40-fold), donkeys (in relation to virus H3N2, up to 32-fold), and horses (in relation to virus B, over 64-fold). The intra-species differences in sensitivity to influenza viruses types A among Leghorn chickens were substantially less (up to 4-fold) than among humans. Human cells also revealed relatively less sensitivity to virus B, that reflects the lower epidemic potential of this virus.

Species and individual diversity in cell sensitivity to the virus' adhesion has been revealed also by tests for attachment of H3N2 virions on leucocytes and tracheal epithelium. Thus, sorption of the virus was especially intensive on chickens' traheal epithelium, but substantially lower on the same tissue of guinea pigs and, in particular, of rabbits. The ability to adsorb the virions was higher on horses' leucocytes than the same cells of sheep and goats. The existence of variation among individuals has been observed also by observation on the attachment of influenza virus H3N2 to the surface of human trachea [95].

Thus, the effectiveness of influenza virus adhesion to cells depends not only on the type and strain of virions, but also on the properties, both species and individual, of the organisms tested, and probably on the molecular ecological congruence between a virus and its victim. Because the attachment of influenza viruses to cell envelopes presents a key step of influenza pathogenesis, the results of relevant tests should be used for evaluation of individual and population immunity as well as the epidemic potential of newly emerged variants of infectious agents. The applied test may have important implications for vaccine design and the strategy of vaccination if it is used in the selection of vaccine candidate strains and the establishment of individual indications to vaccination.

HIV/AIDS

During clinical latency phase of HIV, the infection is active within lymphoid organs, where large amounts of virus become trapped in the follicular dendritic cells. The surrounding tissues that also are rich in CD4+ T-cells also become infected, and viral particles accumulate both in infected cells and as free virus [96].

Rabies

The expressive species and individual differences have been revealed also by in vitro observation of the interaction of the rabies virus with animal and avian red blood cells (Figure 8) [97]. Erythrocytes of two-day-old chicks are more sensitive to rabies virus adhesion than those of adult chickens. When analyzing the data, one can note that the lowest levels of cell susceptibility to adhesion of rabies virus have been revealed in adult chickens, rats, guinea pigs, humans, and sheep, whereas the highest susceptibility have been found in rhesus monkeys, vervet monkeys, and geese. Thus the farther a species has gone away from the ecologically determined possibility of interacting with the source of rabies infection (foxes and wolves), the higher its level of cell susceptibility to the virus, and vice versa.

Figure 8.Effect of the accident with the anthrax bioweapon on the population of Sverdlovsk's citizens [42].

Tick Born Encephalitis

Analogous ecologically understandable differences have been found during testing the animal cells for immunity/susceptibility to adhesion of Tickborne encephalitis virus (Table 7). According to prevalent areal of the ticks inhabitance (the wood) maximum sensitivity to tested virus has been found in donkeys the well known inhabitants of the steppes. Bulls, which pre-descendants lived in the woods, revealed high resistance to adhesion of discussed virus. Goats revealed intermediate features between donkeys and bulls.

Dengue

Once Dengue virus reaches its target cell, a cascade of events generally starting with the interaction of viral envelope glycoproteins with specific entry receptors and co-receptors is necessary in order to trigger the virus–cell membrane fusion. The virus enters the macrophage through a virus receptor [98].

Table 7. Cell immunity to Tick born encephalitis virus in a selection of mammalian species

Cells		Intensity of adhesion (+) at various dilutions of virus suspensions					
Species	No	1:2O	1:4O	1:8O	1:16O	1:32O	1:64O
Donkeys	1st	++++	++++	++++	++++	++++	++++
	2nd	++++	++++	++++	++++	+++	+++
	3rd	++++	++++	++++	++++	+++	+++
	4th	++++	++++	++++	+++	+++	++
Goats	1st	++	++	O	O	O	O
	2nd	++++	++	+	+	O	O
	3rd	++++	+++	+++	++	++	O
	4th	++	++	++	++	+	+
Bulls	1st	O	O	O	O	O	O
	2nd	O	O	O	O	O	O
	3rd	O	O	O	O	O	O
	4th	O	O	O	O	O	O

Gonorrhea

The ability of bacteria to adhere to cells is likely to be advantageous, especially if that initial act of adherence should lead to successful colonization of the particular ecological niche of that organism. Gonococci are able to adhere to human cell but they do not interact with the cells of sheep, guinea pigs, mice, rabbits or goats, all of which are naturally resistant to this infection [99].

Bordetella Infection

In contrast, Bordetella bronchiseptica demonstrates preferential adherence to non-human mammalian ciliated cells of rabbits, mice and hamsters. Bordetella organisms do not attach to chicken ciliated cells. But Bordetella pertussis and Bordetella parapertussis adhere better to human ciliated cells than to those from rabbits, mice, and hamsters [100]. These observations suggest that specificities of microbe attachment may explain the marked predominance of Bordetella pertussis as the cause of whooping cough in humans and of Bordetella bronchiseptica as a respiratory pathogen of many nonhuman mammals.

Diphtheria

Cultured cells of constitutionally sensitive species (guinea pig, human, dog, monkey) are destroyed by diphtheria toxin, whereas cells of hereditary resistant species (rat, mouse) remain intact [101].

Salmonella Infection

The nature of the differences in pathology found between constitutionally immune and susceptible chicken lines in vivo indicates that resistance is expressed at the level of mesenchymal cell, for instance the mononuclear phagocyte system. Macrophages from adult resistant-line birds cleared Salmonella serovar Gallinarum from infected macrophages within

24 h, whereas Salmonella bacteria persisted within macrophages from susceptible-line birds for at least 48 h [59].

Bacteria of genus Salmonella which affect some humans and animals can provoke lysis of victim cells, although this effect is not manifest in every case. Clinical isolates of five Salmonella enterica belonging to serovars of Paratyphy A (strain 50503), Paratyphi B (strain 42), Paratyphi C (strain 1212), Typhi D (strain 4446), and E (strain 337), as classified by Kauffmann [102], were investigated for its influence on the mesenchymal cells (red blood cells) of ten biolological species as well as 1,301 humans, including 150 twins and 264 individuals who either had abdominal typhoid or other salmonelloses. Cells of ten examined biological species were different in their susceptibility to lysis by different Salmonella strains of five serological groups (A, B, C, D, E). All representatives of the four herbivorous species investigated (600 mice, 33 guinea-pigs, 22 horses and 47 rabbits) possessed cells with very high sensitivity to lysis by Salmonella Table 8). Individuals with resistant cells were not seen in these groups of animals. All of the five studied cultures of Salmonella lysed erythrocytes of these four animal species (Table 8) [103].

Table 8. Incidence of constitutional immunity (o) and sensitivity (+) of erythrocytes to lysis by Salmonellae in a selection of animal species [103]

Animal species	Numbers of individuals tested	Salmonella strains				
		A	B	C	D	E
1. Herbivorous animals						
Mice	1st to 600st	+	+	+	+	+
Guinea pigs	1st to 33rd	+	+	+	+	+
Rabbits	1st to 47th	+	+	+	+	+
Horses	1st to 22nd	+	+	+	+	+
2. Omnivorous animals						
Hens	1st to 22nd	+	0	+	0	-
	5th	0	+	0	+	-
	6th	0	0	+	0	-
	7th to 17th	+	0	0	0	-
	18th to 1059th	0	0	0	0	-
Cats	1st to 2nd	0	0	0	0	0
Monkeys	1st to 2nd	0	0	0	0	0
Sheep	1st	+	+	+	+	+
	2nd	0	0	0	0	+
	3rd to 16th	0	0	0	0	0
Dogs	1st to 5th	+	+	+	+	-
	6th	+	0	+	0	-
	7th	0	+	+	+	-

As a rule, the cells of the group of omnivorous species, were not lysed by the Salmonella strains tested, i.e., i.e. they were immune to this cytolytic factor. This group included erythrocytes of cats, monkeys, dogs, sheep, chickens and humans. Only some individuals in the second group demonstrated cells sensitive to cytolysins of some or even all strains studied. This provides evidence for the specific character of determinants of cell immunity as well as of Salmonella cytolytic activity. Most representatives of an examined chicken population (1,042 out of 1,059) demonstrated resistant cells and only 1.6% (17 individuals) had weak sensitivity to lysis by some of the Salmonella strains tested. Only 2 out of 16 sheep had sensitive erythrocytes (one sheep was sensitive to two strains, and another to five ones) and the intensity of lysis varied from very low to pronounced (Table 8).

Analogous individual variability of the sensitivity of human cells was revealed [103] in representatives in urban populations of Saint Petersburg (Russia), Kishinev (Moldova), Tartu (Estonia), and Alma-Ata (Kazakhstan) (Table 9). It has been found that only some individuals of this species had erythrocytes sensitive to the cytolysins of one or more of the Salmonella strains tested (Table 10), and the sizes of the cytolysis zones varied from weak to pronounced [103, 104]. The erythrocytes of 1,037 healthy humans, including 887 healthy blood donors aged 25 to 40 years (Saint-Petersburg, Alma-Ata), pupils aged 14 to 17 years (Tartu, Kishinev), and 150 adult twins were tested. Nearly 90% of the 1,037 healthy individuals had erythrocytes resistant to lysis by one or more Salmonella strains tested. For instance, the incidence of sensitivity to cytolysin of Salmonella serovar D (S. typhi, the abdominal typhoid infecting agents) in the populations of Alma-Ata (Kazakhstan), Tartu (Estonia), Kishinev (Moldova), and Saint Petersburg (Russia) were 4.9, 13.7, 11.6, and 10,7%, respectively (Table 9) [103, 104]. In association with the data on interspecies individual differences in the sensitivity of omnivorous animals, these observations also allow for the supposition that cytolytic molecules produced by different Salmonella strains have differing potential to destroy erythrocytes in the same individual and, consequently, have differing molecular structures. The results allow also the assertion that the victim's target ecological molecule for Salmonella cytolysin is present on the mesenchymal cells of herbivorous but is either absent on the relevant cells of most representatives of omnivorous species or its molecular constitution does not match the ecological effector molecule from the infectious agent [105].

Table 9. Immunity of cells to Salmonella cytolysins in a selection of human populations [103]

Populations	Number of examined persons	Incidence of persons (%) whose cells are immune to destruction by Salmonella strains				
		A	B	C	D	E
Alma-Ata, Kazakhstan	164	93,9	88,5	93,9	95,1	95,7
Kishinev, Moldova	411	90,9	90,4	89,5	88,4	-
Saint-Petersburg, Russia	382	94,0	96,8	91,4	89,3	-
Tartu, Estonia	80	90,0	92,5	88,8	86,3	83,8
Total	1037	92,2	91,6	90,9	89,8	-

Table 10. Diversity in genetic immunity of buccal epithelial cells to adhesion of Neiiseria meningitidis

The peoples tested	The number of cells tested													Indices of adhesion
	In all	Including those with the quantity of meningococces attached to a cell												
		0	1-10	11-20	21-30	31-40	41-50	51-60	61-70	71-80	81-90	91-100	101-200	
1st	25	0	9	8	2	1	3	1	1	0	0	0	0	17,8
2nd	25	0	1	4	7	8	5	0	0	0	0	0	0	31,2
3rd	25	0	1	3	10	5	5	0	0	0	1	0	0	31,9
4th	25	0	0	2	8	8	4	1	2	0	0	0	0	37,0
5th	25	0	0	5	4	4	3	4	2	0	1	2	0	39,1
6th	25	0	1	3	4	5	4	3	0	3	3	1	0	43,3
7th	25	0	1	3	2	5	3	3	3	2	0	1	2	49,3
8th	25	0	1	5	1	3	0	3	2	0	2	3	5	56,3
9th	25	0	0	0	3	0	3	3	3	3	7	1	2	69,3
10th	25	0	0	0	2	1	2	5	3	1	3	1	7	71,4

The genotype dependence of resistance/sensitivity to Salmonella cytolysins was also confirmed by testing for cell immunity/susceptibility in a group of human twins [103, 104]. In the group of human, the incidence of individuals with sensitive erythrocytes correlated to that ascertained for other healthy persons. Out of 62 pairs of monozygotic twins, sensitivity was attributed to 4 pairs (12.9%). Here all 4 sensitive pairs had full double concordance for this characteristic. For instance, in the third and fourth twin pairs in this group, the erythrocytes of both twins were lysed only by Salmonella of types D and E, but were absolutely immune to cytolysins types A, B, and C. The results also demonstrated both the differences between salmolysins of different Salmonella types and the genetic origin of diversity in individual sensitivity to these molecular ecological agents of Salmonella. As for the group of 44 pairs of heterozygotic twins, persons with sensitive erythrocytes were found with the same frequency as other healthy people. The erythrocytes of 37 pairs of dizygotic twins (84.1%) were resistant. Sensitivity to one or more strains was found in 7 pairs (15.9 %), but none of these pairs showed concordance with this characteristic. For instance, in the first pair of dizygotic twins, the erythrocytes of one twin were lysed by Salmonella groups B, D, and E, while the other twin showed sensitivity only to Salmonella group E. In the fourth dizygotic pair, one twin was resistant to all Salmonella cytolysins tested, whereas the other twin showed sensitivity to cytolysins D and E.

The pathogenetic role and epidemic significance of the differences between individual cells immunity was characterized through an examination of 264 people who had abdominal typhoid or other salmonelloses before the testing. These individuals demonstrated lysability of their erythrocytes on a more frequent basis than the other members of the sample population [103]. For example, of the 149 Alma-Ata inhabitants who had abdominal typhoid, sensitive erythrocytes were revealed at a rate of 70.5%, which is 4.4 times more frequent than the same city inhabitants who have never had the disease in their anamnesis, a rate of 15.9%. Analogous data have been revealed by examination of 94 salmonelloses reconvalescents [103].

The high frequency of persons with sensitive erythrocytes among those who have previously had salmonelloses cannot be explained by the influence of infectious disease as a factor that could supposedly change initially resistant erythrocytes, but rather by the fact that people with Salmonella-sensitive erythrocytes have higher receptivity to Salmonella infection. The data can also substantiate the genetic origin of individual sensitivity/immunity of human erythrocytes to Salmonella cytolysins and their correlation with the intensity of the appropriate epidemic process.

While salmonelloses affect their victims by means of nutrition (via alimentary tract), the origin of differences in cell sensitivity to Salmonella cytolysin discovered between mainly herbivorous species (guinea pigs, horses, rabbits, and white mice) and mainly omnivorous ones (humans, dogs, cats, hens, monkeys, and sheep) can be considered as a result of natural selection performed by their regular interaction in this specific "microbe-victim" ecological system. The above data also confirm both the genetic origin of diversity in species and individual sensitivity to this molecular ecological agent of Salmonella virulence and the leading ecological function, i.e., pathogenetic significance of salmolysin, that was only hypotesized by some investigators before [106-108].

The totality of the above-considered data allows for the hypothesis that the genetically determined individual characteristics of cell sensitivity to Salmonella cytolysins can serve as a marker of constitutional predisposition for contamination by the infectious agent. In addition, the detection of resistant erythrocytes can substantiate the genetic immunity to disease. These data suggest a genetic tie to the distinction in sensitivity of erythrocytes to Salmonella cytolysins and also a connection between this phenomenon and some features of these organisms molecular constitution, that ensures genetic immunity against the appropriate infectious agent. One could also suggest that the high frequency of resistance in human and other omnivorous populations can be attributed to the natural selection operating in every epidemic [103, 104]. This puts into question the need for a more detailed study on the genetic origin and epidemic meaning of the differences in cell sensitivity of different organisms to Salmonella cytolytic toxin.

Species and individual diversity in cell sensitivity to Salmonella cytolysin can help in understanding some intriguing pathogenetic, epidemiologic, and clinical observations. The data can serve to orient molecular biological research in the evolution, epidemiology, and pathogenesis of Salmonella infection. Testing for individual differences in resistance/sensitivity to Salmonella cytolysin [103, 104] promotes an approach to compare the coding genome sequences of individuals with high cell sensitivity and sequences of individuals with opposite cell characteristics, thus revealing the molecular determinants of relevant cell quality and creating a way to decipher the molecular target of Salmonella cytolysin on the cell envelope.

This part of the discussed investigation also allows the supposition that cytolytic molecules produced by different Salmonella strains have differing potential for destroying the erythrocytes of the same individual, and, consequently, of different molecular structures. The results also demonstrated that all five Salmonella strains possess cytolytic activity, which can be restricted by species and individual features of erythrocytes [105]. In summary, this data of inter- and intra-specific differences in cell resistance and predisposition to lysis by Salmonellae confirms the genotypic nature of diversity in the incidence of Salmonella infection [4]. The existence of those individual and interspecies differences between cells may explain the origin of diversity in the incidence of Salmonella infection.

During the course of the epizootic swapted a snake farm in Zurich, Switzerland, in vitro interactions were studied between virus and cells of individual animals [89]. Virus replicated efficiently with various degrees of cytopathic effect in each of nine reptilian cell lines tested but the fish cell lines were not susceptible.

Malaria

The constitutional immunity of a cell may change also during the ontogeny of the cell. Malarial parasites demonstrate selectivity towards a particular age group of erythrocytes. For example, young mature chicken erythrocytes are selectively attacked by Plasmodium gallinaceum. Erythrocytes under 12 days of age are preferentially destroyed by the parasites.

Molecular Level of Diversity in the Incidence of Infections

Any infectious agent only causes disease if the both the effecting ecological molecule from the agent and the victim's target ecological molecule match. If the victim does not have compatible ecological molecule, then the victim's molecular constitution will block the infection providing constitutional genetic immunity [109].

HIV/AIDS. HIV - infection can affect and destroy a variety of human cells that express the CD4 receptor molecule on its surface. The virion envelope glycoproteins (gp120) interacts with the CD4 molecule on the target cells but also with its chemokine co-receptors thus performing the first i.e. clue step of intracellular invasion of the virion. Physiologically, the more discovered targets of HIV infection (CD4+ T cells, macrophages and dendritic cells) function as the vital components of reactive production of antibodies performed the lymphatic system. Macrophages and microglial cells are the cells infected by HIV in the central nervous system. In a case when enough CD4+ T cells have been destroyed by HIV, the production of antibodies by the lymphatic immunogenic system decreases leading to the syndrome known as AIDS.

The intensity of interplay between virus and its receptors on the cell surface receptors and ligands has fundamental role both in cell penetration by HIV-1 and in immunity to this clue step of the infection thus determining all levels of diversity characteristic of HIV infection. The structure of chemokine co-receptor is coded by CCR5 gene has been mapped to chromosome [110]. The mutant allele CCR5-[Delta]32, which is characterized by a 32 bp deletion in the single coding exon of the gene, was identified as responsive for coding the receptor structure not compatible to relevant molecular ecological agent of the virus [111].

The modified co-receptor along with CD4+ receptor for HIV-1 is incapable to promote cell penetration by HIV. Individuals homozygous for CCR5-[Delta]32 display no clinical symptoms and appear to be healthy. They possess structural (constitutional) cell immunity to the infection. Heterozygous individuals also exhibit slower progression to AIDS after seroconversion, and protection from AIDS-related lymphoma. Thus, the risk of acquiring HIV infection is individually modulated by genetic polymorphisms in the chemokine receptor/ ligand and the antigen-processing/presentation systems.

This mutation occurs at an allele frequency of 9% and a carriage frequency of 15%–18%, among white European individuals, but the frequencies in other major racial groups are negligible. Additional variants, most of which are codon-altering, have also been identified. Some of these variants may protect against HIV-1 infection as a result of severe alteration in the conformation of the molecule. The effects of host genetic variation on acquiring HIV infection are inextricably bound to the well-established and powerful genetic variants that can function at different stages of infection [111, 112]. The separate effects of these variants can be tested at cellular and molecular level.

Infection with H.pylori

This infectious agent produces cytotoxin VacA that causes gastric tissue damage, leading to gastric ulcers when administered intragastrically. This effect is associated with the ability of the toxin to cause vacuolation and death of gastric epithelial cells both in vivo and in vitro. In contrast, mice deficient in protein tyrosine phosphatase receptor type Z Ptprz gene do not show mucosal damage by cytotoxin VacA, although VacA is incorporated into the gastric epithelial cells to the same extent as in wild-type mice [65].

Primary cultures of gastric epithelial cells from Ptprz+/+ and Ptprz-/- mice also showed similar incorporation of VacA, cellular vacuolation and reduction in cellular proliferation, but only Ptprz+/+ cells showed marked detachment from a reconstituted basement membrane 24 h after treatment with VacA. This toxin bounds to Ptprz-receptor, and the levels of tyrosine phosphorylation of the G protein-coupled receptor kinase-interactor 1 (Git1), a Ptprz substrate, were higher after treatment with VacA, indicating that VacA behaves as a ligand for Ptprz. Furthermore, pleiotrophin (PTN), an endogenous ligand of Ptprz, also induced gastritis specifically in Ptprz+/+ mice when administered orally. Taken together, these data indicate that gastric ulcers are induced by erroneous Ptprz signaling [65].

Malaria. Malarial parasites demonstrate selectivity towards a particular age group of erythrocytes. The factor involved in such a phenomenon may be that the young erythrocytes contain more lipids in their membrane than the older cells. The factor involved in such a phenomenon may be that the young erythrocytes contain more lipids in their membrane than the older cells. [113]. Age-related changes in resistance of cells to mengo virus and encephalomyocarditis are conditioned by corresponding changes in maturation of the cell membrane molecular composition, namely the structure of the molecular receptor [12]. Moreover, there is a possibility that the large ribosomal RNA component in malarial ribosomes is provided in part or entirely from the host cell ribosomes. In the young red cell, the total volume of ribosomal RNA is greater than in the older red cell and this volume decreases with age [113]. In addition, there may be a factor that prevents the plasmodium from entering cells of a certain age due to changes in the red cell surface receptor sites that are known to bind merozoites.

Age-related changes in resistance of cells to mengo virus and encephalomyocarditis are conditioned by corresponding changes in maturation of the cell membrane molecular composition, namely the structure of the molecular receptor [12]. Therefore, certain immunogenic aspects of the molecular constitution of the organism are controlled by regulated changes expressed by the genetic program of individual development.

The Origin of Diversity in the Incidence of Infections

Some species of mosquitoes are extremely susceptible to the virus of western equine encephalitis while others are not infected with this pathogen. The species belonging to the genus Culex are very susceptible to Bacillus sphericus toxin while the Aedes species are immune to the this action.

In August of 1972 a viral respiratory epizootic swept a snake farm in Zurich, Switzerland. During the course of the epizootics , 128 (29%) of 431 snakes of Bothrops atrox species died, whereas Naja naja species or Elaphe longissima species did not. Human personal of the farm have not been affected [89].

For centuries it was claimed that over the course of human history infectious diseases functioned as the crucial ministries of human life and death, taking away hundreds of millions of lives. The intrusion of conquistadors exposed the Native Americans for the first time to smallpox, tuberculosis, measles, influenza and some other diseases that were unusual for America's aborigines before. As a result, during the period of three hundred years, between 1492 and 1800, the aboriginal Native American population plunged from 72 million to 600,000 [114] i.e. population was reduced to 0.8 percent. It is the most impressive example of mass annihilation produced by infectious agents. But the population of European conquistadors and colonists increased sharply over the same period. The same annihilating agents did not touch them so remarkably.

It has been shown that susceptibility to many infectious diseases has a genetic basis. Variations often associated with polymorphism in the genome, can now be detected [98]. We discuss special categories of genetic diversity namely genetic variations related to infectious diseases, for instance in humans,

There are both direct and indirect reasons to suppose that this is a case of genetic immunity to influenza virus. The set of the reasons includes selective intra-individual, individual (intraspecies) and interspecies differences in the incidence and severity of the disease with resultant differences in recovering and mortality, as well as discovery of the influenza resistance gene Mx [115] and observations of individual and species diversity in immunity of cells to the infection [95].

Species and individual diversity in genetic immunity to many infections is already a widely known fact. For a long time, however, this phenomenon has been neglected by both immunologists and epidemiologists, although it is regarded as essential [2, 7]. Near the end of the last century, that situation began to change [1, 4, 13, 14, 37]. It was discovered that this kind of immunity is determined by genetically programmed and very specific constitutional features of the individuals, populations and species under consideration.

Being obligate intracellular parasites, influenza viruses enter and affect victim cells by means of their specific chemical (molecular ecological) substances (adhesins, enzymes, cytolysins, polynucleotides). Infectious disease develops as a result of the interaction between the molecular ecological agents of the virus, on the one hand, and the affected body molecular structures on the other. Specificity in victim-parasite interaction involves mutual molecular recognition of the co-actants, mediated by chemically complementary molecules. The first step is the firm adhesion of the micro-organism to the victim cell membrane. Adhesion is mediated by glycoprotein molecules called adhesins. Adhesins are the molecular substances produced and used by microbes to adhere to target cells. Adhesion of influenza virus is a highly specific phenomenon [94].

Pathogenic effects of microbes and microbial molecules on molecular targets in the victim body are provided also by the genetically determined molecular structure of the targets, thereby defining constitutional susceptibility or constitutional immunity to infection. As a result of this property of molecular constitution, the viability and propagation of intruders can be either allowed or restricted. This kind of immunity is provided by constitutional incongruence between acting molecules of infectious agents and their molecular targets in the attacked organism. The success or failure of microbial aggression depends on the precise molecular congruence of infectious agents with the molecular make-up of the attacked organism, especially that of its receptors. Structural modifications of relevant molecular attributes of a victim are able to reduce or even eliminate the success of microbial intrusion [1]. The incongruence arises as a result of mutations and is propagated in a population, thanks to natural selection, as an evolutionary adaptation to life-threatening environmental agents.

Innate immunity to influenza can be considered as a phenomenon that encompasses specific structures possessed by the attacked organisms and functions in defense against the molecular ecological aggression of the virus. These structures can be presented with a lack or modification, for instance, of specific cell receptors for viral adhesion (hemagglutinin). Certain stereochemical modifications, as well as chemical modular compositional variations, can arise as a result of mutations. These may diminish, even completely exclude, the possibility of intermolecular interaction between microbes and their victims, thus lessening the success of microbial aggression.

The organisms possessing this modified molecular constitution are incapable of interacting with the microbe; they are not susceptible to this disease but are constitutionally resistant to it. These organisms are not infected by the corresponding microbe, but, in contrast, continue to develop normally and produce healthy progeny while other individuals of the same species which do not have this molecular structure become ill and die. On repeated exposure to a given pathogen, the progeny of resistant mutants will finally predominate in a population; an individual protective mutation becomes a property of a group, then of a population, and finally of the species (Figure 3). The selective principle of the appearance and spread of antimicrobial resistance lies at the base of natural constitutional immunity, which is characterised by specific molecular constitutional factors that exclude the possibility of interaction between parasite and victim.

The possibility and intensity of evolving constitutional immunity is determined by ecological prerequisites for regular interaction of potential victims with populations of infectious agents [3, 4, 116]. According to their different ecological features, living beings have been differentially subjected to interaction with different infectious agents. Populations of domestic poultry do not regularly contact the sources of influenza virus H5N1 in its artificial environment and have not experienced intensive natural selection for genetic immunity to the agent (Figure 4, a). In contrast, the presence of genetic immunity to influenza in most ducks, geese and other aquatic birds shows that they had regular interaction with the sources of relevant infectious agents in their natural environments, especially in the hydrosphere [117], and had been subjected to natural selection for genetic immunity to influenza (Figure 4, b, c).

Like the evolution of human genetic immunity to most of the existent infections [3, 14], the formation of an analogous immunity to influenza could have been induced by humans' continual, inevitable and intensive carnivorous contact with various potentially infected animals, including aquatic birds. It is known, for example, that diseased animals are caught by hunters more easily than healthy ones. Innate resistance to influenza infection in aquatic birds can be considered to be the result of their evolutionary adaptation to inevitable coexistence with this virus over a great many generations [45]. The adaptation could have been performed by means of natural selection among both victim species, leading to formation of its genetic immunity, and populations of the parasites, leading to formation of new pathogenic variants being able to break down the victims' immunity [3, 4, 14].

The selection has continued over millions of years, and species living today presumably possess genes for natural defense against relevant infectious agents. As a consequence, constitutional immunity currently seems to be a leading factor in the immunological structure of the population that determines the main parameters of epidemic processes. Nevertheless, one should take into account that, in fact, all "viral" and "bacterial" cell receptors exist not to interact with microbial pathogens, but to provide important physiological functions of relevant physiologically active molecules. The evolutionary adaptation of harmful microbes to a parasitic way of life could have been performed by the exploitation of pre-existent physiological receptors. The subsequent evolution and spread of genetic immunity to infections could have created a constitutional genetic immunity to the molecular physiological agents, thus providing the spread of relevant forms of pathology [4, 41]. The regularities in genetics, together with the data on the phenomena of evolutionary resonance and conjugated evolution, reveal the possibility of changes in the strength ratio in parasite–victim systems, thus entailing a change in the sensitivity to infectious agents and next intensification of epidemic processes [{Rumyantsev SN, 1983 122 /id].

Modern knowledge of the origin of diversity in susceptibility to infections has come from above discussed observations on the species, population, individual, cellular, subcellular and molecular levels of the specific kind of ecological interactions resulted in infectious diseases The initial contact of infectious agents with the attacked organism and their subsequent penetration inside of its body are carried out by means of a victim's natural communications, with which the especially regular physiological functions of feeding (alimentary mechanism of transfer), and breathing (respiratory mechanism of transfer) are provided. Of the two, alimentary transfer of infectious agents functions more effectively. The intensity and efficiency of respiratory transfer depends on the density of the population under consideration.

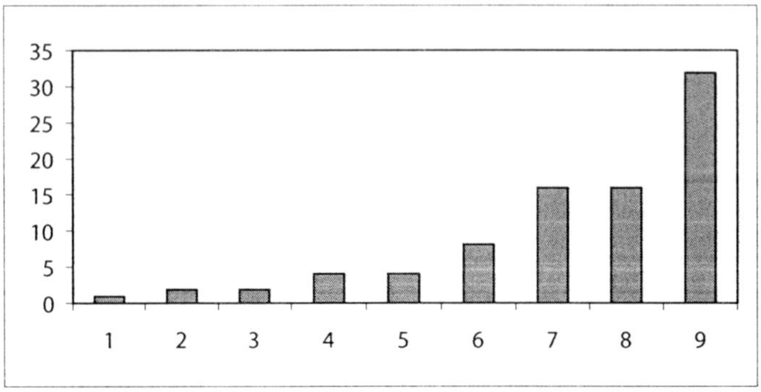

Figure 9. Interspecies differences in cell susceptibility to adhesion of rabies virus (according to [97]. 1. Adult chickens; 2. Rats; 3. Guinea pigs; 4. Humans; 5. Sheep, 6. 2-day-old chickens; 7. Rhesus monkeys; 8. Vervet monkeys; 9. Goose.

Clinical Observations

Interspecies Diversity in the Severity of Infectious Diseases

The course and severity of any infectious disease and the disease signs associated with it can vary with the victim's species or individuality, from completely asymptomatic, lengthy carrying of the potentially high pathogenic agent, to generalized systemic infection accompanied by digestive, respiratory, motor and central nervous system involvement, with death occurring within couple of hours or days.

Influenza

Human influenza is mostly located onto the upper respiratory tract or conjunctival mucosa and v9rus is excreted mainly via the respiratory tract. But in waterfowl regular portal of entry and a target organ for this virus is gastrointestinal tract and virus is excreted in high concentrations in the feces [{Webster RG, 1978 563 /id}]. H5N1 is present in the meat and eggs of infected birds, and animals have become infected by eating diseased birds. The cats became infected with H5N1 virus after being fed infected chickens. In those cats, virus-associated ganglioneuritis occurred in the submucosal and myenteric plexi of the small intestine, suggesting direct infection from the intestinal lumen. All cats excreted virus not only via the respiratory tract but also via the digestive tract. In wild aquatic birds (ducks, geese, swans and screamers, shorebirds, gulls), influenza viruses replicate preferentially in the intestinal tract and are excreted in the feces. The viruses have been isolated from freshly deposited avian fecal material and from unconcentrated lake water [118]. Thus infected birds share water with the uninfected ones. Aquatic birds mostly reveal no disease signs, and the relevant morbidity is very low among them (Figure 4, b, c).

In contrast, non-aquatic birds (chickens and turkeys) develop typical signs of respiratory influenza infection, which include respiratory signs; rales; excessive lacrimation; sinusitis; cyanosis of unfeathered skin, especially of the combs and wattles; edema of the head and face; ruffled feathers; diarrhea; and nervous disorders, decreased egg production and a high level of mortality.

Above mentioned harbor seal also died of a severe respiratory infection with consolidation of the lungs, typical of primary viral pneumonia [45]. In experiments with squirrel monkeys, the seal influenza virus replicated in the lungs and nasopharynx after intratracheal administration, and in the conjunctiva after administration into the eye. In one monkey that died of pneumonia, the virus was recovered from the spleen, liver, muscles, and lungs [119], indicating that in this kind of primate the virus has potential capability for generalized lesions.

The signs of expressive disease in non-aquatic mammals, for instance, in horses, pigs and humans, are the same as in their aquatic mammalian relatives and consisted of nasal discharge, dry hacking coughing, fever, labored breathing, loss of appetite, muscular soreness, tracheobronchitis, conjunctivitis, and secondary bronchial pneumonia.

Parainfluenza

A study of parainfluenza virus 3 (PIV3) infection in two species of cotton rats (Sigmodon hispidus and Sigmodon fulviventer) demonstrated that, after intranasal application of human PIV3, both species sustained a high level of viral replication in nasal tissues, although neither developed substantial nasal histopathology (rhinitis). This contrasts sharply with the pattern seen in respiratory syncytial virus infection of S.hispidus, where slightly lower levels of viral replication caused moderate to severe rhinitis. PIV3 replicates well in the lungs of S.hispidus and the pathology consists primarily of bronchiolitis with most lesions appearing six or seven days after infection. In contrast, S.fulviventer develops maximal pathology at the same time post infection, but the virus causes a severe pneumonia, with only minor components of bronchiolitis. The levels of viral replication and the duration of the infection were virtually identical. In all localisations, the clinical course of infections ranged from acute to chronic and persisted in the victim in latent forms [120]. The fact that two distinctly different histopathological patterns were seen in two different rat species tested suggests that genetics of the victims are the major determinants of disease manifestation.

Infection with Sendai virus of hamsters or parainfluenza viruses (PIV) of sheep typically results in both bronchitis and interstitial pneumonia in the same individual animal. PIV3 infection may also be associated with the croup syndrome but more often causes bronchitis, bronchiolitis, or interstitial pneumonia. PIV3 infection in hamsters may result in mild bronchiolitis but generally does not cause interstitial pneumonia. The same infection can cause a severe bronchiolitis and interstitial pneumonia in sheep [120].

Inter-individual Diversity in the Severity of Infectious Diseases

All infections are characterized by the manifestation of individually different spectrum of disease clinical signs, with resultant clinical courses, which range from acute to chronic and possible persistence in a latent form.

Infection with N.meningitidis

There is a massive clinical evidence, that persons have extraordinary different signs, courses and severity of meningococcal infection [66, 80]. Meningococci are carried as harmless commensals in the nasopharynx in about 1% of the population. At the same time the meningococcal carriadge is highly variable. It peaks at 15-20 years of age and decreases with age. In some individuals, namely in only a small proportion of carriers, however, the same microbe causes devastating and fatal illness. Meningitis and sepsis are the most dangerous forms or meningococcal infection especially while they are accompanied by massive cutaneous hemorrages (Figure 10). Sometimes the infection is going in the form of arthritis, otitis, tonsillitis, pneumonia, pericarditis, vaginitis, urethritis, peritonitis.

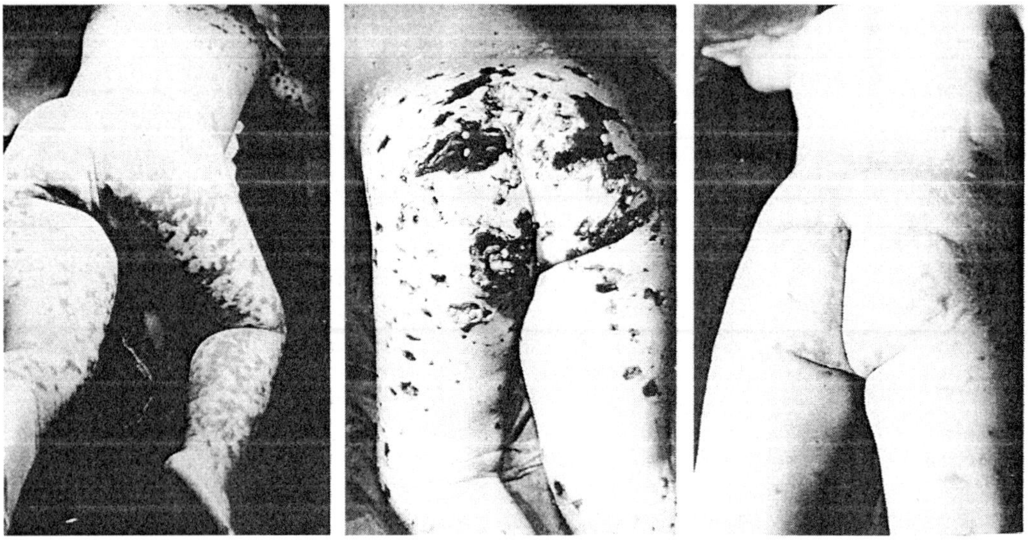

Figure 10. Large petechiae (left) and massive cutaneous hemorrages 10 days later (center) in a three – year-old boy with group A meningococcal disease at Aurora Hospital, Helsinky, Finland, in 1974. The photograph on the right shows the same patient six months later, after several skin grafts [147].

Infection with H.pylori

In most individuals *Helicobacter pylori* infection is asymptomatic. In other people the infection induce various intragastral affections provoking symptoms of heartburn, epigastric pain and nausea. Individually, including ulcers and gastritis, the infection can exist either in the form of acute but transient gastritis or at least in seven different chronic forms. In some patients acute gastritis resulted in spontaneous healing where as the majority of acutely infected individuals go on to develop some of six chronic clinical forms (pangastritis, antral gastritis, gastric ulcer, duodenal ulcer, gastric canacer, duodenal cancer). These can develop either pangastritis or antral gastritis, which is located in the pillory end of the stomach and is most closely associated the development of duodenal ulcer. Alternatively a pangastritis is associated with gastritic atrophy, solitary or multiple gastric ulcer and gastric cancer (Figure 11). The mechanisms which determine how and why this various clinical forms develop are not clear.

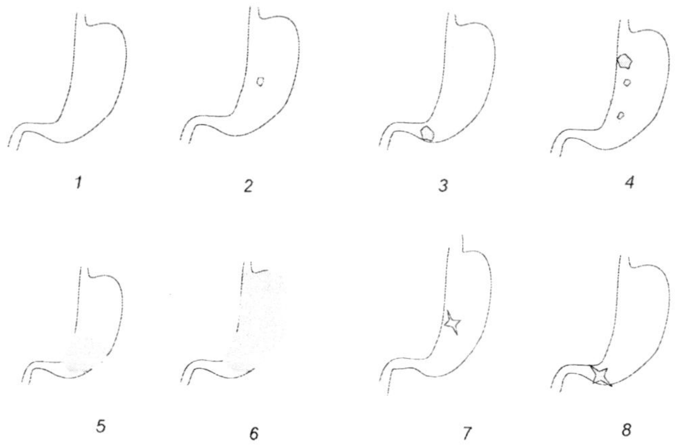

Figure 11. Variations in the locations and forms of gastric lesions by H.pylori. (1) Healthy stomach, (2) Solitary gastric ulcer (3) Duodenal ulcer (4) Multiple gastric ulcers, (5) Antral gastritis, (6) Pangastritis (7) Gastric cancer (8) Duodenal cancer.

Anthrax

The infection can exists in hyperacute, acute (pulmonary, septic and enteral) or subacute (cutaneous) form. Hyperacute and acute forms are characteristic of anthrax disease in herbivorous animals. Subacute (cutaneous, local) form accounts for well over 95% of anthrax cases in humans. In contrast to 100% fatality of untreated acute human anthrax, less then 20% of untreated cutaneous cases were fatal. [121].

Dengue

Dengue infection can develop three distinct syndromes characterized by different degree of severity: the most frequent dengue fever, more severe dengue hemorrhagic fever (DHF), and dengue shock syndrome. Although caused by the same viruses, dengue and dengue hemorrhagic fever are clinically distinct. The classical features of DHF are increased capillary permeability without morphological damage to the capillary endothelium, altered number and functions of leucocytes, and increased haematocrit and thrombocytopoenia (Halstead, 1993; Chaturvedi et al., 1997).

Dengue fever is characterized by sudden onset of systemic toxicity, fever, headache, vomiting, and severe myalgia or bone pain of escalating intensity. Either coincident with or following remittance of fever on days 3 to 5 of the illness, there appears a maculopapular or morbilliform rash on the trunk which spreads to the limbs and face. This phase of the illness is often accompanied by recrudescence of fever, lymphadenopathy, granulocytopenia, and thrombocytopenia. Minor mucocutaneous bleeding is occasionally manifested by petechiae, epistaxis, menorrhagia, and a positive tourniquet test. Dengue fever lasts 3 to 9 days, is self-limiting, and is rarely associated with serious sequelae.

The clinical course and signs of dengue hemorrhagic fever is characterized by an initial stage of fever, rash, and anorexia (lasting 3 to 5 days) followed by a shock phase in which hepatomegaly, hypotension, and a hemorrhagic diathesis occur. Dengue hemorrhagic fever results from additional pathogenetic processes not present in classical dengue fever; the most important of which are diffuse capillary leak with hemoconcentration, thrombocytopenia, and disseminated intravascular coagulation.

In dengue shock syndrome, the decreased plasma volume which results from increased vascular permeability causes clinical shock that, if uncorrected, may lead to acidosis, hyperkalemia, and death. Ninety percent of dengue hemorrhagic fever cases occur in children experiencing multiple infections with dengue.

The list of such examples could be broadened up to size of a fundamental handbook. Nevertheless, until recently, the origin of this kind of biodiversity has been poorly understood. Recent advances in immunology — especially in identifying the constitutional mechanisms of immunity [1, 4, 14] have contributed to our understanding of the origins of this kind of biodiversity in infectious diseases.

Age Diversity in the Severity of Infectious Diseases

Immunogenic characteristics of the molecular constitution of the organism are controlled by genetic program of its individual development. According to the program, constitutional immunity may change during ontogeny of the organism. That is why newborn and 3-week-old mice are totally defenseless against the virus of yellow fever, but by adulthood they have become completely immune. Newborn mice are also very susceptible to the herpes virus, but by 4 weeks of age this sensitivity is replaced by constitutional immunity [86, 87].

Dengue

Dengue infection mainly affects adults in Brazil but predominant children in Asia (Phuong et al., 2004; Siqueira et al., 2005). In Asia, risk of severe Dengue disease is greater in children than in adults, and outcome is worse in younger children. As with many viruses, neonates and children tend to be particularly susceptible to dengue. By contrast, in America the disease largely affects the adult population and produce milder disease. [98] .

Tick Born Encephalitis

Tick born encephalitis and myelitis dominate in the age group > 50 years. With increasing age the clinical course of the disease is more severe. Also this age group suffers significantly more sequelae, has a need of longer rehabilitation and a higher case fatality [83].

Meningococcal Infection

A general characteristic of most infections in animals and plants is the changed response to disease, which develops in the course of growth and maturation. The changes are particularly dramatic in the neonatal period. Meningococcal infection presents one of the rare exceptions to this common law. About 15% of all cases of childhood meningitis is due to the meningococcus [80]; however, children under 1 month of age rarely fall ill with a meningococcal infection. In a series of 211 cases of neonatal meningitis, Neisseria meningitidis was an uncommon organism, with only 2 of the 211 due to N. meningitides [79]. In accordance with these epidemic and clinical observations, cells of human newborns are absolutely immune to the attachment of meningococci [94]. However, in the neonatal and following periods, the cells become far different and individually sensitive to meningococcal adhesion. Thus, the function of meningococcal receptors on human cells develops during postnatal ontogenesis. In contrast, the influenza viral infection is characterized by different individual cell sensitivity in all age groups tested [94]. Erythrocytes of 2-day-old chicks are more sensitive to rabies virus adhesion than those of adult chickens [97, 122].

Senescence increases the frequency and severity of infections, which cannot be explained only by the decline in efficiency of the lymphoid system immunogenic response. The increase in susceptibility of the elderly to certain infectious diseases can be explained by the changes in molecular constitution of organs, tissues, cell clones, cells, and extracellular structures that occur with age. There are developmental changes in the molecular constitution of cells that are correlated with developmental differences in their physiological and ecological characteristic, and this may cause the age-related differences in susceptibility to relevant agents.

Varicella Zoster Infection

Varicella zoster infection causes chickenpox (Varicella), a generally mild infection of childhood but shingles (Herpes Zoster) of old age, a painful unilateral vesicular rash which follows reactivation of virus latent in dorsal root and some cranial nerve ganglia. The incidence of Herpes Zoster increases with age, particularly in those over 50 years. Two thirds of zoster cases occur in those under 65 years whereas the severity of the disease is lower in this age group. Between 13 and 26% of patients with Herpes Zoster develop persistent neuropathic pain known as post herpetic neuralgia. [123].

Malaria

Malaria demonstrates selectivity towards a particular age group of erythrocytes. Young mature chicken erythrocytes are selectively attacked by Plasmodium gallinaceum. The erythrocytes under 12 days of age are being preferentially destroyed by the parasites [113]. Factor involved in such a phenomenon may be that the young erythrocytes contain more lipids in their membrane than the old cells. There might be factors that prevent the plasmodium from entering cells of a certain age due to change of the red blood cell surface receptor sites that are known to bind merozoites [113].

Several explanations of age-dependent change in age variation of immunity have been advanced: maturation of lymphatic immunogenic system; development of "barriers" to the spread of microbe from the periphery to highly susceptible target organs; augment interferon productivity; maturation of macrophages; higher and more stable body temperature [79, 80]; inherent variation in the complement pathway [124].

The extreme rarity of meningococcal infection in the new-born period is not inconsistent with transplacental immunity [80]. On the other hand, there are assumptions that the nature of discussed phenomenon is associated with some age peculiarities of the organism's molecular constitution, that not involve the reactive production of immunoglobulines [12, 14, 113], e.g. inherent differences in cellular receptor structures should be responsible for the agents adhesion which in most cases is the key stage of infectious processes [{Petri W.A.Mann B.J., 1995 76 /id}].

Age related changes in resistance of cells to mengo virus and encephalomyocarditis virus are conditioned by corresponding changes in maturation of the cell membrane molecular composition, namely the structure of molecular receptor [12]. Erythrocytes of 2-day-old chicks are more sensitive to rabies virus adhesion than those of adult chickens [97].

Inter-organ Diversity in the Location of Infectious Affections

When any kind of infection is developing, it affects far not all parts of the organism. As a rule, infectious lesions are found in strictly definite organs, tissues and cell populations. Usually, infectious lesions are found only in strictly definite organs and tissues, whereas others appear to be immune. Thus, grippe chiefly affects the upper respiratory tract, whereas small pox is manifested mainly with skin eruptions. The like specific distribution of susceptible and immune parts of the body is characteristic also of intestinal infections (dysentery, cholera) and the ones, affecting chiefly internal organs (tuberculosis, glanders, syphilis, leprosy, etc.). Bacterial dysentery affects mucosa of large intestine, while cholera - the one of small. Clinical pictures of tetanus and botulism are dependent on the damage of nervous system synaptic elements. Bacteria of the genus Bordetella adhere predominantly to ciliated respiratory epithelial cells. The multiplication of bacteria attached to the ciliated respiratory epithelium is a unique attribute of the genus Bordetella.

HIV-infection can directly attacks organs, such as the kidneys, the heart and the brain leading to acute renal failure, cardiomyopathy, dementia and encephalopathy. Nearly every organ or system can be affected at this stage of HIV. People with AIDS have also an increased risk of developing various cancers such as Kaposi sarcoma, cervical cancer and cancers of the immune system known as lymphomas [125]. Additionally, people with AIDS often have common symptoms of many infection like fever, sweat, swollen glands, chill, weakness, and weight loss [126]. Most AIDS patients die from pulmonary, gastro-intestinal, neurological and other opportunistic infections or malignancies associated with the progressive failure of the lymphatic immunogenic system [127]. Meanwhile, AIDS can directly affect brain cells, intestinal cells, spinal cord cells, and many other.

Streptococcus mutans is abundant in dental surface but is found to a lesser extent, if at all, on the epithelial cells of the tongue. The reverse is true for Streptococcus salivarius, normally found in large numbers on the tongue [128]. E.coli colonizes the gastrointestinal tract, while streptococci colonize the skin and esophagus in humans. Any microbial aggression is regularly revealed in infected organisms according to hereditary heterogeneity of the tissues, differing for each type of infection. Bacterial dysentery affects mucosa of large intestine, while cholera - the one of small.

Investigation of the cause of the localization of brucellae infection in certain tissues in some animal species began with cattle. A survey of heir organs showed that over 90 % of brucellae were in the foetal placenta, chorion and foetal fluids. In regard to brucellosis in other species, erythritol was found in the placentae of goats, sheep and pigs, but not in the placentae of humans and quinea-pigs, which explains why, in cases of human brucellosis, an undulating fever occurs with no overwhelming reproduction or localisation of brucellae in either male or female; abortion is rare. The same is true for guinea-pigs, the preferred animal model for human brucellosis. Furthermore, the testes and seminal vesicles of cattle, sheep and pigs contained erythritol, but the seminal vesicles of humans did not [23]. Thus, a component of the molecular constitution of some species and their tissues is responsible for the restriction of brucellae reproduction. This is a major factor of constitutional immunity to brucellosis. The localisation of erythritol absence explained the localisation of immune and susceptible tissues in cattle, goats sheep and pigs, and why Brucella infection in man, guinea pig and other species tested does not show such tissue predilections.

It has been demonstrated just recently [{Shinya K. 2006 574 /id}] an molecular anatomical difference in the distribution in the human airway of the different binding molecules preferred by the avian and human influenza viruses. The respective molecules are sialic acid linked to galactose by an α-2,3 linkage (SAα2,3Gal) and by an α-2,6 linkage (SAα2,6Gal). Human flu viruses preferentially bind to what are known as α 2,6 galactose receptors, which populate the human respiratory tract from the nose to the lungs. Avian viruses prefer α 2,3 galactose receptors, which are common in birds but were thought to be nearly absent in humans. Humans also have α 2,3 galactose receptors, but only in and around the alveoli. In human 2,6 receptors were common in the nose and throat, but 2,3 receptors – H5N1's preferred site – were common in the alveoli. The virus can also affect the gut and most worryingly, the brain.

Avian influenza virus (H5N1) may cause severe lower respiratory tract (LRT) disease (lethal pneumonia) in some humans. This pattern is consistent with autopsies that have shown heavy damage to the lungs but little involvement of the upper respiratory tract. However, the LRT cells to which the virus attaches are unknown for both majority of humans and other mammals. H5N1 virus attached predominantly to type II pneumocytes, alveolar macrophages, and non-ciliated bronchiolar cells in human LRT. This pattern was most closely mirrored in cat and ferret tissues. These findings may explain, at least in part, the localization and severity of H5N1 viral pneumonia in humans. Among experimental animals cats and ferrets more closely match the human pattern of infection and are considered as more suitable experimental animals than do mice and macaques [129].

Individual Locations and Sizes of Infectious Affections

Every infectious disease affects only focal areas in the infected organism. Multiple foci of infectious damages are even more clearly demonstrated in infectious exanthemas. Focal vascular affections i.e. a rapid development of petechial rash or purpura fulminans is characteristic of sepsis and septic shock induced by Neisseria meningitidis. Intra-individual variation can be observed in the distribution of H3N2 virus particles on the surface of human trachea.

Smallpox

That focal (mosaic) exanthema is seen, for instance, during poxvirus infection in human and animals. The focal lesion of cutis is seen, for instance, during poxvirus infection. It consists of a number of skin lesions scattered over the surface of the body, each of which is the seat of inflammation so intense as to lead to the formation of small abscesses in the course of 4 to 5 days (Figure 12). Beyond the edge of the lesions, the skin is normal. The number of lesions present may be less than a dozen in a minor case of illness, or they may number in the thousands. In a severe course of disease, lesions may be set so closely as to conceal almost the whole cutaneous surface[130]. The smallpox rash is most severe on the face, then the hands and upper extremity. From the hands upwards, the scabs diminished in density. On the front part of the trunk they are scarce, especially on the abdomen. The areas adjacent to the specific damage are found almost intact, though all the skin is presumably homogeneous [130-132].

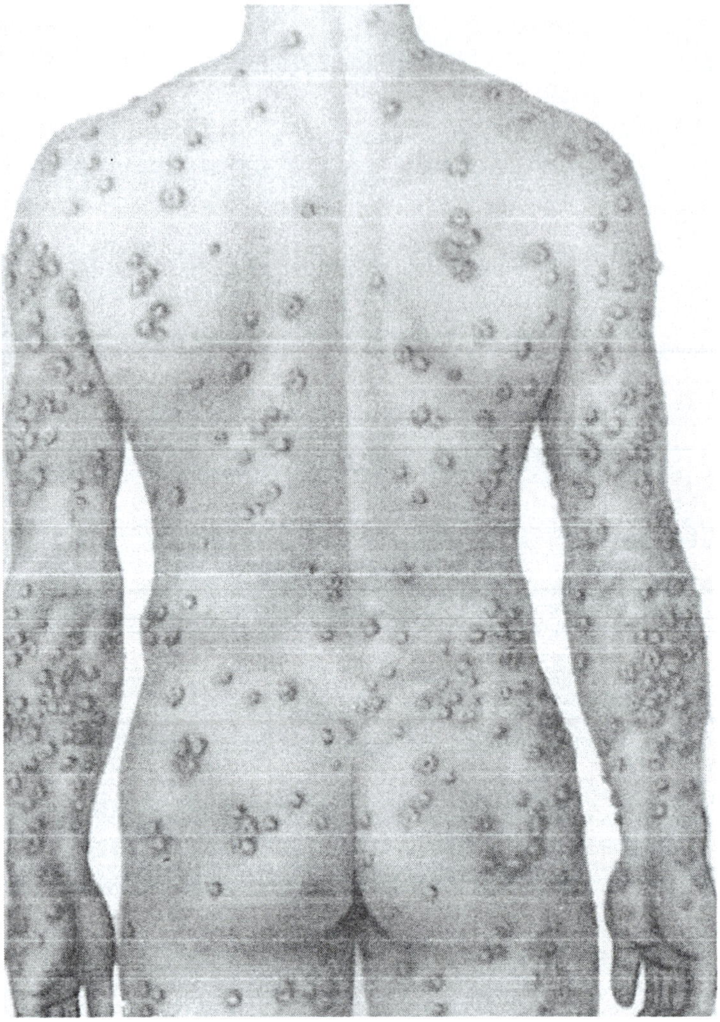

Figure 12. Dispersion of the smallpox exanthema

Anthrax

The name of the disease has been given following by its focal (cutaneous) form, that accounts for over 95% of cases on records. The focal lesions vary greatly in size from about 2 cm to several centimetrs across and begin to resolve about 10 days after the appearance. The resolution takes from 2 to 6 weeks living minor scarring

Polio

Poliovirus infection targets nerve cells at random although the virus is present throughout the nervous system [133]. The brain and spinal cord are involved in every case of poliomyelitis, but not every area of these organs is susceptible to the virus. This can only mean that nerve cells in certain areas are susceptible to viral damage, whereas those in other areas are not. In the worst cases, nearly every skeletal muscle may be paralysed. In the mildest cases, paralysis may be limited to part of one muscle. All degrees and combinations of paralyses can be observed [131].

Mosaic distribution of damages is also characteristic of hepatitis, tuberculosis, and many other infections that chiefly affect internal organs [131, 134]. The major recognized forms of rickettsial Q fever are pneumonia, hepatitis, meningoencephalitis, vascular infection, bone infection, osteomyelitis, endocarditis, unexplained fever, and flu-like form [135]. Typhoid and paratyphoid fever produce inflammatory foci in many organs and, in some areas, necrosis and sloughing of tissues. Necrotic or inflammatory foci are commonly seen in the spleen and the liver, and sometimes in the kidney, the meninges, the bones, the endocardium, the joints, and many other portions of the body[131].

Dysentery

In bacillary dysentery, the specific pathological changes are located in the colonic mucosa and consist of random patches of greyish membrane composed of thickened, necrotic mucosa. In severe cases, the disease may produce the destruction of the whole of the large bowel. In less severe cases, the areas of necrosis of the mucous membrane are less extensive, with small, scattered ulcers or nodular crusted lesions appearing on the bowel wall [131].

The disease was characterized by either circumscribed papular skin lesions or spreading grey skin patches. Examination of the skin of animal with an exanthema will reveal that the areas adjacent to the specific damage are found almost intact, though all the skin is presumably homogeneous.

Mosaic distribution of susceptible and immune areas is also characteristic of infections that affect chiefly internal organs (hepatitis, pneumonia, tuberculosis and so all) [134]. Every infectious disease affects specific areas in the infected organism with focal lesions of characteristic morphology and distribution.

The same kind of diversity is characteristic of infectious diseases in animals. Focal bone lesions were characteristic of some infections in Pleistocene bears which lived over 10000 years B.C [136]. Such focal damage is regularly observed in infectious diseases of plants and invertebrates. A viral skin disease of green sea turtles (Chelonia midas) was described at a commercial sea turtle farm in the British West Indies. Turtles developed spontaneously "grey-patch disease". The disease was characterized by either circumscribed papular skin lesions or spreading grey skin patches [89]. Analogous mosaicism are regularly observed in infectious diseases of plants [132, 137].

Each infectious disease is expressed in the infected organism by at least two categories of the same tissue, outwardly identical and differing only in their relationship to a given micro-organism. Constitutional immunity explains why parts of one category are affected by a given microbe, while at the same time other morphologically identical components of the organism remain uninvolved - although both exist under the same conditions and may be equally attacked by the infectious agents. These morphologically identical parts, which vary in resistance, appear as a nidus, and the distribution of such foci are characteristically distinct for each type of infection.

When cell cultures are affected by some microbial pathogenic factor, most cells are killed. At the same time, however, a number of outwardly undistinguishable cells fully preserve their basic functions despite the unfavourable conditions. Such constitutional resistance may involve only a few cells or the majority of cells in a population. In all cases, the culture contains cells which are not susceptible to that infection along with those cells which are susceptible to contamination and thus killed.

This phenomenon is highly specific: the cell may be immune to microbial agent of one type, but susceptible to another one. The number of immune cells in the cultured population may be extremely small or may form the majority. In fact, in an affected animal every infection manifests at least two populations of homogeneous cells which differ sharply from each other with regard to one highly specific feature: cells of the first population are susceptible to agents, whereas those of the second are immune to the infection.

Influenza

Human influenza is usually located onto the upper respiratory tract or conjunctival mucosa [138, 139]. Besides, some cases of influenza infections may be initiated by gastrointestinal inoculation of virus and continued by its intra-intestinal reproduction [140]. In some people the diarrhoea cases of influenza had no respiratory symptoms. Thus influenza virus could be entered the food chain of ancient human.

H.pylori Infection

Different sensitivity to H.pylori infection which forms the base of gastritis, peptic ulcer and other manifestations of the disease is characteristic not of entire epithelial layer of the stomach, but of its dispersed areas of various sizes and locations. The affections are always patchy. Gastric antrum is predominant site of colonizaion by H.pylori, but the whole body of the stomach (Figure 11) can be involved in severe cases 141. In patients with antral gastritis and pangastritis the infection forms great number of smallest ulcerations of stomach mucosa whereas in other patients far more large but usually solitary zones of stomach and duodenum affection áre observed (Figure 11).

The ulcerations of the stomach are located presumably near the lesser curvature of stomach and are rare at or near the greater curvature. But in any case, the areas adjacent to the ulcers are left unaffected by initial destructive process but are subjected only to reactive inflammation 142. The factors influencing the site and number of ulceration are of interest and could shed light on the pathogenesis of this group of diseases 142.

Hence, the infection with H.pylori reveal the presence of at least two cell clones in gastrointestinal mucosa, located in the closest neighborhood and distinguished in the range of their sensitivity towards this specific microbial factors. Up to now nobody could answer, why the interaction between H.pylori and the mucosa can result in either multiple damages of smallest size (Figure 13) [143] characteristic of chronic gastritis, single crater of middle size (ulcer of stomach or duodenum), few craters (multiple gastric ulcers) or gastric adenocarcinoma [144].

The size of affections and its location in the stomach is of crucial importance for the course and severity of the diseases that can result from *Helicobacter pylori* infection. About 10-15% of infected individuals experience ulcer disease. Such ulcers are more common in the duodenum than in the stomach itself. Severe complications include intragastric bleeding and perforation of the gastric wall. In some individuals *Helicobacter pylori* also infects the corpus region of the stomach . This results in a more widespread inflammation that predisposes not only to ulcer but also to cancer in the same stomach region,. This cancer has decreased in incidence in many countries during the last half-century but still ranks as number two in the world in terms of cancer deaths. Inflammation in the stomach mucosa is also a risk factor for gastric lymphoma.a special type of lymphatic neoplasm in the stomach. Since such lymphomas may regress when *Helicobacter pylori* is eradicated by antibiotics, the bacterium plays an important role in perpetuating this tumour. Only a minority of infected individuals develop stomach disease. After Marshall's and Warren's discovery, research has been intense. Details underlying the exact pathogenetic mechanisms are continuously being unravelled.

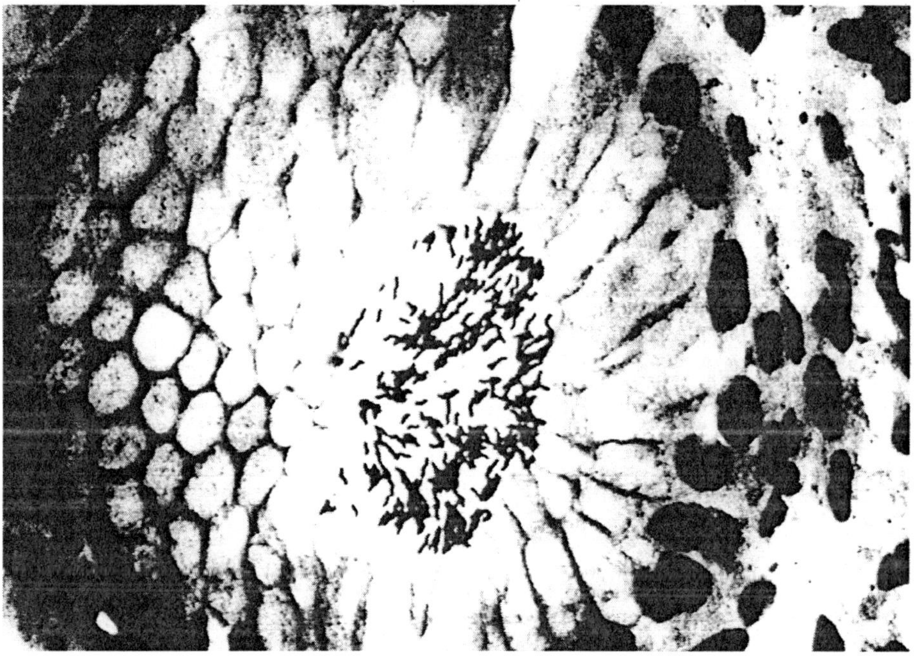

Figure 13. The microulceration of the stomach mucosa induced by H.pylori [143].

Cellular and Molecular Levels of Intra-individual Diversity in Infections

The focal nature of infectious damages can not be explained by the lack of uniform predilection of microbes for the organism. Moreover in such cases, the infectious agent is found within both the typical damaged elements and the adjacent areas which have been left intact. The virus of German measles, for example, reveals this both in the typical elements of the rash and in the neighboring parts of the cutis which remain uninfected [145]. By the way, in the course of rubella exanthem investigation, it was demonstrated, that the reason of lesions focalization had nothing in common with the germs absence or presence in the undamaged parts of the organism. Rubella virus was found within the typical rash elements, as well as in adjacent areas, left absolutely intact [145]. All these features are observsd although in infectious diseases of plants {Vavilov N.I. 1964 ID: 38}.

When cell culture was affected by some microbial pathogenic factor, many cells have been killed. At the same time, however, a number of outwardly indistinguishable cells fully preserved their basic functions despite the conditions unfavourable for other cells of the same kind [146]. Such immunity may involve only few cells or a majority of cells in a culture. In all cases, the culture consists of cells which are not susceptible to that infection along with those cells which are susceptible to contamination and thus damaged.

We studied intra-individual variations in sensitivity of human epithelial (buccal) cells to adhesion of Neisseria meningitidis, counting under microscope the number of microbes attached to every of 25 cells derived from every of tested humans. While the mean quantities of attached cells have varied among tested persons from 12 to 86 microbes per cell, some cells were fare more susceptible to the adhesion. They attached up to 200-250 microbes whereas each of the most resistant cells has attached less than 10 microbes (Table 10).

Qualitatively analogous diversity was observed in experiments with human erythrocytes, although in this case the quantity of microbes attached on a cell was substantially lesser (from 1 to 20). Besides, in contrary to epitheliocytes, every of these cell populations contained over 30% of erythrocytes that were absolutely unsusceptible to the attachment of this kind of microbes. One could suppose, that this difference has arisen because most susceptible erythrocytes have been agglutinated by attached microbes and thus excluded from observation. Thus each of observed individual cell populations, both epitheliocytic and erythrocytic ones, represents by itself an innate mixture of cells with different grades of genetic susceptibility to adhesion of N.meningitidis.

The diversity of red blood cell populations and its grade can be observed by in vitro quantitative measuring of the reaction of haemagglutination induced by by N.meningitidis, influenza viruses or any other hemagglutinating microorganism [94, 97, 148]. When the quantitative microassay has been performed in U-bottom wells of polysterene plates the results can be observed as follows:

0 – agglutination did not happen: all erythrocytes are settled on the U-bottom of the well in a form of a compact small round. That means the tested cells are resistant the agent.

+ - The minority of erythrocytes are agglutinated and formed light cloudiness of the sample whereas the majority of cells has not been agglutinated but are settled on the bottom of the well in a form of compact small ring.

++ - Many erythrocytes are agglutinated and formed medium cloudiness of the sample whereas the minority of unagglutinated cells are settled on the bottom of the well in a form of a loose small ring.

+++ - The majority of erythrocytes are agglutinated and formed dense cloudiness of the sample but on the bottom of the well in a form a small weak ring is seen..

++++ - All erythrocytes are agglutinated and formed a dense cloudiness of the sample. The tested sample contains of only susceptible cells.

Heterogeneity of cell populations, which are sensitive to an infectious agent, has been reported frequently by many investigators. Sensitivity may involve only a few cells or the majority of cells in a population. When a susceptible cell culture is affected by some microbial pathogenic factor, most cells are killed. At the same time, however, a number of outwardly indistinguishable cells fully preserve their basic functions despite unfavourable conditions.

This phenomenon is highly specific: a cell may be immune to microbial agents of one kind, but susceptible to another. The number of sensitive cells in a cell population may be extremely small or may form the majority. Analogous situation is usually observed in a hole organism. Some of its cells may be sensitive to an infectious agent, whereas another clones of the same kind reveals immunity to the same agent. Immune polymorphism of cell populations contributes to the topography of the affected tissue areas and the favourite focus' localisation typical of certain infectious diseases. Focal character of infectious affection is due to the resistance of certain cell clones in their sensitivity to an infectious agent and their focal distribution.

Diphteria

Cell strains that are spontaneously immune to the lethal effect of diphtheria toxin have been selectively isolated from highly sensitive human cell lines. The toxin-immune cells were found to be 100,000 to 1,000,000 times more resistant to toxin than the toxin-sensitive cells of the parent cell population [50, 149]. Thus, the resistance of these cell strains was comparable to the resistance of cells from animal species that are naturally immune to the toxin, such as rat and mouse species. This phenomenon is highly specific: the cell may be immune to microbial agents of one type, but susceptible to another type.

According to [150], the number and size of infectious lesions depend on the patient reactivity and the potency of the infecting strain. The examples considered above illustrate the manifestation of the mosaic configuration within affected homogenous structures.

Evidently, the ubiquitous phagocytes and immunoglobulins are not functional in this phenomenon. The focal nature of infection is not explained by the lack of uniform predilection of micro-organisms for the organism. Every infection in its pre-eruptive stage is a septicaemia, and the causative agents are circulated everywhere in the affected tissue but they induce specific eruption only in susceptible areas of the tissue. In such cases, the infectious agent is found within both the typical damaged elements and the adjacent areas that have been left intact [131, 132, 145].

Origin of Intra-individual Diversity in Infectious Diseases

Transgenic technology entails the insertion of foreign genetic material into the germ. The transgenically received himeric organism consists of both innate and foreign components of its set up. Genes responsible for resistance or sensitivity to particular infectious diseases might be introduced into otherwise vulnerable or susceptible individuals to change their immune status.

Poliovirus infects only primates and does not infect mice who have total constitutive immunity. The species specificity of poliovirus is primarily determined by a unique cell surface receptor. An important determinant in the development of disease relies on the characteristic cell tropism which appears to be governed by the accessibility of specific cellular receptors. Transgenic mice carrying poliovirus receptor genes transferred from the human genome, unlike innate mice, are sensitive to poliovirus inoculation [33].

One of the main reasons for introducing new genes into plants and animals is to increase their immunity. Genetic engineers can introduce increased resistance to drought, salinity, pests, herbicides and infections into the plant genome. The innate presence of Mx1 protein gene in some strains of mice specifically prevents reproduction of the influenza A virus and other orthomyxovirus family members both in vivo and in vitro. Mx1 on its own, as a constitutive transgenic product, is an efficient anti-influenza virus effector immunogenic in the transgenic mouse. Genetic engineers are now able to remove and transfer Mx1 protein gene from constitutionally resistant to a naturally sensitive organism with extraordinary precision [151]. Transgenic animals received the mosaic structure of their sensitivity to infection.

The examples considered above illustrate the manifestation of the mosaic configuration within affected homogenous structures. Evidently, the ubiquitous phagocytes and immunoglobulins are not functional in this phenomenon. The focal nature of infection is not explained by the lack of uniform predilection of micro-organisms for the organism. In such cases, the infectious agent is found within both the typical damaged elements and the adjacent areas which have been left intact [132, 145].

In a population reliant to inherent constitutional immunity to an infectious agent, individuals can be conveniently divided into three categories, as follows: 1) totally resistant organisms (no susceptible structures), 2) mildly susceptible organisms in which a few foci appear and the infection runs a benign course, 3) organisms in which the number of susceptible structures is high and the infectious process develops in a severe form with formation of many foci of specific infection [4, 78].

The mating of resistant and susceptible individuals gives rise to progeny with intermediate degrees of infection and extent of infectious foci thus the hybrids can never be more resistant than their parents. These genetically predetermined differences are proof of heterozygosity: one of the constitutionally immune parents passes on a lack of receptivity to a certain infection to the descendant in some parts of organs , while the susceptible parts of a different molecular configuration are inherited from the second parent (Figure 14).

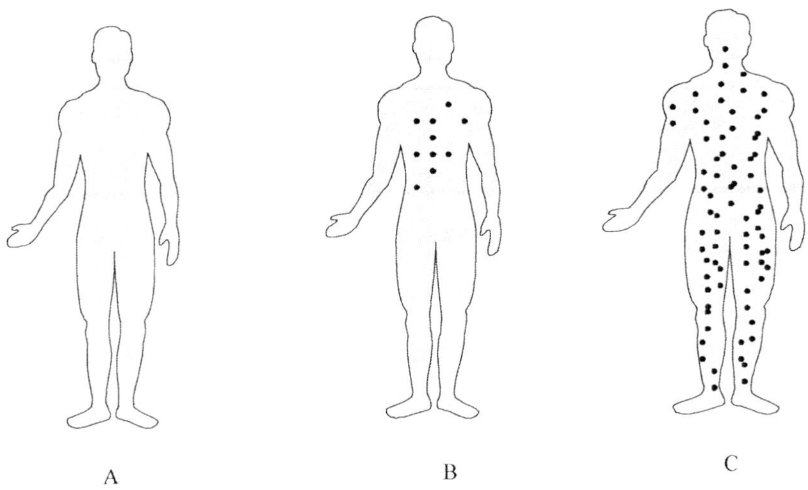

Figure 14 . Structural predestination of individual diversity in the severity of a disease. A, totally resistant organism which contains no susceptible structures; B, mildly susceptible organism in which a few sensitive foci exists and the disease runs a benign course; C, organism in which the number of constitutionally susceptible structures is high and the disease develops in a severe form [78].

To study these problems, the molecular aspects of ecology and physiology must be explored. The clonal histochemical architecture of the small intestine, oesophagus, breast, and thyroid was demonstrated in normal, as opposed to mosaic animals, using a diversity of the X-linked enzyme glucose-6-phosphate dehydrogenase (G6PD), as a clonal marker. The results were found to depend upon the simultaneous study of tissue from normal mice, mice homozygous for abnormal G6PD levels, and heterozygous mice. A dual population of cells was found in the heterozygous animals in all tissues studied. The oesophageal epithelium showed sharply separated alternating paths of positive and negative cells. Small intestinal crypts were always monophenotypic while villi were polyphenotypic, indicating that crypts are monoclonal and villi derive from more than one crypt. In contrast, thyroid follicles and breast acini showed a mixture of mono and polyphenotypia, consistent with a polyclonal origin [152].

Any infectious disease affects only focal areas in the affected organisms, and the amounts and distribution of infectious lesions vary from patient to patient. The diversity highlights both individual and intra-individual differences in susceptibility of various parts of the organism under consideration.

Observation of Biodiversity in Non-infectious Diseases

During the last decades of XX Century, researchers have proposed bacterial and viral triggers for some diseases that has been accounted as non-infectious before [153]. Many bacteria and viruses have been linked to "non-infectious' diseases. Human papillomavirus (HPV) get an uncontested role in cervical cancer, hepatitis B and C viruses – in hepatic cancer, Epstein–Barr virus - in lymphoma and schistosomes - in bladder cancer.

Infectious agents have also been suggested to contribute to leukaemia, Crohn's disease, arthritis and rheumatic fever. However, despite the growing evidence, the role of infectious agents in a wide range of diseases is still contested. For decades, scientists have unsuccessfully attempted to link infection with atherosclerotic cardiovascular disease as well as .to find the link between herpes simplex virus type 1 (HSV1) and the Alzheimer's disease. The triggering role of microbial agents in obesity, diabetes and other metabolic diseases continue to be very controversial too [41].

Nevertheless although the above mentioned links are preliminary and not supported we can continue to consider these group of disease as non-infectious ones. From a vast number of noninfectious diseases only some of them (atherosclerosis, obesity, senescence, schizophrenia, and autoimmune aggression) will be considered below in the try to get new understanding of this kind of pathology to be gained from the view-point of biodiversity.

Atherosclerosis

Atherosclerosis is accounted as the world's biggest killer. Since 1990 more people have died worldwide from only one clinical form of atherosclerosis – the coronary artery disease than any other disease. Cardiovascular diseases now represent more than 30% of all deaths worldwide. [154].

The disease can show itself in different forms form of the affection of blood circulation: coronary artery disease (CAD), myocardial infarction, stroke, high blood pressure or other. Any associations of above forms has been observed too. The etiology of atherosclerosis remains uncertain. The "response-to-endothelial injury" theory is most widely accepted, but nobody can name which ecological or physiological factor is responsible for both development of this "injury" ane its epidemic spread around the World.

Epidemiological Observations

Racial and Other Genetic Differences in Atherosclerosis

Studies of atherogenic vascular pathology have shown a racial differences in the predilection to atherosclerosis. In the Caucasian population, the pathology is likely to be in a proximal location such as the bifurcation of the carotid arteries in the neck. This type of lesion is accessible to surgical correction. Among the Orientals the vascular degeneration is more likely to occur within the brain. The Orientals are also more likely to develop small vessel diseases that may cause different types of stroke including the hemorrhagic one [155]. Orientals are generally known to have more intracranial vascular stenosis and a lower incidence of carotid stenosis than Caucasians [156].

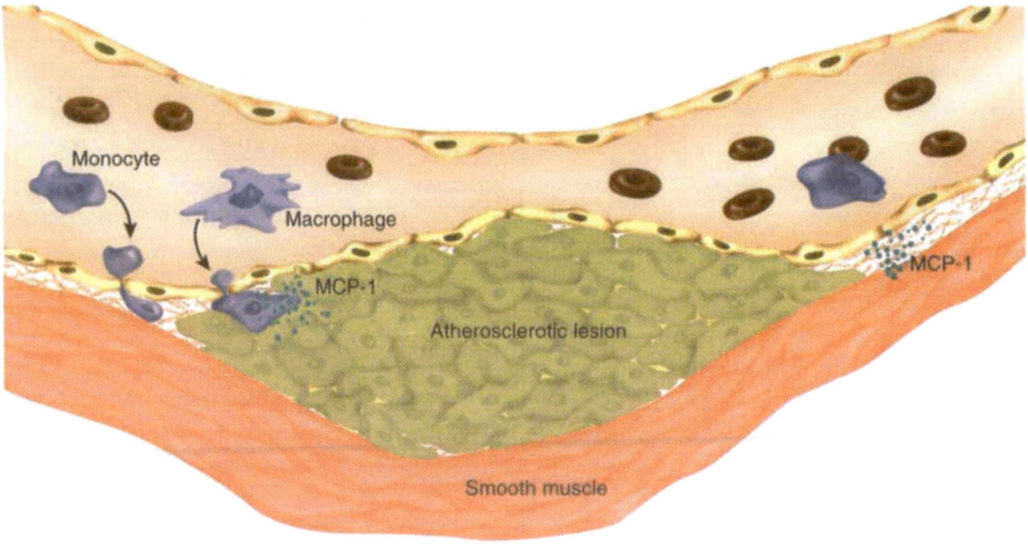

Figure 15. The scheme of an artery with atherosclerous plaque, buildup of waxy deposits inside the blood vessels.

Beside, some observations indicate allelic heterogeneity for genetic susceptibility to atherosclerosis [154]. The Icelandic HapA risk haplotype was associated with increased neutrophil release of leukotriene B4 (LTB4), which is synthesized from LTA4, indicating that the ALOX5AP variants might be pro-inflammatory. LTB4 activates monocytes, which migrate through arterial endothelium and differentiate into macrophages that take up lipid to become foam cells, a key early event in atheroma development (Figure 15). Blockade of the LTB4 receptor has been shown to reduce the amount of lipid accumulation, monocyte infiltration and lesion size in Ldlr–/– or ApoE–/– (apolipoprotein E) knockout mouse models of atherosclerosis (which are susceptible to atherosclerosis owing to dyslipidaemia). There is still a large gap in our knowledge of the genes that are involved in susceptibility for most patients with multifactorial CAD [154]. But the most gap is formed by phenomenal diversity in the location and size of atherosclerotic affections.

Inter-population Diversity in the Incidence of Atherosclerosis

Atherosclerosis is the most common cause of death not everywhere in the World but among populations of "industrialized" countries especially in the USA, England, Scotland, Finland, Scandinavia, and Russia that is likely considered as the result of diet and other environmental factors (widespread economic hardship and social upheaval, a high prevalence of cigarette habituation, and a diet high in saturated fats). In contrast, the frequency of the disease among "non-industrialized" populations of the Far East is significantly lower than that in the West. Atherosclerosis is also rare among the native populations of African continent with traditionally rural and agrarian way of life. This kind of pathology it is more characteristic of the more mixed populations. Recently its prevalence in developing countries has rapidly increased also. The epidemic of cardiovascular diseases has taken on a global dimension.

Age Diversity in the Incidence of Atherosclerosis

Atherosclerosis may begin as early as in childhood, but it is the advanced stages of this condition that are the most dangerous later in life. It is a slow developmental disorder. It usually takes decades for the signs and symptoms of the disease to be apparent. The age-adjusted mortality that is due to CAD in North America and Western Europe is falling each year, but as these populations age the prevalence of CAD is actually climbing [154]. Some people experience rapidly progressing atherosclerosis in their 30s or later. Before age 60, men are more likely to have heart attacks than women are.

Clinical Observations

The clinical entity of atherosclerosis results from a particularly complex disease process, with scope for a large degree of heterogeneity among patients who have apparently similar clinical presentations and, as a corollary, extensive genetic heterogeneity. This disease is characteristic of especially variable manifestations. Early lesions appear as fatty streaks,. These later develop into atherosclerotic plaques, which consist of a lipid core and an overlying fibrous cap.

Individual Diversity in the Signs of Atherosclerosis

Individual manifestations and course of atherosclerosis are highly variable. Many people with structurally advanced atherogenesis may have no symptoms and experience no functional impairment. On the other hand, initially asymptomatic patients may be subjected to sudden cardiac or aortal death, whereas many other people present for a long time with clinically important symptoms and signs of relevant chronic diseases (coronary heart disease, high blood pressure, renal deficiency or other local dysfunction of blood circulation) or suffer from various consequences of former myocardial infarction or stroke. Initially thought to be a chronic, slowly progressive, degenerative disease, it is now apparent that atherosclerosis is a disease with periods of activity and quiescence.

Cellular and Molecular Diversity in Atherosclerosis

Atherosclerosis is an intra-arterial pathological process, which induces inflammatory endothelial lesions with appearance of lipid-containing foam cells. The earliest pathologic lesions of atherosclerosis are the fatty streaks that can be found in the arteries of many individuals by age 20 years. The fatty streaks are formed by focal accumulation of serum lipoproteins within the intima of the vessel wall. The followed cholesterol metabolic pathways have a key role in atherosclerosis development [154]. They continue in the development and further progression of specific patched formations (plaques) on the innermost of arterial wall leading to reduced or blocked blood flow. The plaques can be either single or multiplel.

It was hypothesized that formation of plaques may be triggered by some infectious agents [157]. There are a few privileged individuals that are resistant when challenged by risk factors. Knowledge of the resistant factor(s) and their mechanism(s) of action would lead to a solution of the etiology. This leads to interest in genetic immunity to atherosclerosis.

Diversity in the Location and Size of Atherosclerotic Affections

Although a systemic disease, atherosclerosis manifests in a focal manner and affects stochastically different organ and systems in different patients for reasons that remain unclear. Any individual case of atherosclerosis begins when some initial injury of arterial intima induces the fatty molecular material to deposit at the injured place in the form of plaques, resulting in focal narrowing of the arterial lumen and eventual impairment of blood flow. Atherosclerotic plaques can either remain silent for a long time, or progressively narrow the lumen, which restricts flow (e.g. producing angina pectoris). Rupture or erosion of a plaque can precipitate an acute occlusion owing to the formation of a thrombus or blood clot. In some cases the progression of the plaques can lead to rupture of arterial wall and appropriate bleeding.

Plaques (either single or multiple) can be individually stochastically localized on the wall of either one or some of arterial vessels. Plaque buildup in the vessels that supply an organ with blood may cause relevant dysfunctions of the organ. Symptoms of any atherosclerotic affection depend on the size of the affection's location as well as on the danger of the focal blood stream blockage. The most clinically important manifestations of the disease are focal and tend to occur at individually located sites.

The symptoms of atherosclerosis differ depending upon the size and location of plaques and the grade of its development. In some patients the intensity of affection and its progression is rather high, whereas in other ones the manifestations are weak. The focal character of the affection is the principal feature in this kind of pathology.

Any artery in the body as well as any part of the artery can be involved, but only severe local narrowing or obstruction of some arteries, those that supply more critically-important organs first of all heart, brain, kidney are revealed itself in life threatening form. Plaques may also build up in more than one place in the artery. According to location and degree of specific affections this can result in relevant local ischemic accidents which manifests appropriate clinical syndromes and outcomes [154].

Coronary (heart) arteries atherosclerosis leads to coronary artery disease i.e. to chest pain, heart attacks, myocardial infarction and sudden death. Thoracal aorta atherosclerosis leads to the aorta aneurism. Atheroma in arm or leg arteries may cause occlusive disease of relevant peripheral artery. In carotid arteries of the brain the atherosclerosis leads to sudden dizziness, weakness, loss of hearing or speech, and blindness. Cerebral artery atherosclerosis may lead to a stroke on one side of the body. Hepatic artery atherosclerosis leads to insufficiency of relevant lobe of the liver. Affection of pancreatic artery leads to insufficiency of the pancreas. Abdominal aorta atherosclerosis leads to abdominal aorta aneurism. Location of plaques in a mesenterial artery can induce the ischemic bowel disease. The atherosclerosis of a renal artery can leads to insufficiency of relevant kidney, for instance high blood pressure resistant to treatment. Peripheral artery atherosclerosis is caused by plaque buildup in the arteries that supply the extremities of the body (such as the hands and feet) with blood. In the femoral artery of the legs the atherosclerosis leads to cramping and fatigue in the calves of the legs when walking and finally to the necrosis of the leggs or fingers.

Moreover the unequal affection of different parts of organism circulatory system is uneven, too [158]. The progression of the plaques leads to narrowing of the affected arteries and relevant local limitation of blood flow. Atherosclerosis usually does not cause symptoms until it severely narrows an artery, or totally blocks an artery. Usually symptoms begin when there is at least a 75 percent blockage in the locality of affected artery. As a plaque encroaches on the luminal area blood flow becomes limiting and flow cannot increase on demand.

The Origin of Diversity in Atherosclerosis

Altough the identification of susceptibility genes for these common, complex diseases is a vital goal the real progress in unearthing novel genes has been slow. The optimistic predictions of the impact of genetics on the diagnosis and prevention of atherosclerosis [159] can not be realized because it is impossible to image the existence of genes responsible for any individual localization, size or intencity of development of atherosclerotic affection. Classical genetics can not explain why some person revealsonly one asymptomatic small plaque on the wall of a peripheral artery whereas other person suffer of multiple plaques located in many arteries which provide very important organs with blood. In other words, which genetic factors are able to determine while some small parts of arterial wall are receptive to atherogenic influence, whereas far more extensive neighboring parts of the same vessels are not injured.

Thus, it become obvious that humans, i.e. different parts of their individual aortal circulatory systems reveal extraordinary genetic diversity in their predisposition to atherosclerosis. Affected people have many differences in manifestations of their atherosclerosis especially in its location, size, and the grade of expression. Individuals of an observed human population can be conveniently divided at least into four principal categories, as follows:

1) The totally unaffected (homozygous) organisms that have no signs of atherosclerosis. Those people formed the majority of ancient and current naïve populations. They also form the substantial quote of some modern populations.
2) Mildly affected (heterozygous) organisms in which few small signs have been appeared whereas other areas of the body arterial vessels are left unaffected.
3) Mostly serious form of the disease with intermediate grade of atherosclerosis formed by location of solitary plaques in the artery of especially important organs.
4) Extraordinary rare but exceptionally dangerous disease with dissemination of multiple atherosclerotic lesions in many parts of arterial basin.

The future identification of specific susceptibility genes will add to our knowledge of the molecular pathophysiology of atherosclerosis, [154] but will not explain the phenomenal diversity in the location and size of atherosclerotic affections that is the main factor which determine the location, the course, the severity and the exit of any case of disease. It will not refine the identification of high-risk individuals and suggest areas of research for drug discovery. Escape from atherosclerosis can be achieved on the basis of its prophylaxis, which should be elaborated by medical genetics.

Obesity

Epidemiological Observations

Obesity is far more than a cosmetic problem. The global epidemic of obesity is associated closely with a plethora of dangerous chronic illnesses such as heart disease, diabetes, high blood pressure, stroke, different kinds of cancer, liver disease, gout, disease of the joints, and so on. More than 300,000 adult excess deaths in the United States each year are related to obesity. Obesity is now the second leading cause of excess deaths in the USA [160]. and may reduce the life expectancy for the most affected populations [161].

Obesity takes its toll in death and disability, as well as in the considerable effort required to understand the molecular and genetic aspects of how it develops, and many questions about the nature of the disease do not receive the answers they deserve. Nothing, for example, is known about the factors and mechanisms of the epidemic spread of obesity in populations, as well as the origin of the extraordinary diversity in the clinical manifestations of the disease.

Racial Disparity in Predisposition to Obesity

Epidemiologic data demonstrate a positive association between obesity and certain measures related to the metabolic syndrome that differ among individuals of different ethnic/racial backgrounds. Although the effect of genetic factors on these differences is not completely deciphered, more insightful information have been obtained by taking into account the degree of ancestral admixture in populations [162].

Inter-population Diversity in Obesity

Obesity is referred to as a disease of civilization, but it is not a recent phenomenon. Its history can be traced back at least 25,000 years, thanks to the evidence of corpulent women sculptures made by Stone Age artists. But by the end of the twentieth century, the number of obese people in some countries had begun to increase in a genuine epidemic. In the USA the epidemic dated from1960, and continues to progress [163] In just two decades, the incidence of adult obesity has increased by 80 percent: from 15 percent in 1980, to over 27 percent in 2000 [160, 164]. The childhood obesity epidemic is bourgeoning too [165].

Today, obese peoples form more than 60 percent of Americans aged 20 years and older. Of these, obesity is more prevalent in African-American women than in their white counterparts [166]. Differences among ethnic and racial groups in obesity-related traits have been clearly established in the scientific literature [167]. More than half of non-Hispanic, black women aged 40 years or older are obese. Thus this kind of pathology it is more characteristic of the more mixed populations. This kind of pathology it is more characteristic of the more mixed populations.

Genetic Predilection to Obesity

Heritability is the less disputable paradigm about obesity. Many compelling evidence confirms that development of obesity as well as most variance in its incidence is attributable to some genetic factors [168-170]. Consequently, one can maintain that main features of obesity are determined by genetic predestination. The most evidence comes from classic genetic analyses of pedigree, twin, adoption and familial aggregation studies which estimated of obesity heritability of 0.7-0.8. Obesity tends to run in the affecting families, suggesting a genetic cause of the disease. Adults who were adopted as children were found to have weights closer to their biological parents than to their adoptive parents. Twin studies suggest a heritability of fat mass of between 40 and 70% with a concordance of 0.7-0.9 between monozygotic twins compared to 0.35-0.45 between dizygotic twins [169, 171]. Nevertheless, many aspects of obesity inheritance are considered as extraordinary complex and still poorly understood, despite active investigations [172].

Several genes have been shown to contribute to obesity predilection. Most of them is now discovered up to the level of molecular regulation of food intake and expenditure [173, 174]. Like any other genetically predetermined attribute of human body, obesity can not be corrected by restrictions in eating and intensification of physical exercises. Any voluntary efforts to reduce fat deposition are resisted by very potent response of abnormal regulation of fat-storehouses [173]. Really, "Can a Cush'ite change his skin? or a leopard its spots?" [{1984 523 /id}].

Clinical Observations

Individual Diversity in Manifestations of Obesity

Humans are extraordinarily diverse in their predisposition to obesity. Affected people have many differences in manifestations of their obesity as well as in the grade of its expression. Individuals of an observed human population can be conveniently divided at least into four principal categories, as follows:

1) The totally unaffected organisms that have no signs of obesity. Those people formed the majority of ancient and current native populations. They also form the majority of some modern populations.
2) Mildly affected organisms in which few small signs have been appeared whereas other areas of the body are left unaffected.
3) The organisms with intermediate grade of obesity. Most expressive are locations of fat-storehouses around belly ("apple" shape) and hips and buttocks ("pear" shape).
4) Most serious form of the disease with development of nearly total obesity of entire body, formed by combination of several locations, shapes and sizes.

Diversity in the Locations and Sizes of the Fat-storehouses

The shape, location, size and rate of obesity progression may be different in different individuals or different organs as well. Equally with total affection of almost all the body, there are several key areas of restricted focal fat-storehouses locations around the body. The areas of a human body where solitary focal obesity can be observed include facial jowls, neck, upper back of the arms, upper back underneath the bra straps, areas near the underarms, waist, love handles, hips, breast, buttocks, abdomen, belly, thighs, knees, calves, and ankles. In central obesity (or 'apple-shaped' or 'masculine' obesity) the main deposits of body fat are localized around the abdomen and the upper body. In mild and intermediate forms of the disease total affection of almost all the body can include all the locations whereas the locations in the form of solitary bulges also can be observed in various individual combinations. In most cases, obesity is presented by focal manifestations of individually different grade.

The locations and sizes of fat-storehouses are different for each individual (Figure 16). Someone has only one small bulge whereas another person has three or four local areas where excess fat exists. The neck alone also can be composed of an excess amount of fat. Some people may have sagging jowls right underneath their jaw line. The whole belly or its upper and lower parts are the typical areas where extra fat is stored. A couple folds are occurred when increased fat accumulates in the upper back. Some people have bugles of fat that begins below the underarm and then comes down before the waist starts. Some affected women have fat in the back of the upper arms. The thigh, especially its outer part is very commonly affected. Some patients have a significant amount of fat in the inner low quadrants and not much in the outer areas of the buttocks. For some people, the fat is stored in the upper buttock whereas for others, it is in the lover outer buttock. Extra fat bulges may be symmetric or asymmetric. Someone can be heavier on the right side, but another one on the left. Not much fat has been observed in the calf or ankle. Fatty calves and ankles are often an innate condition that patients have had from childhood, even when they have no weight problems. The sizes of obese masses and their dislocation around the body predestine individual diversity in the course and severity of obesity.

Figure 16. Some variants of the sizes and locations of the fat storehouses in obesity.

Thus, one can conclude that within every obese human body there are at least two co-existing parts of fett tissue with unequal predisposition to obesity. The parts exist in a form of separate patches of different sizes and locations and are stochastically dispersed around the body [41].

Cellular and Molecular Diversity in Obesity

Every kind of living being is provided a constitutive physiological system that maintains normal body structure within a genetically predetermined shape and size. A part of the physiological system that regulates body fat has now been discovered to the molecular level [173]. It acts through the molecular physiological mediators of energy intake and energy expenditure. The ability to burn fat for energy, and the achievement of satiety are performed by the functioning of a particular set of hormones and other molecular physiological agents. The system includes the hormonal regulation of food intake, first by leptin and insulin. In a case of obesity, the work of this precise and powerful system becomes disturbed.

In a normal organism, leptin performs the role of inhibiting excess food intake [175]. Disturbance of the leptin production or function determines the development of obesity. Deficiency of leptin production results in an insatiable appetite leading to severe early-onset obesity. The same result is achieved by a mutant modification of the leptin structure which forms an incongruence between leptin and its receptor. Both states can be corrected by leptin treatment [176].

On the other hand, in an obese body, the leptin function can be normal or its content increased by leptin treatment, but its performance can be stifled by a derangement of receptor cells, i.e., by the body's constitutional immunity to leptin's influence because of incongruent receptors [170, 176].. This blocking effect of mutant modification of leptin receptors cannot be corrected by leptin treatment.

Analogous dramatic immunity of cells to the specific action of relevant physiological agents functions in many kinds of hormonal pathology [41]. For example, the immunity of cells to insulin is a major determinant of the decline of glucose tolerance. Similarly, non-insulin-dependent diabetes mellitus is characterized by pathological hyperglycaemia in the presence of higher-than normal levels of plasma-insulin. Likewise, a pathogenic decrease in cell sensitivity to vitamin D3 determines the familiar forms of rachitic, and the immunity of cells to androgens causes the phenomenon of testicular feminization. An analogous resistance of cells to corticosteroids determines the pathogenesis of Cushing's disease. In yet another example, the general and peripheral immunity of cells to thyroid hormones determines the genesis of a range of relevant disturbances (RTH-syndrome). This resistance is an inherited inability to respond appropriately to the T3 hormone linked to mutations in the thyroid hormone receptor (TR)-beta [177]. The principles of cell immunity to physiological agents are analogous to those in the constitutional (genetic) immunity to infections [13, 41].

There are some commonalities in the structures and functions of molecular physiological and infectious agents [4, 13, 41]. The B chain of the cholera toxin and the beta-subunits of thyrotropin, luteinizing hormone, human chorionic gonadotropin, and follicle-stimulating hormone are shown to have a region of sequence analogy believed to correlate with their ability to bind to receptors on cell membranes. A sequence analogy is also defined in the alpha-subunits of these glycoprotein hormones and the region of the cholera toxin A1 chain believed to be responsible for adenylate cyclase activation [178]. Like hormones, cholera toxin increases adenylate cyclase activity, raising levels of cyclic adenosine 3',5'-monophosphate, which activates a cellular lipase [179].

What is more, many hormones share their specific targets with the acting molecules of infectious agents. Tetanus toxin binds specifically with the thyroid thyrotropin receptor [180]. Similarly, the rabies virus receptor appears to be closely associated with the acetylcholine receptor. Epstein-Barr virus has a strong tropism for lymphocytes and those of the complement C3 receptor [181]. The human T-cell leukemia virus preferentially infects the cells of the helper/inducer class [182]. Likewise, the cardioviruses and influenza A and B viruses bind to the glycophorin A molecule on human erythrocytes and this glycoprotein also carries the M and N blood group antigens [183]. Virus-induced obesity in mice alters their response to leptin [184]. Human adenoviruses increase adiposity in humans [185] and other animal models [186, 187].

Taken together, these data make it permissible to suppose that in fact all "viral" and "bacterial" specific receptors have not been evolved to interact with microbial pathogens, but to provide the physiological functions of hormones or other physiologically active molecules. The evolutionary adaptation of harmful microbes to a parasitic way of life could be performed by the exploitation of pre-existent physiological receptors or other cell structures and functions [185]. The subsequent evolution and spread of genetic immunity to infections (Figure 3) could create a constitutional genetic immunity to the molecular physiological agents responsible for regulation of fat intake and metabolism, thus providing the spread of relevant forms of pathology [41].

Although the spread of obesity is progressing as in a worldwide pandemic, the possibility that infections might be etiologic agents in obesity has received relatively little confirmation. Experiments link obesity to four animal viruses (rabies virus, Rous-associated virus, Borna virus, and SMAM-1). A human virus, adenovirus 36, increases adiposity in experimentally infected chickens, mice, and marmosets. But direct correlation of human obesity with an epidemic of any infection was not found [153, 185-189]. Nevertheless the indirect action of infections on obesity genesis and spread cannot be denied today [41].

Age Diversity in the Onset of Obesity

Different parts of a body can be undertaken by obesity at different ages. The question arises immediately about genetic determination of individual and intra-individual diversity in the age of obesity development. How can we account for the fact the individual differences in the age of obesity development? Should one accept that in some individuals the genes of obesity begin to work in the period of childhood whereas in other bodies the time to become obese begins at the adulthood? Can one suppose that some human genomes can contain different genes of obesity which work at different age? A lot of observations about constitutional immunity to infectious agents and senescence confirms that correlation of sensitive and constitutionally immune cell clones of an organism can change with age [4, 94, 190] and allow us to accept this working hypothesis for the discussed cases of analogous immunity to leptin or other molecular physiological agents involved in the development of obesity.

The Origin of Diversity in Obesity

The mutant formation of the life-saving genetic immunity to infection could create a life-threatening resistance to the relevant hormone. This kind of individual molecular constitution could spread in a population as a result of natural selection for genetic immunity to infections that has been performed differently in ecologically distinct populations [14, 191], thus forming ethnic differences in the molecular constitution of both hormones and their targets. Thus, molecular physiological agents of one ethnic group become xenobiotic to the analogous agents of another group, and vice versa [41]. On the other hand, any change in one molecular physiological agent should entail relevant evolutionary changes in all other molecules connected to it [191].Thus, the evolutionary process removes the molecular physiological incongruence within an ethnic group but creates an incongruence between groups.

Humans are extraordinarily diverse in their manifestations of obesity. Affected people have many differences in manifestations of their obesity as well as in the grade of its expression. The shape, location, size and rate of obesity progression may be different in different individuals or in different organs as well. Equally, with total affection on almost all the body, there are several key areas of restricted, focal, fat-storehouse locations around the body. In most cases, obesity is presented by focal manifestations of individually different grades and locations (Figure 16). One person has only one small bulge, whereas another person has one, two, three or four local areas of excess fat. Additionally, these extra fat bulges may be symmetric or asymmetric. Someone can be heavier on the right side, another on the left.

Thus, one can conclude that focality of affection is the principal feature of obesity. Within every obese body there are at least two co-existing clones of homogenous fat cells with opposite predisposition to the development of obesity. These clones exist in a form of separate populations stochastically dispersed around the body. The sizes of obese cell masses and their dislocation around the body predestine individual diversity in the course and severity of obesity.

The origin of this very important feature has been ignored by previous systematic investigations. Heretofore, the origin of individual and intra-individual differences in location, shape and size of fat-storehouse has formed an unsolved riddle of obesity. Should one suppose that the physiological functions of fat metabolism and their regulation in obesity are disturbed differently in different populations, in different generations, as well as in different body areas of the same person?

How can we account for the fact that the identical cells of an organism differ from one another when all contain the same genome complex? Up to the end of the twentieth century, it was a riddle for both pathology and genetics [78, 190]

Although it is now an obvious truism that a person's genetic makeup has a principal influence on the development of obesity, the special characteristics of the genome that determine the zonal distribution of obesity around the body has been unknown. In contrast, the origins of the analogous features of infectious diseases and many noninfectious ones have been deciphered [4, 190]

The phenomenon of dapple distribution, as well as the diversity in the grade of focal diseased affections, is not new for the general paradigm of pathology, but its origin and significance were discovered only at the end of twentieth century [78, 190, 192]. All diseases (heart, circulatory, pulmonary, mental, skeletal, renal, gastrointestinal, endocrine or metabolic) are characteristic of the diverse grade and focal dislocation of specific lesions.

The genesis of these intra-individual diversities in the case of obesity can also be explained by the hybridization of persons with different genetic immunity/susceptibility to internal factors determining obesity [41]. The mating of resistant and susceptible individuals gives rise to progeny with intermediate degrees of obesity and extent of obese foci. One can suppose that focal diversity in the rate and kind of dispersion of obesity manifestations is such as a result of hybridisation between two genetically different organisms: one of them was constitutionally immune to leptin and insulin whereas its mating partner was normally sensitive to this molecular physiological agents [41]. This conclusion is supported by observations that subjects who are heterozygous for the leptin mutation develop obesity which is not as severe as in homozygotes [193], and that variation in obesity-related traits is associated with interethnic genetic admixture [194].

The modern increase of obesity incidence in some countries can be associated with the spread of constitutional immunity to a lot of infectious diseases performed by natural selection, followed by increased hybridization between representatives of populations with different genetic make-up which in turn could provide incomplementariness between a hormone and its receptor, thus determining the state of leptin deficiency. These data provide a starting point for the clinical, molecular and genetic analysis of the obesity cases affecting different populations. At least four variants of the incomplementariness could be considered here:

Aboriginal hormone meets mutant aboriginal receptor; 2) Mutant aboriginal hormone meets aboriginal receptor; 3) 'Alien' hormone meets aboriginal receptor; 4) Aboriginal hormone meets 'alien' receptor.

The existing set of observations allows us to conclude that the spread of obesity is determined by a rise of heterozygosity in the population under observation. Extraordinary diversity in the obesity course, the manifestations and severity of specific affections, and their sizes and focal disposition around the body could be created by inter-ethnic hybridization (xenogamy) of persons with different grades of hormone-receptor heterozygosity. Prerequisites for the intensive performance of this mechanism arise today everywhere in the world, especially among more mixed populations.

Unlike the populations of many other countries most of the ethnic groups in the United States have resulted mainly from the intensive intermixing of European, African, and Native-American populations during the colonization and continued mutual habitation of the New World. Genetic variants or alleles from these previously isolated parental populations were brought together in new combinations, establishing the gene pools of the various contemporary European-, African-, Hispanic-, and Native-American resident populations. Heterozygous individuals of these populations who inherit variants that predispose them to disease-related traits have a greater chance of acquiring the disease of obesity[167, 194].

The mutual exposure, analysis and evolutionary comprehension of epidemiological, clinical, immunological, genetic and experimental data concerning principal characteristics of obesity showed that many of them are also common to those of some other kinds of diseases, especially infectious ones. As with genetic immunity to infections, obesity is determined by a constitutional incongruence between relevant acting agents and their molecular targets. Thus, according to the hypothesis that was discussed above, the genesis of obesity and its spread in a population are associated with evolutionary formation and the spread of genetic immunity to infections followed by inter-ethnic genetic admixture. The number of obese people first began to increase in a genuine epidemic in territories with ethnically-mixed populations. In these circumstances, thinking about the eradication of obesity, should we pause to remember the commandment of Abraham: "... you will not take a wife...from the daughters of the Ca'naan'ites in among whom " we are dwelling [33]? What are the tests for genetic congruence that should be elaborated and used today [195]?

SENESCENCE

Senescence is one of the most important topics of human biology and medicine. It is the result of a productive form of aging leading to organism's programmed death. In analyzing the origin of senescence the entire body of facts is considered. Scientists are using observations in populations and clinic, cells in culture, yeast, nematodes, and mice, and the human genome, to gain a better understanding of why and how we age. Although the aging field is full of theories, the phenomenon continues to remain mysterious. The pages immediately following are devoted to what we have to say about the phenomenon today. The author tried to analyze the nature and importance of this kind of human body' genetic diversity, using examples from epidemiology, clinical observations, genetics, immunology, and molecular biology.

Epidemiological Observations

Diversity in Human Life Span

The human life span has varied historically from generation to generation, from population to population and from individual to individual. According to [196], during the patriarchic period the average life span consisted of 912 years of age, after Flood – 484 years and at the end of biblical times – 185 years. Today the average life span of man is approximately 70 years of age (Figure 17). In the 20th century, most industrialized countries have experienced large increases in mean life expectancy, from around 50 years to around 75–80 years. This has been due to a marked reduction in early life mortality during the first half of the twentieth century, followed by a less recognized almost twofold reduction in mortality at ages above 70 in the second half of the century [197].

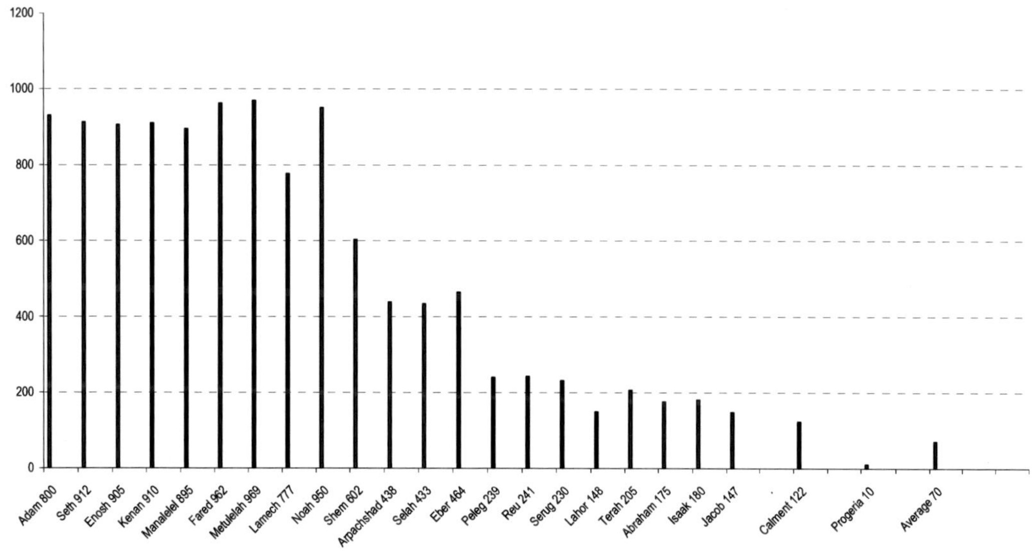

Figure 17. The biblical account of the life span for the men living either before or after the Noah's flood.

Meanwhile, there are individual differences in life span. Senescence changes are not synchronous in any population. Different people do not grown old simultaneously. If we look at the current diapason of individual life spans, we would see a definite contrast between them. Jeanne Calment was the oldest person whose exceptionally long lifespan, had been verified by official document. She died at 122 in 1997 [198]. Some scientists discuss on whether today's humans can live to 130 or 150 years.

In contrast, there are premature ageing syndromes (Werner Syndrome, Hutchinson-Gilford Progeria Syndrome, and some other) that mimic many signs associated with ageing and result in very early death. This kind of human very short lifespan phenotype is associated with early-onset of atherosclerosis and other diseases of aging. For example, Hutchinson-Gilford Progeria Syndrome is a genetic state that accelerates the ageing process to about seven times median rate or about seven times lifespan of Jeanne Calment that . Because of this accelerated ageing, a child of ten years will have similar respiratory, cardiovascular, and arthritic conditions that a 70-year-old would have. If they have reproduced, their offspring would be expected to inherit the same genetic status, Werner Syndrome has its onset between the age of 15 and 20 [199].

Modern human populations are very different in their age structure as well as in intensity of their senescence. Old-timers may make only a smallest part of a population, for instance in Cuwait, or fare more large part of it, for instance, in Monako Figure 18). Human individuals of an young generation can be conveniently divided into three main categories, as follows: 1) the majority of totally unaging organisms that have no senile signs; 2) mildly senile organisms in which a few senile signs have been appeared; 3) the minority of organisms in which the number of senile structures is high and the senescence develops in a severe form with formation of many foci of senile regression (e.g. the premature ageing syndromes). In a more mature and especially in an old generation one can observe the change of correlation between the above groups.

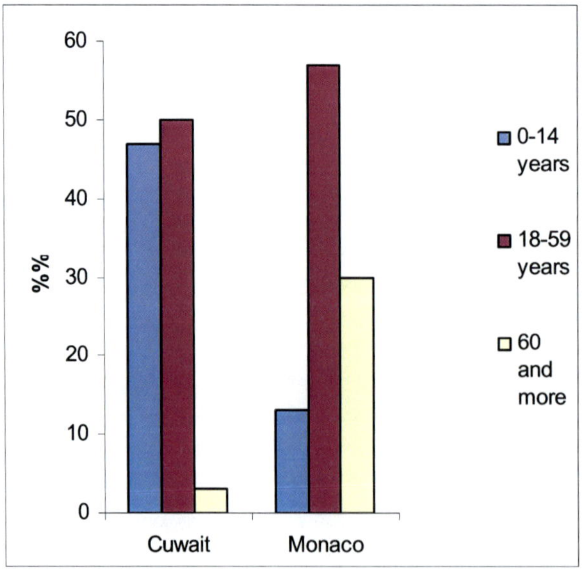

Figure 18. Different age structure of two human populations (according to [208]).

Genetic Determination of Aging

It was believed for a long time, that our senescence is programmed in our genes and it is true. Firstly, genetic regulation determined essential life span of every species. Secondly, specific genes that increase life span were first found in the nematode roundworm. Mutations in the *age-1* gene in *C. elegans* are particularly interesting. Findings from animal studies have provided evidence that individual genes can have a significant effect on lifespan. Some alleles of *age-1* can confer a near doubling of lifespan. This mutation results in a 65% extension of mean life span and a 110% extension of maximum life span [200]. Thirdly, there are genetic differences between populations and families. Extreme longevity is often clustered in families, with the percent of very old siblings greater than expected. According to [201], Werner's gene [WRN] is a longevity assurance gene. If WRN is missing, then we see accelerated ageing [199].

Findings from animal studies have provided evidence that individual genes can have a significant effect on lifespan. Furthermore, human genetic studies have shown that common polymorphisms in some genes influence lifespan, probably mainly through their association with diseases [{Christensen K, 2006 635 /id}] that arose with the development of human society [190].

Twin studies show that genetic differences account for about a quarter of the variance in adult human lifespan. Common polymorphisms that have a modest effect on lifespan have been identified in one gene, APOE [{Christensen K, 2006 635 /id}]. Gene variants with substantial beneficial effects have been proposed to exist. Human studies of longevity face numerous theoretical and logistical challenges, as the determinants of lifespan are extraordinarily complex. These challenges should be solved before other genetic determinants can be uncovered.

Molecular Regulation of Aging

The modern knowledge about the mechanism of aging has been achieved just recently in the try to decipher the Hutchinson-Gilford Progeria Syndrome (HGPS). From biological point of view, this syndrome can be considered as a legal genetic state that is characterized by acceleration of ageing process to about seven times the normal rate. The syndrome is attributed to aberrant structure of a gene known as lamin A, which leads the body to produce a shortened version of a protein molecule by the same name that is a part of the covering of the cell nucleus. The relevant cells share some of the aberrations that occur in cell nuclei in people with the syndrome, such as histone modifications and increased DNA damage [202]. Accordingly to the appearing of the 'mutation' around the body, children with the devastating Hutchinson-Gilford Progeria Syndrome age at seven times the normal rate, so a 10-year-old patient could have health conditions typical of a 70-year-old people.

Just recently, it was found that the same molecular process responsible for HGPS is active in healthy cells. Cell nuclei from 'normal' old individuals acquire similar defects as 'diseased' patient cells including changes in age-related nuclear defects are caused by sporadic use in healthy individuals of the same cryptic splice site in lamin A whose constitutive activation causes HGPS. The same molecular transformations occur legitimately in the cells of physiologically aged people [202].The finding suggests that lamin A participates in the aging process both in healthy and in 'diseased' individuals, bat this mechanism is "exaggerated" in the progeria people. Thus certain premature-aging disorders like progeria , which can end children's lives by their early teens, are somehow accelerated versions of current physiological aging [202].

Clinical Observations

Diversity in the Locations and Sizes of Senescence Affections

When any individual begin to grow old only some places in his/her body are involved in the process. The entire body do not undergo the senile regression simultaneously [203]. The forms and rate of ageing progression are different: in different species, individuals, organs, systems, tissues, organelles and macromolecules. At the same time, individually no tissue, organ or system becomes functionally exhausted even in very old organism [204]. That means while some parts of a considered organ are affected by senescence and exhausted others ones continue to function. They have longer life span.

All tissues can undergo regressive structural changes with age but not simultaneously, and not uniformly in their different locations, sizes and dispersion. Different parts of entire human body age differently. Moreover, the forms and rate of ageing progression are usually different not only in different organs but also in different parts of the same organ. Also they may be different not only in different tissues but in different places of the same tissue too. For instance, if in one person the senile wrinkles are displayed mainly on the face but in another person they are located mainly on one's back of his neck. When the smooth skin of the face turns into the wrinkles, it happens just owing to the local inequality of senile regression. At the same time, however, a number of neighbouring parts of the skin fully preserve their basic structure and function.

Another of the most evident examples of such senescence inequality is the difference in the rate and distribution of hair turning grey. Some people can begin to gray before 30. Most people over 40 have some graying of scalp heir. Although the tendency to graying is inherited as is baldness [205], different parts of hair are becoming an object of senile regression not simultaneously even being located in immediate proximity to unaffected loci of the same skin. On the early steps of senescence the grey hair flocks are observed on the temple (Figure 19a) or in the beard (Figure 19b). Usually they are located symmetrically being surrounded by a normally coloured hair. Asymmetrical location both of solitary and multiple grey hair flock also can be observed. In contrast, on the late steps of senescence a plethora of the grey hair surrounds the dispersed flocks of a normally coloured hair.

Also hair growth does not uniformly decrease after adult life. On the other hand, most women past 65, especially those of Mediterranian origin, have excess of hair localized over the lip and shin [206]. Conglomerates of melanocytes in the skin are also characteristic of oldest white people. Nevertheless, in every age group, even among the oldest, individuals can be found whose performance on many structures and functions does not decline over the total age span. For example, some people's performance of mental tasks is indistinguishable from that of younger adults [207]. No one died of absolute (total) senescence.

An impressive diversity of the adjacent areas of a blood vessel wall is characteristic of sclerotic manifestations of senile regression. Mosaic distribution of senile and intact areas is also characteristic of structure and function of bones, brain and all internal organs, such as lungs, liver, pancreas, stomach, heart, kidneys. This peculiarity of senile regression is demonstrated not only in form of senile changes but also in their fantastically mosaic dislocation, making impression, that aging is stohastic and nondeterministic process. At the same time those focal differences form basis for shaping the body structure during senescence.

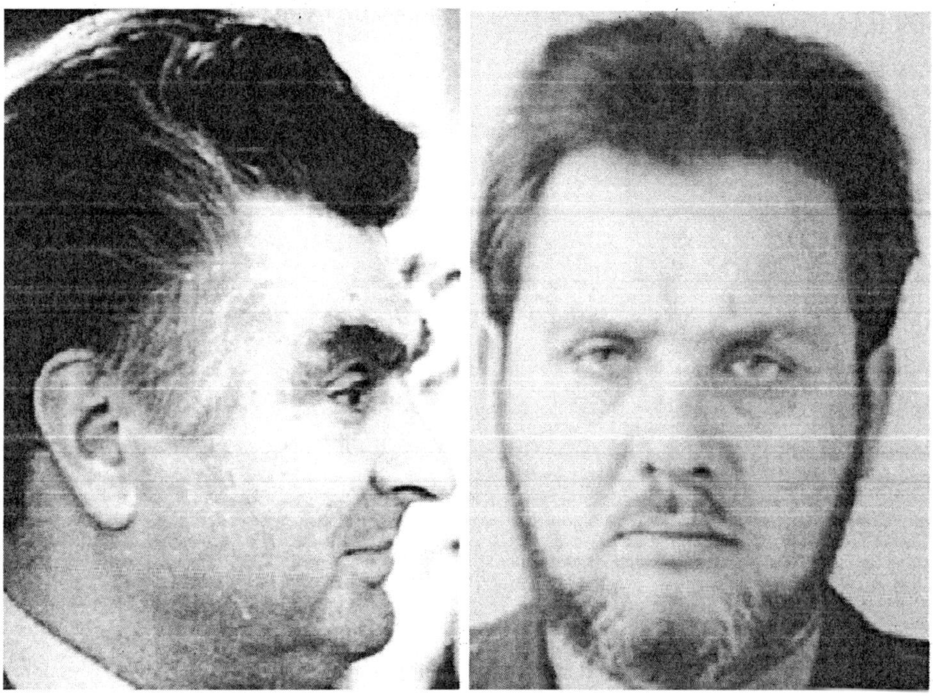

Figure 19. Some variants of dapple location of grey hair flocks in a temple (a) and a beard (b) [189].

Different places of an entire organ or tissue has been undertaken by senescence at different age. Thus, although ageing is usually named as a generalized process, in fact, ageing is a diffuse process with stochastically disseminated focal manifestations. Every senile affection is dispersed in disseminated focal areas of the aged organism. Without any exception a relevant tissue is affected irregularly and duration of life is determined not only by a degree of senescence dissemination but in some cases by location of relevant changes. This is a widely known phenomenon to which nobody called special research attention before. The nature of dapple ageing manifestation has completely eluded investigation and interpretation.

One could suppose that those foci do not appear simultaneously but according to their own programs of ageing. The early aged foci may form only small solitary places of a tissue, an organ, a system, a region or an organism or the large areas of it. One can suppose that the proportions and the features of dislocation change according to features of individual genetic makeup thus determine individual variations in the course of senescence as well as in the life span of an organism under consideration.

Senile regression reveals heterogeneity patterns in cell clones too. Cells age also differently. There is now *in situ* evidence that senescent cells accumulate with age in human tissue. For instance the share of fibroblasts that express a marker of senescent cells are readily apparent in the dermis of skin from old donors, but much less so in skin from young donors [208]. Recent studies on longevity of organelles and molecules also indicated a large range their of lifetimes [202, 209]. Such local sensitivity to senescence revealed by tissues, cells, subcellular and molecular structures may involve only a few parts of the body or majority of them.

In all cases, a considered body consists of parts which are not predisposed to early age along with those parts which are susceptible to senile regression and thus are grown old. One can conclude that within every human body there are co-existing pools (clones) of homogeneous cells and subcellular structures with unequal predisposition to ageing. The quantity, dispersion and localization of more susceptible structures are individually diverse and variable. It is the reason for the organism's varied individual predisposition to ageing in respect of it course, degree and time of development. Thus one may assume, that different parts of a human body can have unequal limit of life duration; each of them has its own `biological watch' telling different time. Why do they age differently? How could different parts of a human body receive so different watch?

The Origin of Diversity in Senescence

Thus one of the most characteristic and widely observable features of senescence, usually ignored by scientific investigation, is very variable dapple distribution of its manifestations seen within systems, organs and tissues of the same individual. Different parts of an organ or tissue has been undertaken by senescence at different calendar age. The early aged foci may form only small solitary places in a tissue, an organ, a region or an organism or the large ones of it. The proportion of early aged parts to later aged ones as well as their localization is predestined by individual genetic makeup thus determining diversity in the course and forms of senile regression and the life span of organisms under consideration. It is a kind of biodiversity that is known least. To clear up what we can to say about the nature of the phenomenon today, the features of this kind of a human body diversity were interpreted, using the data and conceptions of immunology, molecular biology and genetics. The results of the interpretation allow to suppose that different parts of a human body usually have unequal genetic limits of life duration and their own programs of ageing: each of them has its own 'biological watch' which tell different time. The genesis of this diversity can be explained by hybridization of persons with different genetic susceptibility to internal or/and environmental factors determining senescence.

What is the nature of this kind of biodiversity? Although different late-acting genes influencing ageing can be present in all organs and tissues [210], changes of the age vary widely amongst different tissues as well as within a tissue. Cultured cells from people with hereditary premature ageing syndromes grow old prematurely. According to data of many experiments and observations, those unequal quality of analogous structures of the same genetic origin in an entire organism arises as a result of intra-species hybridization [14, 38, 78, 132, 152, 211]. This analogy allows to suppose the same genetic origin in the case of intra-individual diversity in senescence. It is such genetic example that challenged Mendel's first low. That is codominance, a kind of relationship between two alleles of a gene pair in a heterozygote at which the alleles do not express a typical dominance/recessive relationship but are both fully expressed.

One can suppose that the mosaic dapple dispersion of senile changes is also caused by different local predisposition of some parts of outwardly homogenous tissue to senile regression. The focal resistance to ageing can explain why any parts of a tissue are affected by the process, while at the same time the morphologically identical components of the same organism remain uninvolved although both exist under the same conditions and may be equally attacked by relevant agent. These morphologically identical parts, which vary in resistance/sensitivity to ageing, appear as a nidus, and the distribution of such foci are distinct for different people. The genesis of this intra-individual diversity in senescence can be explained by hybridization of persons with different genetic susceptibility to internal or/and environmental factors determining aging. It is possible that another explanation can be proposed but as for us the above interpretation is more acceptable today.

Schizophrenia

Schizophrenia is a devastating psychiatric disorder characterized mainly by 'positive symptoms' that include delusions and hallucinations, 'negative symptoms' that include blunted emotions and social isolation, and cognitive deficits that include impairments in executive function, attention and working memory.[212]. Although unprecedented advances in schizophrenia research have been achieved over recent years especially by improved neuroimaging technologies, little touched about the disease's origins or its mechanisms.

The causes of schizophrenia are considered as a mosaic of multiple factors including changes in the biochemistry of the brain function, changes in the structure of the brain, and genetic predilection. Viral infections and head injuries may also play a role. Finally, schizophrenia is probably a group of related diseases, some of which are caused by one factor and some by another [213].

Epidemiological Observations

Inter-population Diversity in the Incidence of Schizophrenia

Although schizophrenia occurs worldwide at an average frequency of about 1 percent (i.e, nearly 65,000,000 around the World), there is about tenfold inter-population variation in its prevalence [214]. The lower frequencies of shizophrenia are seen in "nonindustrialized" societies, the "industrial" societies have many more chronic schizophrenics. It has been revealed positive correlation between the "complexity" of a society and the number of schizophrenics it allows [215]. The aboriginal Taiwanese groups are found to have much lower rates of schizophrenia than those found in ethnic Chinese population also living on the island [216]. Those living in the centers in developing countries had, on the average, considerably better course and outcome than schizophrenic patients in the centers in developed countries.

The inhabitants of Micronesia live on nearly identical islands dispersed over 2760 miles of ocean in the east-west direction. There are eight distinct language, cultural and racial groups in these islands. A study of the prevalence of schizophrenia in Micronesia in these eight cultures showed that there were no cases in the islands settled in pre-historic times by the Polynesians, only a few cases and a low prevalence (one case per thousand adults) in the remote eastern islands [217] , and a high prevalence (6.8 case per thousand adults) of schizophrenia in the western islands [218]. The study demonstrates that in island populations, separated one from the other both by distance and by cultural attributes, schizophrenia is in some islands a rare occurrence—on some it may not occur at all; but on other islands in the same region, the disease is prevalent. Therefore, schizophrenia is not a universal affliction on mankind.

In the Hawaii state, 1,689 schizophrenic patients have been followed from 1962 to 1972. This cohort included persons of Japanese, Caucasian, part-Hawaiian, Filipino, Chinese and other ancestries. Those of Japanese and Chinese descent had the highest rates for schizophrenia, Filippinos intermediate and Caucasians the lowest [219]. The highest rates for schizophrenia is characteristic of the more mixed populations.

Age Diversity in the Incidence of Schizophrenia

Schizophrenia occurs in all races, in all cultures, in all social classes, in all ages, and in both sexes. The risk for anyone in the population becoming schizophrenic is generally accepted as one percent. Young people are commonly attacked for the first time between 15 and 30 years of age, but the illness may develop as late as age 40. The disease onsets usually in late adolescence or early adulthood and follows an episodic and deteriorative course where the prognosis becomes worse with each episode [220]. Individual differences in liability are largely genetic, with heritability estimates of around 80%.

Clinical Observations

Inter-individual Diversity in the Severity of Schizophrenia

It does not belong to very rare disease. 1% of the population may suffer from schizophrenia worldwide, but the disorder is more prevalent in families where schizophrenia has previously been diagnosed. The bench psychiatrists feel justified in strongly suspecting that our species is divided into a very large majority of individuals who do not possess the schizophrenic genotype, and a small minority (perhaps 5 percent) who do [215].

Beside with nonpsychotic carriers of the schizophrenic genotype the schizophrenics patients demonstrated very marked variations of course and outcomeof the disease. In some cases schizophrenia may develop very gradually. This slow development is referred to as gradual-onset or insidious schizophrenia. A gradual build-up of symptoms may or may not lead to an acute or crisis episode of schizophrenia which involves short period of intense hallucinations, delusions, thought disorder, and an altered sense of self.

Sometimes schizophrenia has a rapid or sudden onset with very dramatic changes in behaviour that occur over a few weeks or even a few days. Sudden onset usually leads fairly quickly to an acute episode. Some people have very few such attacks in a lifetime; others have more. Some people lead relatively normal lives between episodes. Others find that they are very listless, depressed, and unable to function well. Besides, the cases of chronic schizophrenia are known. This is a severe, long-lasting disability characterized by social withdrawal, lack of motivation, depression, and blunted feelings. In some patients with chronic disorders moderate versions of acute symptoms such as delusions and thought disorder may be happen too. Clinical diagnosis on initial evaluation appeared to be a good predictor of subsequent symptomatology, but not of the length of the episodes, the total time during which the patient would be psychotic, pattern of course, or the degree of social impairment [221].

Intra-individual Diversity in Schizophrenic Affections

The manifestations of schizophrenia are extraordinary diverse. Only initial signs of the disease onset accounts over 20 very impressive manifestations including excessive fatigue and sleepiness or an inability to sleep, inability to concentrate or cope with minor problems, dropping out of activities (skipping classes), decline in academic and athletic performance, deterioration of personal hygiene; eccentric dress, forgetfulness, apparent indifference, even in highly important situations, bizarre behaviour, frequent moves or trips or long walks leading nowhere, excessive writing without apparent meaning, inappropriate laughter, peculiar use of words or language structure, strange posturing, staring; vagueness, social withdrawal, isolation and reclusiveness, deterioration of social relationships, undue preoccupation with spiritual or religious matters, drug or alcohol abuse, unusual sensitivity to noise or light, low tolerance to irritation, inability to express emotion, irrational statements, conversation that seems deep but is not logical or coherent,

The following development of schizophrenia also does not annihilated initial abundance of the manifestations but makes it more expressive and diversely. The above discussed initial signs begin to be supplemented by some of symptoms symptoms including delusions (false beliefs), paranoia, hallucinations (hearing or seeing things), extremely disconnected speech or thinking, lack of speech, lack of feelings, feelings that don't match the situation, social isolation, being unable to function at home, work or school, lack of interest or initiative, lack of energy, problems with hygiene or grooming, odd behavior, overreactive senses, pacing, rocking, catatonia (being unable to move) or very disorganized behavior, strong feelings of depression, anxiety, anger, or unreality, excessive concerns about one's body.

Each of above manifestations, both initial and developed ones, can appear either lonely or in association with any quantity of any other signs of the disease. This indicate that very different parts of the mind can be involved where as the neighboring once continue to function normally although before the onset of the dysfunction both the affected and unaffected parts of the brain had the equal chances to be damaged. One can propose this is a result of unequal constitutional susceptibility of the discussed structures and their functions to relevant etiologic agents responsible for onset of the disease.

Cellular and Molecular Diversity in Schizophrenia

Each nerve cell has branches that transmit and receive messages from other nerve cells. The branches release chemicals, called neurotransmitters, which carry the messages from the end of one nerve branch to the cell body of another. The mind function of human brain consists of myriads transmissions performed inside and between neurons by a plethora of specialized biomolecules. In the brain afflicted with schizophrenia, something goes wrong in this system of communications between nerve cells. Molecular origin of focal character of schizophrenic dysfunctions has been confirmed by experiments with special neuroplegic chemicals.

Individual human are known to differ genetically with respect to amphetamine sensitivity and other dopamine agonists [222]. Increased sensitivity to amphetamines is characteristic of schizophrenics and such test have been suggested to identifying those individuals who possess the schizophrenia genotype [215]. In patients with schizophrenia, D2 dopamine receptors in the basal ganglia have been observed to be more numerous than in patients with no history of neurological or psychiatric disease. Because most patients with schizophrenia are treated with neuroleptic drugs that block D2 dopamine receptors in the caudate nucleus, it has been suggested that this increase in the number of receptors is a result a biochemical abnormality intrinsic to schizophrenia [223].

With positron emission tomography D2 dopamine receptor density in the caudate nucleus of was measured in normal volunteers and in patients with schizophrenia. D2 dopamine receptor densities in the caudate nucleus in patients were higher than in the normal people. Thus, schizophrenia itself is associated with an increase in brain D2 dopamine receptor density [223].

Only some schizophrenic patients (13 of 45) showed an acute improvement in schizophrenic symptoms following d-amphetamine infusion that confirms individual features of the dopamine receptors distribution. The 18 patients who worsened tended to have higher CSF 3-methoxy-4-hydroxyphenylglycol levels than did those who improved. Patients who improved had been significantly more psychotic before the infusion0f amphetamine. Patients who worsened had been more psychotic than those who did not change. The authors suggest that sensitivity to dopamine stimulation in schizophrenia is state-dependent rather than trait-dependent and that the simple, undirectional hypothesis of schizophrenia needs to be reformulated [224].

The Origin of Diversity in Schizophrenia

Asthenic somatotype is the basis for continuous-progressive forms of schizophrenia; the pyknic type of the build-up is based on the recurrent and shift-like forms with circular features, whereas the normosthenic somatic type forms the basis for the attack-like progressive form. The data obtained permit a conclusion that formation of various schizophrenia forms is associated with the individual typological peculiarities characteristic of different morpho-phenotypes of the body constitution [225]. The increase of the population heterogeneity as a result of migration processes is one of the factors contributing to greater variation of schizophrenia forms [226]. Structural investigations have suggested that schizophrenia is associated with various cerebro-craniofacial dysmorphogenesis including cerebral asymmetry [227] usually induced by heterozygosity.

Amid the future findings one can predict the disclosure of the phenomenon of functional cytoecological dysfunction, dependent upon constitutional incongruence between molecular neurotransmitters and their receptors possessed by outwardly homogeneous clones of different phenotypes being distinguished by tiny but very important features of their molecular constitution. Those incongruence is supposed to arise as a result of mating between peoples whose previous molecular evolution had performed differently in ecologically different environments [41].

Molecular composition of such heterozygotes in some parts of a body is formed by genotype of one parent whereas in neighboring parts of the same body its formation is controlled by genotype of another parent. According to this opinion, such leading symptom of schizophrenia like mental splitting, as well as a positive correlation between the "complexity" of a society (mosaic heterozygosity of human populations) and the number of schizophrenics it allows witnessing the vital necessity for specialized investigations in this direction [215].

AUTOIMMUNE DISEASES

Epidemiological Observations

Inter-population Diversity in Predisposition to Autoimmunity

There are important inter-population's differences between various localizations of autoimmune aggression. Incidence rates vary among them from less than one newly-diagnosed case of systemic sclerosis to more than 20 cases of adult-onset rheumatoid arthritis per 100 000 person-years. Prevalence rates range from less than 5 per 100 000 (e.g. chronic active hepatitis, uveitis) to more than 500 per 100 000 (Grave disease, rheumatoid arthritis, thyroiditis). [228]..

Ethnic and geographic differences in incidence of autoimmune diseases have been documented, but this effect is non-specific and some groups may be at higher risk for some diseases and lower risk for other diseases [228]. The incidence of type 1 diabetes increased but the rates of rheumatoid arthritis declined over the past 40 years. Besides, the frequency of autoimmune diseases in multiethnic India parallels that of Western countries [229]. This kind of pathology is more characteristic of the more mixed populations.

Individual Diversity in Predisposition to Autoimmunity

In spite of ethiological commonality, there are also important individual (intra-population) differences between various localizations of autoimmune aggression. Autoimmunity affects today many millions of individuals. Thus, autoimmunity involves practically every field in medicine. E.g. 5% of the US population is affected by autoimmune diseases. Other 95% are not affected. Autoimmune diseases are more common among women and often increase in incidence following reproductive years [230]. This kind of diseases forms one of the leading causes of death among young and middle-aged women in the United States. At least 85% of thyroiditis, systemic sclerosis, systemic lupus erythematosus, and Sjögren disease patients are female [228]..

Age Diversity in Predisposition to Autoimmunity

Although most autoimmune diseases can occur at any age, some diseases primarily occur in childhood and adolescence (e.g. type 1 diabetes), in the mid-adult years (e.g. myasthenia gravis, multiple sclerosis), or among older adults (e.g. rheumatoid arthritis, primary systemic vasculitis) [228]. Rheumatoid arthritis can start at any age but in the 20-40 age group it appears most frequently.There are no evidence of the connection between the establishment of any autoimmune disease and the known effect of aging on the lymphatic immunogenic system.

Cell and Molecular Diversity in Autoimmunity

Although it has been clearly documented that B and T lymphocytes lose the ability to respond to antigenic stimulation with age, it has nonetheless been noted that the frequency of auto-reactive antibodies is higher in older individuals. Given that the majority of the age-associated defects in immune regulation target the naïve T and B lymphocyte subsets, it has been presumed that this paradoxical increase in antibodies specific for self antigens was due to changes in the B cell repertoire and/or to differences in the mechanisms responsible for generating immune tolerance in primary responses. Alternatively, it may be possible that memory immune responses, first generated when the individual was young, may play a critical role in the appearance of serum autoantibodies by reactivation later in life (recall memory). Really, memory immunity can be maintained over the lifetime of the organism. Thus, memory B cells which are self-reactive may be harbored within an organism as it ages and the potential exists that they become re-activated at a later time, resulting in a vigorous autoreactive recall response [231] .

This may occur preferentially in older individuals due to several factors, including deficiencies in immune tolerance with age, progressive age-associated loss of tissue integrity yielding neo-self antigens, and possible re-exposure to an infectious agent which induces an autoimmune memory response through molecular mimicry. Thus, we propose that some of the autoantibodies seen in elderly patients may have been produced by memory lymphocytes originally generated against antigens encountered during one's youth, but maintained in a tolerant (non reactive) state until a subsequent triggering event occurs [231].

The origin and development of non-self parts inside of a body during its postnatal ontogenesis forms the main question of autoimmunity pathogenesis. Some opinions are devoted to the attempts to find the answer. Recent studies indicate cell traffic occurs from the fetus into its mother during pregnancy and that low numbers of fetal cells commonly persist in the maternal circulation for years thereafter. Microchimerism refers to a small number of cells from one individual harbored in another individual. These observations led to the hypothesis that microchimerism of host and non-host cells are involved in autoimmune disease.

The hypothesis is applicable to men, children and women without pregnancies because there are other sources of microchimerism, including from a twin, the mother or a blood transfusion. Microchimerism has now been investigated in a number of different diseases with some results supporting a potential role in disease pathogenesis. However, fetal and maternal microchimerism are also found in organs affected by non-autoimmune conditions. Moreover, microchimerism is commonly detected in the peripheral blood of healthy individuals raising the intriguing question of whether these cells are simple remnants of pregnancy or whether they might also have beneficial effects for the host [230].

Clinical Observations

Diversity in the Locations and Sizes of Autoimmune Damages

Autimmunity can affect either blood (autoimmune hemolytic anemia, pernicious anemia) blood vessels (polyarteritis nodosa, Wegener's granulomatosis), digestive tract (autoimmune hepatitis, Behçet's disease, Crohn's disease, primary bilary cirrhosis, ulcerative colitis), eyes (Sjögren's syndrome, uveitis), glands (Graves' disease, thyroiditis, type 1 diabetes mellitus), heart (myocarditis, rheumatic fever), joints (ankylosing spondylitis, rheumatoid arthritis), kidneys (glomerulonephritis), lungs (sarcoidosis, scleroderma), muscles (dermatomyositis, myasthenia gravis, polymyositis), nerves and brain (Guillain-Barré syndrome, multiple sclerosis), or skin (pemphigus, psoriasis, scleroderma) separately or in association with other organs. Most of autoimmune affections presents focal syndromes. Lupus erithematous, Sjögren's syndrome, scleroderma, rheumatoid erthritis, and polymyositis are systemic autoimmune disease. They often also affects many organs throughout the body including the skin, blood vessels, heart, lungs, and muscles.

Autoimmune aggression can affect almost any part of the body i.e. any tissue of any organ or body system. Sometimes it can affect even many parts of the body. But it never can affect all parts of a tissue at once. Extraordinary diversity oin the locations and sizes of the affections is characteristic of every kind of autoimmune diseases.

The manifestations of autoimmune diseases are extraordinary diverse. This depends both on which tissue of the body and which part of the body became 'non-self' for it and thus is targeted by autoimmunity. Three of some dozens of possible examples (myasthenia gravis, ankylosing spondilitis and psoriasis) are briefly considered below to illustrate clinical diversity of this kind of human pathology

Myasthenia gravis, i.e. localized muscular weakness, e.g. ocular motor disturbances, ptosis or diplopia, are the initial symptom of myasthenia gravis in two-thirds of patients; almost all had both symptoms within 2 years. Weakness is restricted to the ocular muscles in about 10% of cases. Oropharyngeal muscle weakness, difficulty chewing, swallowing, or talking, is the initial symptom in one-sixth of patients, and limb weakness in only 10%. Initial weakness can be limited to single muscle groups such as neck or finger extensors or hip flexors. The rest have progressive weakness during the first 2 years that involves oropharyngeal and limb muscles. Maximum weakness occurs during the first year in two-thirds of patients. Approximately one-third of patients improved spontaneously, one-third became worse, and one-third died of the disease [232].

Ankylosing spondylitis primarily affects the spinal joints. In advanced cases this inflammation can lead to new bone formation on the spine inducing kyphosis (Figure 20). Other joints of the body such as the shoulders, hips, ribs, heels and small joints of the hands and feet can become involved too. Sometimes the eyes, lungs and heart can be affected too [233]. Each organ can be involved either separately or in association with other ones.

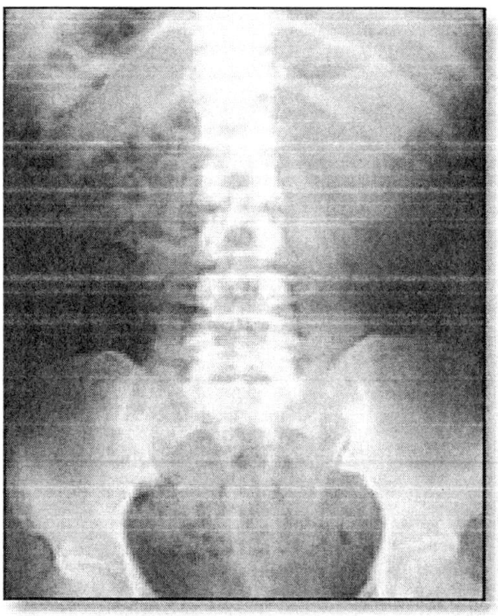

Fig.20. Dapple location of bones lesions in a case of ankylosing spondilitis. [232]

Psoriasis is a lifelong disease that can occur on any part of the body skin. Sometimes it appears on the eyelids, ears, mouth and lips, as well as on skin folds, the hands and feet, and nails. Psoriasis can vary widely among individuals. There are five morphological types of psoriasis: plaque (most common form of the disease), guttate (appears as small red spots on the skin), inverse (occurs in armpits, groin and skin folds), pustular (white blisters surrounded by red skin), and erythrodermic (intense redness over large areas). The most common form, plaque psoriasis, appears as raised, red patches or lesions covered with a silvery white buildup of dead skin cells, called scale.

The Origin of Diversity in Autoimmunity

The exact mechanisms that cause autoimmune disorders are not clear, and the past two decades of research on the lymphatic immune system have not seen substantial progress in our understanding of autoimmunity. Autoimmune diseases are considered as multigenic once. Meanwhile the evidence for genetic involvement in the pathogenesis of autoimmune diseases comes only from the studies of twins: really, in the case of monozygotic twins, both twins are more frequently affected than in the case of dizygotic twins. It is also accepted to suppose that autoimmune diseases arise as a result of aberrant misrecognition of some self-epitopes of a body as non-self by the lymphatic immunogenic system, which begins mistakenly attack the body's own tissue and kill some body parts like germs inducing inflammation and tissue damage.

It is proposed, that in the case of autoimmune aggression the lymphatic immunogenic system in afflicted individuals selectively loses the ability to distinguish self from nonself and switchs "modes" - from protection to attacking. Paradoxically, the inability to distinguish self from nonself is not manifested in all parts of afflicted individual but only in some of them. One question arose: why some parts of the same cellular category are feeling by lymphatic system as a self while another part of the same kind are feeling as "non-self". Those hypothesis can not explain the spotty distribution of autoimmune lesions. This is one of the profound delusions of the 20^{tth} century's immunology.

Opposite, one can suppose, that some focally dispersed parts of afflicted organism become antigenically "non-self" at the latest postnatal ontogenesis whereas the lymphatic system continues to perform its innate functions that has been formed at prenatal ontogenesis. Autoimmune diseases arise as a result of the attack of lymphatic system against aberrant epitopes emerged in a body during the latest postnatal ontogenesi

Every human body consists of coexisting phenotypically dissimilar mosaic populations of homogenous cells of unequal molecular constitution. The phenotypic manifestations of the mosaicism are being formed during all individual life along with realization of individual program of ontogenesis. One can suppose that some subpopulations of cells being formed in the course of postnatal ontogenesis are appearing to be non-self for the lymphatic system of immunogenesis, thus provoking the immune self-aggression. In a case of hormone adaptive systems are of normal function the selfaggression is restrained. In the contrary, in consequence of stresses, age transformation and other immune depressive factors the defensive potential of hormone adaptive systems diminishes allowing immune self-aggression.

CONCLUSION

The investigation has been performed above presented initial perspective of a special kind of biodiversity which is characteristic of both infectious and noninfectious pathology. In the light of data considered and generalized above it becomes clear that manifestations of this multilevel kind of biodiversity may be observed beginning from species to populations, individuals, organs, tissues, cells and finally to their organelles and molecular structures. Besides, one should strongly accentuate, that the beginning of this biodiversity is rooted on the molecular level. Any disease either infectious or non-infectious ones arise as a result of intermolecular interactions. Both kind of diseases possess at least 9 similar traits (Table 11).

Both infectious and non-infectious kinds of biodiversity are of genetic origin being strongly connected with molecular features of constitutional immunity that are exceptionally specific for each disease. At the same time the functions of this immunity are diametrically opposite. At the case of any infectious disease constitutional immunity provides the attacked body with life-saving resistance while at the case of undoubtedly non-infectious diseases (obesity, diabetes, other hormonal dysfunctions and probably the diseases of mind) constitutional immunity forms principal prerequisites for their development. Atherosclerosis, senescence and some other diseases of undetermined etiology may be accounted as exceptions of this gradations.

Table 11. Similarity and differences between infectious and non-infectious diseases

Features	Kind of pathology	
	Infectious	Non-inectious
1.Infectious trigger	+	o
2.Constitutional (genetic) predilection	+	+
3.Individual variations in susceptibility	+	+
4.Stochastic distribution of individual cases amongst a population	+	+
5.Mosaic (focal) affections	+	+
6.Stochastic distribution of affections amongst a body, organ or tissue	+	+
7.Intra-individual diversity in predilection	+	+
8.Molecular bases of pathogenesis	+	+
9.The possibility of spontaneous recovery	+	o

Both infectious and non-infectious kinds of biodiversity arise as a result of sexual self-reproduction that forms inevitably this or that grade of heterozygosity. Any case of biodiversity that has been considered above is a result of hybridisation between two genetically different organisms: one of them was constitutionally immune to relevant ecological or physiological agent whereas its mating partner was constitutionally sensitive to it. As a result the descendant's body cells are formed under control of two codominant allelomorphic genes. The first step of those hybridizations is revealed itself at the dissimilarity of molecular constitution of homologous cells and their subcellular structures.

There is a popular fallacy that all individual's somatic cells are genetically identical (with a few possible exceptions) and so they do not have conflicting interests. In reality, they may have different phenotypes, and therefore different interests. Such a symbiotic coexistence of genomes forms the main source of biodiversity both in infectious and non infectious diseases,

The basis for biodiversity in both infectious and non-infectious diseases is formed by the organism's heterozygosity as a result of the coexistence in the body of two active allelomorphic genes and of two heterogeneous parts of the body, for instance of its organs, tissues or cell clones. One of these heterogenous parts possesses the feature of one parent (e.g. genetic immunity to an infection) where as the other possesses the alternative feature of the other parent (i.e. genetic susceptibility to the infection).

The individual grades of heterozygosity may be extraordinarily variable. In the simplest cases of hybridization the quotes of between-spouse admixture can consist either of 50% to 50%, 75% to 25%, 87,5% to 12,5%, 93,75% to 6, 25% of the whole genome or so all. Besides, each of these quotes may be stochastically dispersed around the body making extraordinary different forms of the admixture. How can we account for the fact that identical cells of an organism differ from each other when all contain the same genome complex? It is both a genetic and gerontologic puzzle.

The phenomenon of heterozygosity can not be considered as something unusual. Even in a case of inbreeding both male and female genomes are not identical 234. Males and females differ greatly in their parental investment. Each sexual co-operator has had a different evolutionary history, and has a different both genotypic and phenotypic make-up. That is why each descendant of any couple is heterozygotic even in the case of inbreeding and thus can be considered at list as bi--genome organism. This may lead to the differences in the way and sometimes to conflicts inside the descendant's phenome. Heterozygosity and relevant intra-individual biodiversity are intrinsic of sexual reproduction.

In a case of outbreeding the grades of heterozygosity are especially variable. Each heterozygotic body is formed and functions under control of two genomes mixed in individually different proportions. As a result of heterozygosity, within many multicellular organisms, phenotypically dissimilar clones of homogeneous cells and extracellular structures with unequal molecular constitutions coexist. The clones vary in their liability to relevant exogenous and endogenous factors. They exist in a body in the form of separated populations dispersed either stochastically or legitimately around the body in mosaic patches.

As a result of heterozygosity, within many multicellular organisms, phenotypically dissimilar clones of homologous cells and extracellular structures with unequal molecular constitutions coexist. Such cells are varied in their liability to this or that exogenous or endogenous factor; moreover, the correlation between the dissimilar clones within the specimens of one species is very changeable. Probably, it is the reason for the organism's varied individual predisposition to different diseased states in respect of their course and severity. Human populations are unprecedented in the number of differences of their genetic and phenotypic make-up. The genome of the modern civilized human can be considered as a mosaic composed of genetic segments inherited from a long row of different ancestors.

Thus, within every living body there are at least two co-existing homologous parts with unequal make-up and relation to either ecological (infectious) or physiological agents. The parts exist in a form of separate patches of different sizes and locations being stochastically dispersed around the body. The quantitative correlations between the patches within the specimens of a species are variable in size and location. According to individual variation in the grade, location and form of heterozygous admixture the organism under consideration can be succumbed by a disease of extraordinarily different forms, danger, severity and location.

This is the reason for the organism's varied individual predisposition to different diseases, their spread, relevance, courses, and severity.

The biodiversity in infectious diseases is predestined by the organism's heterozygosity resulted in coexistence in the body of two active allelomorphic genes and of two allelic cell clones. Both of these alleles function dominantly. The heterozygous individual shows both alleles expressed equally although in different locations of the body. One of these cell clones possesses the genetic immunity to an infection where as the other possesses the alternative feature of the other parent, i.e. genetic susceptibility to the same infection. The diversity in the danger and severity of infectious diseases depends exclusively on the differences in the genetically predetermined variations in the quantity of susceptible cells in the body under consideration.

Because parasitic microbes can not exist out of the results of biosynthetic processes performed by affected living organisms, any epidemic is a process of continuous transmission of infectious agents between susceptible members of relevant species. The intensity of any epidemic process, i.e. its ebb or flow is predestined by the number of susceptible members in the relevant species as well as by the grades of their susceptibility.

Intra-population differences in susceptibility to an infection determine the variations in the spread of its epidemics. Like the immune individual, the immune population does not allow the infectious agent to enter into it and spread around it. Immunity of a population, especially of genetic origin forms the mightiest barrier to the spread of relevant epidemics around this population. In contrast, the existence in a population of a relevant number of partially susceptible persons predisposes the population to rapid spreading of relevant epidemics. The existence of genetic immunity in a population performs the mightiest anti-epidemic defense whereas the biodiversity in the susceptibility to infections functions serves as a savior of epidemic processes.

The biodiversity in different kinds of non-infectious diseases can be of different genesis. Firstly, some diseases" that are supposed as non-infectious", e.g. atherosclerosis do not show principal pathogenetic differences from infectious ones. But in contrast to many infectious diseases the diversity in grades of danger and severity of atherosclerosis depends not on the quantity of specific affections but on its location. One small atheromatous patch located in a relevant myocardial or brain artery may lead to deadly heart attack or stroke, whereas some dozens of the same patches located along abdominal aorta can be asymptomatic for a life time.

Secondly, the senescence (aging) should be considered as a a result of normal physiological performing of genetically programmed process of individual ontogenesis. The biodiversity in senescence begins from species, populations, individual and intra-individual differences in the level of the age where and when this process has been programmed to be finished. Besides, nobody died of absolute senescence being free of some diseases and their consequences.

Thirdly, the epidemiological and clinical diversity in metabolic diseases like obesity, diabetes, rickets, testicular feminization, Cushing's disease, the syndrome of cell immunity to thyroid hormone is associated with genetic resistance to relevant molecular physiological agents i.e. hormones or mediators. The principles of cell immunity to physiological agents are analogous to those in the constitutional (genetic) immunity to infections which is formed also by structural incongruence between the relevant agent and its cell receptors. One could suppose this immunity could arise thanks to mutation. But it is absolutely impossible to image that these pathogenic mutations can be distributed around populations by direct natural selection for immunity to relevant molecular physiological agents.

Because many hormones share their specific targets with the acting molecules of some infectious agents, the evolutionary formation of constitutional immunity against an infection forms immunity to relevant physiological agents. Thus the susceptibility to metabolic non-infectious diseases is formed by insusceptibility to some infections. It arises and spreads in populations as a result of the spread of genetic immunity to relevant infectious agent. This kind of intermolecular incongruence can be reinforced by interethnic admixture. Thus the mutant formation of the life-saving genetic immunity to infection can create a life-threatening immunity to the relevant hormone. The individual diversity functions as a savior of populations whereas the intra-individual diversity functions as a savior of individual life.

Analogical ways of pathogenesis can be discussed also in relation to schizophrenia. It can be induced by cell immunity to neurotransmitters. This kind of pathology could arise and spread in a population as it has been hypothesized in the relation to obesity's pathogenesis. Nevertheless the direct post-infectious origin of the disease also cannot be excluded for a while. In this case one can propose innate differences in cell immunity to an infection resulted in the death of a susceptible clone. Thus diversity presents the most general feature of any kind of pathology.

In analogy to infectious diseases the individual diversity in the grades of danger and severity of metabolic diseases depends directly on the quantity of specific affections and its location. The biodiversity in epidemiologic and clinical manifestations of metabolic diseases depends on the existence of their not deadly but chronic forms that provide the preservation of this kind of non-infectious pathology.

The integrated view on different levels of biodiversity in both infectious and non-infectious diseases shows new direction toward the deciphering of many actual aspects of epidemiology, etiology, pathogenesis, prophylaxis, diagnostics and treatments unopened until now. Whereas the possibility to be affected by an infection depends on both infectious agents and the attacked organism, the severity of disease and its outcome depends exclusively on the genetic make-up namely on the grade of the attacked organism diversity.

Human reproductive ecology is changing, and these changes have manifestations in the terms of diseases and their distribution. High rates of current outbreeding, migration, and cross-ethnic mating will re-assort our genes at unprecedented rates. The rates of biodiversity and the number of non-infectious diseases will arise. Genetic admixture forms the main moving force for the worldwide epidemics of non-infectious diseases. Besides, the growth of biodiversity will provide the World population with defense against any new or re-emerged infectious disease. The existing grade of infectious biodiversity ensures that there is enough variations among individuals to provide the World population with resistance to any newly emerged or re-emerged infectious disease.

The anti-infectious prevention programs can be made far well more effective by targeting them towards susceptible individuals. In agriculture, infectious biodiversity is a key tool in breeding new crop, fowl and cattle varieties A recent revival of research on genetic immunity has been stimulated by the recognition that both farming and human health care must become more effective. In contrast the programs of prevention of non-infectious diseases should be targeted on the restriction of relevant biodiversity.

The biological functions of biodiversity in diseases disperse far out of the interests of medicine, agriculture and biotechnology. Infectious diseases and their diversity are becoming increasingly recognized as having a major influence on the evolution of biological populations, species and all life as a whole. Although the systematic application of medicine to evolutionary biology is a very new and unaccustomed enterprise, medicine begins to inform biology of previously unknown principles of living beings' constitution and evolution.

REFERENCES

[1] Bieniasz PD Intrinsic immunity: a front-line defense against viral attack . *Nature Immunology 5*, 1109-1115 (2004).

[2] Boyd,W.C. *Fundamentals of immunology.*(Interscience publishers, New York: 1966).

[3] Burgasov PN and Rumyantsev SN *Evolution of Clostridiosis [in Russian]*(Meditsine, Moscow, 1974).

[4] Rumyantsev SN *Constitutional immunity and its molecular-ecological principles [in Russian]*(Nauka, Leningrad, 1983).

[5] Cui Z *et al.* Spontaneous regression of advanced cancer: Identification of a unique genetically determined, age-dependent trait in mice. *Proc Natl Acad Sci U S A 100*, 6682-6687 (2003).

[6] Kimman TG *Genetics of infectious disease susceptibility*(Kluwer Academic Publishers, Dordrecht, Netherland, 2001).

[7] Kaufmann SHE, Sher A, and Ahmed R Introduction in *Immunology of Infectious Diseases* (eds. Kaufmann SHE, Sher A and Ahmed R) 1-2 (ASM PRESS, Washington, DC, 2002).

[8] Petri WA and Mann BJ Microbial adherence in *Principle and practice of infectious diseases.* 11-19 (Churchill Livingston, 1995).

[9] Zinkernagel R.M. and Kelly J. How antigen influences immunity. *The Immunologist 5*, 114-120 (1997).

[10] Yuen KY, Chan PK, Peiris M, and et al Clinical features and rapid viral diagnosis of human disease associated with avian influenza A H5N1 virus. *Lancet 351*, 467-471 (1998).

[11] Soosaar JLM, Burch-Smith TM, and Dinesh-Kumar SP Mechanisms of plant resistance to viruses . *Nature Reviews Microbiology 3*, 789-798 (2005).

[12] Morishime T, McClintock PR, Aulakh GS, Billups LC, and Notkins AL Genomic and receptor attachment differences between mengovirus and encephalomyocarditis virus. *Virology 122*, 461-465 (1982).

[13] Rumyantsev SN Observations on constitutional resistance to infection. *Immunology Today 13*, 184-187 (1992).

[14] Rumyantsev,S.N. Constitutional and non-specific immunity to infection. *Rev. sci. tech. Off. int. Epiz. 17*, 26-42 (1998).

[15] Rumyantsev S.N., Avrova N.F., Pospelov V.F., and Denisova N.A. Influence of gangliosides on adhesive interaction of Neisseria meningitidis with human cells.[in Russian]. *ZhMEI*29-32 (1990).

[16] Foster,J.W. and Kinney,D.M. ADP-ribosilating microbial toxins. In. CRC critical reviews in microbiol., 11, 273-298. *CRC critical reviewsin microbiol. 11*, 273-298 (1984).

[17] Middlebrook JL and Dorland RB Responce of cultured mammalian cells to the exotoxins of Pseudomonas aeruginosa and Corinebacterium diphtheriae: differential cytotoxocity . *Canad J Microbiol 23*, 183-189 (1977).

[18] Cheney CP, Schad PA, Formal CB, and Boedeker EC Species specificity of in vitro Escherichia coli adherence to host intestinal cell membranes and its correlation with in vivo colonization and infectivity . *Infect Immun 28*, 1019-1027 (1980).

[19] Haywood AM Characteristics of Sendai virus receptors in a model membrane . *J Molec Biol 83*, 427-436 (1974).

[20] Strobel G.A. A mechanism of disease resistance in plants. *Sci. Am. 232*, 81-88 (1975).

[21] Wyler DJ Malaria: Host-Pathogen Biology . *Rev Inf Dis 4*, 785-797 (1982).

[22] Kasarov LB Degradation of the erythrocyte phospholipids and haemolysis of the erythrocytes of different animal species by leptospirae . *J Med Microbiol 3*, 29-37 (1970).

[23] Smith H The influence of the host on microbes that cause disease . *Proc R Soc Lond 246*, 97-105 (1991).

[24] Kruger DH and Bickle TA Bacteriophage survival: multiple mechanisms for avoiding the deoxiribonucleic acid restriction systems of their hosts. *Microbiol Rew 1983 47*, 345-360 (1983).

[25] Pavlovic J, Haller O, and Staeheli P Human and mouse Mx proteins inhibit different steps of the influenza virus multiplication cycle. *J Virol. 66*, 2564-2569 (1992).

[26] Jones D Cyanogenesis in animal-plant interactions in *Cyanide compounds in biology* 151-170 (Wiley, Chichester, 1988).

[27] Currier WW Molecular controls in the resistance of potato to late blight. *TIBS 6*, 191-194 (1981).

[28] Rowland MW Fitness of insecticide resistance. *Nature 327*, 194 (1987).

[29] Miller ME Natural defence mechanisms: development and characterization of innate immunity in *Immunologic disorders in infants and children* (eds. Stiehm ER and Fulginiti VA) 127-131 (Saunders WB, Philadelphia , 1973).

[30] Hutt F.B. *Genetic resistance to disease in domestic animals*(Comstock Publishing Associates, Ithaca, 1958).

[31] Webster L.T. Inheritance and resistance of mice to enteric bacterial and neurotic virus infections. *J. exptl. Med. 65*, 261-286 (1937).

[32] Darnell MB, Koprowsky H, and Lagerspetz K Genetically determined resistance to infection with group B arboviruses.I. Distribution of the resistance gene among various mouse populations and characteristic of gene expression in vivo. *JInf Dis 129*, 240-247 (1974).

[33] Koike S., Ise I., and Nomoto I. Functional domains of the polyovirus receptor. 11-15. *Proc Natl Acad Sci USA 88*, 11-15 (1991).

[34] Hormaeche C.E Natural resistance to Salmonella typhimurium in different inbred mouse strains. *Immunology 37*, 311-318 (1979).

[35] Abel L. and Dessein A.J. The impact of host genetics on susceptibility to human infectious diseases. *Current Opinion in Immunology 9*, 509-516 (1997).

[36] Hill AVS Genetics of infectious disease resistance. *Current Opinion in Genetics and Development 6*, 348-353 (1996).

[37] O'Brien SJ and Dean M In search of AIDS-resistance genes. *Scientific American 277*, 44-51 (1997).

[38] Paxton W.A. *et al.* Relative resistance to HIV-1 infection of CD4 lymphocytes from persons who remain uninfected dispite multiple high-risk sexual exposure. *Nat. Med. 2*, 412-417 (1996).

[39] Nagaraj AN and Black LM Hereditary variation in the ability of a leafhopper to transmit two unrelated plant viruses. *Virology 16*, 152-162 (1962).

[40] Anderson R.M., Crombie J.A., and May R.M. Predisposition to helminth infection in man. *Nature 320*, 195-196 (1986).

[41] Rumyantsev SN Obesity: a reckoning both for genetic immunity to infection and xenogamy. *Medical Hypothesis 66*, 535-540 (2006).

[42] Rumyantsev SN *Biological Weapon. A Terrible Reality? Profound Delusion? Skillful Swindling?*(Vantage Press, New York, 2006).

[43] Monto AS The Threat of an Avian Influenza Pandemic. *N Engl J Med 352*, 323-324 (2005).

[44] Stohr K Avian influenza and pandemics--research needs and opportunities. *N Engl J Med. 352*, 405-407 (2005).

[45] Webster RG, Bean WJ, Gorman OT, Chambers TM, and Kawaoka Y Evolution and ecology of influenza A viruses. *Microbiol Rev 56*, 152-159 (1992).

[46] Brachman PS and Friedlander A Anthrax in *Vaccines* (eds. PlotkinSA and Orenstein WA) 629-637 (WB Saunders Co , Philadelphia, Pa , 1999).

[47] Kohout E, Sehat A, and Ashraf M Anthrax: a continuous problem in south west Iran. *Am J Med Sci. 247*, 565 (1964).

[48] Lew D. *Bacillus anthracis* (anthrax) (ed. Mandell GL,B.J.D.R.) 1885-1889 (: Churchill Livingstone Inc; New York, NY, 1995).

[49] Meselson M *et al.* The Sverdlovsk anthrax outbreak of 1979. *Science 266*, 1202-1208 (1994).

[50] Gabliks J. and Falkone M. Interaction of diphtheria toxin with cell cultures from susceptible and resistant animals. *J. Exp. Med. 123*, 723-732 (1966).

[51] Korotich AS, Vasilchenko AA, and Sobko A.I et al. Concerning footh-and-mouth disease in man. *Zh. Mikrobiol. Epidemiol. Immunobiol.*132-135 (1974).

[52] Allison A.C. Coevolution between hosts and infectious disease agents and its effects on virulence in *Population biology of infectious diseases* (eds. Anderson R.M. and May R.M.) 245-267 (Springer Verlag, Heidelberg, 1982).

[53] Marshall BJ, Armstrong JA, McGechie DB, and Glancy RJ . Attempt to fulfill Koch's postulates for pyloric Campylobacter. *Med J Aust 142*, 436-439 (1985).

[54] Chagnon NA and Melancon TF Reproduction, numbers of kin, and epidemics in tribal populations: a case study. in *Population and biology* (ed. Kreryfitz N) 147-168 (Ordina Editions, Liege, 1980).

[55] Stohr K Influenza - WHO cares. *Lancet Infectious Diseases 2*, 517 (2002).

[56] Hoffman SL, Punjabi NH, Kumala S, and et al/ Reduction of mortality in chloramphenicol-threated severe typhoid fever in high-dose dexamethasone. *N. Engl. J. Med. 310*, 82-88 (1984).

[57] Rogerson SI, Spooner VJ, Smith TA, and Richens J Hydrocortisone in chloramphenicol-treated severe typhoid fever in Papua New Guinea. *Trans R Soc Trop Med Hyg 85*, 113-116 (1985).

[58] Wain J, Hous D, Parkhill J, Parry C, and Dougan G Unlocking the genome of the human typhoid bacillus. *THE LANCET Infectious Diseases 2*, 163-170 (2002).

[59] Wigley P, Hulme SD, Bumstead N, and Barrow PA In vivo and in vitro studies of genetic resistance to systemic salmonellosis in the chicken encoded by the SAL1 locus. *Microbes and Infection 4*, 1111-1120 (2002).

[60] McNeil W.H. *Plagues and people* (Blackwell Scientific Publishers, Oxford, 1976).

[61] Stearn EW and Stearn AE *The effect of smallpox on the destiny of the Amerindian*(B. Humphries, Inc., Boston, 1945).

[62] Thomson D The ebb and flow of infection. *JAMA 235*, 269-272 (1976).

[63] Vries RR de, Khan PM, Bernini LF, Loghem E, and Rood JJ Genetic control of survival in epidemics. Rood JJ. *J Immunogenet 6*, 271-287 (1979).

[64] Windsor HM *et al.* Prevalence of Helicobacter pylori in Indigenous Western Australians: comparison between urban and remote rural populations. *Med J Aust. 182*, 210-213 (2005).

[65] Fujikawa A *et al.* Mice deficient in protein tyrosine phosphatase receptor type Z are resistant to gastric ulcer induction by VacA of Helicobacter pylori. Nat Genetics 2003; 33: 375-381. *Nat Genetics 33*, 375-381 (2003).

[66] Cartwright K. Meningococcal carriage and disease, in *Meningococcal disease* (ed. John Wiley) 159-177 Chichester, 1995).

[67] Johnson N P and Mueller J Updating the accounts: global mortality of the 1918-1920 "Spanish" influenza pandemic. *Bull. Hist. Med 76*, 105-115 (2002).

[68] Carr EA and Renfrew RR Recovery of Bacillus anthracis from the nose snd throat of apparently healthy workers. *J Infec Dis 100*, 169-171 (1957).

[69] Brachman PS *Inhalation anthrax* in *Airborn contagion* (ed. Kundsir RB) 83-93 (New York Academy of Sciences, New York, 1980).

[70] Myenye KS, Siziya S, and Peterson D Factors associated with human anthrax outbreak in the Chikupo and Ngandu villages of Murewa district in Mashonaland East Province, Zimbabwe . *Cent Afr J Med 42*, 312-315 (1996).

[71] Tekin A, Bulut N, and Unal T Acute abdomen due to anthrax . *Br J Surg 84*, 813 (1997).

[72] Sirisanthana T, Nelson KE, Ezzell JW, and Abshire TG Serological studies of patients with cutaneous and oralpharyngeal anthrax from northern Thailand. *AmJ TropMed Hyg 39*, 575-581 (1988).

[73] Regis E Evaluating the threat. Does mass biopanic portend mass destruction? *Sci. Am.*(2001).

[74] Guillemin J *Anthrax. The investigation of a deadly outbreak.*(University of California Press, Berkeley, 1999).

[75] Kahn JO and Walker BD Acute Human Immunodeficiency Virus type 1 infection. *N Engl J Med 331*, 33-39 (1998).

[76] Hornick R.B. *et al.* Typhoid fever: pathogenesis and immunologic control. *N. Engl. J. Med. 283*, 686-739 (1970).

[77] Pang T, Levine MM, Ivanoff B, and Finlay BB Typhoid fever - important issues still remain. *Trends Microbiol*131-138 (1998).

[78] Rumyantsev S.N. Structural predestination of individual diversity in the course and severity of infectious diseases. *TheScientificWorldJOURNAL 2*, 205-216 (2002).

[79] Groover R.V.Sutherland J.M.Landing,B.H. Purulent meningitis of new-born infants: eleven year experience in the antibiotic era. *New England J. Med. 264*, 1115-1118 (1961).

[80] Klein J.O.Marcy S.M. Bacterial sepsis and meningitis in *Infectious diseases of the fetus and new-born infant.* (ed. Remington J.S.Klein J.O.) 843-844 (W.B.Saunders Company, 1995).

[81] Arvin AM and Maldonado YA Other viral infections of the fetus and newborn. in *Infectious diseases of the fetus and newborn infant* (eds. Remington JS and Klein JO) 749 (W.B.Saunders Company, 1995).

[82] Berger U, Saeftel S, and Schlez K Untersuchungen uber die Meningokokkentragerquote im Hinblick auf Alter und Geschlecht. *Z Med Mikrobiol Immunol 155*, 192-202 (1970).

[83] Kunze U, Baumhackl U, Bretschneider R, and et al. Tick-borne encephalitis in the golden agers: A conference report of the International Scientific Working Group on Tick-Borne Encephalitis (ISW TBE). *Vaccine 24*, 1236-1237 (2006).

[84] Prince GA and Porter DD The pathogenesis of respiratory syncytial virus infection in infant ferrets. *Amer J Pathol 82*, 339-350 (1976).

[85] Maitland K, Williams TN, and Newbold CI Plasmodium vivax and P. falciparum: Biological interactions and the possibility of cross-species immunity interactions and the possibility of cross-species immunity. *Parasitol Today 13*, 227-231 (1997).

[86] Mogensen S.C. Role of macrophages in natural resistance to vius infections. *Microbiol. Rev. 43*, 1-23 (1979).

[87] Mogensen S.C. Viral interaction with phagocyte functions. in *Microbial perturbation of host defence* (ed. O'Grady,S.H.) 165-184 (Academic Press, New York, 1981).

[88] Eggers HJ Notes on the pathogenesis of enterovirus infections. Observations, experiments and speculations. *MedMicrobiol Immunol 179*, 297-306 (1990).

[89] Clark HF and Lunger PD Viruses in *Diseases of reptilia* (eds. Cooper JE and JacksonOF) 135-164 (Acad Press, New York, 1981).

[90] Gengenbach BG, Green CE, and Donovan CM Inheritance of selected pathotoxin resistance in maize plants regenerated from cell cultures . *Proc Natl Acad Sci USA 74*, 5113-5117 (1977).

[91] McGee Z.A.Stephens D.S. Common pathways of invasion of mucosal barriers by Neisseria gonorrhoeae and Neisseria meningitidis. *Surv Synth Path Res 3*, 1-10 (1984).

[92] Scheuerpflug I, Rudel T, Ryll R, Pandit J, and Meyer TF Roles of PilC and PilE Proteins in Pilus-Mediated Adherence of Neisseria gonorrhoeae and Neisseria meningitidis to Human Erythrocytes and Endothelial and Epithelial Cells. *Infect. Immun. 67*, 834-835 (1999).

[93] Virji M, Alexandrescu C, Ferguson DJ, Saunders JR, and Moxon ER Variations in the expression of pili: the effect on adherence of Neisseria meningitidis to human epithelial and endothelial cells. *Mol Microbiol 10*, 1271-1279 (1992).

[94] Rumyantsev S.N., Shabalow N.P., Pyasetskaya M.F., Rogacheva N.M., and Bardakova L.I. Species, population and age diversity in cell resistance to adhesion of Neisseria meningitidis serogroups A, B and C. *Microbes and Infection 2*, 447-453 (2000).

[95] Rumyantsev SN, Aron RA, and Beljakova IV. Adsorbing interaction of influenza viruses with human and animal cells.[in Russian] . Genetic ingeneering of immunomodulators and vaccinepreparations , 99-104. 1989. Leningrad, All-union Research Institute oflnfluenza. Ref Type: Magazine Article

[96] Levy JA HIV pathogenesis and long-term survival . *AIDS 7*, 1401-1410 (1993).

[97] Halonen PE, Murphy FA, and Fields BN et al Hemagglutinin of rabies and some other bullet shaped viruses. *Proc Soc ExpBiol and Med 127*, 1037-1042 (1968).

[98] Chaturvedi UC, Nagar R, and Shrivastava R Dengue and dengue haemorrhagic fever: implications of host genetics. *FEMS Immunol Med Microbiol*(2006).

[99] Koransky JR, Scales RW, and Kraus SJ Bacterial hemagglutination by Neisseria gonorrhoeae. *Infec Immun 123*, 495-498 (1975).

[100] Tuomanen EI, Nedelman J, Hendley JO, and Hewlett EL Species specificity of Bordetella adherence to human and animal ciliated respiratory epithelial cells. *Infec Immun 42*, 692-695 (1983).

[101] Eidelis L, Proia RL, and Hart DA Membrane receptor for bacterial toxins. *Microbiol Rev 47*, 596-620 (1983).

[102] Kauffmann F. *The bacterology of Enterobacteriaceae.* , .(Munksgaard, Copenchague: 1969).

[103] Rumyantsev SN, Bobrakova LV, and Perkus LV Genetic limitations of erythrocyte lysis by Salmonella genus bacteria.[in Russian]. *Genetika*2037-2041 (1978).

[104] Bobrakova LV. Epidemiological aspects of the populations heterogeneity in the resistance of erythrocytes to Salmonella haemolysines.[in Russian]. 1-24. 1986. Central Scientific Research Institute of Epidemiology.
Ref Type: Thesis/Dissertation

[105] Rumyantsev SN Toward Molecular Level of the "*Salmonella*-Victim" Ecology, Genetics, and Evolution . *TheScientificWorldJOURNAL 1*, 193-199 (2004).

[106] Falkow S. What is a pathogen? *ASM News 63*, 359-365 (1997).

[107] Libby SJ *et al.* A cytolysin encoded by Salmonella is required for survival within macrophages. *Proc. Natl. Acad. Sci. USA 91*, 489-493 (1994).

[108] Simonian RV Hemolytic activity of Salmonella typhi murium. *Zh. Exptl. Clin. Med. 1*, 372-374 (1988).

[109] Burgasov,P.N. and Rumyantsev S.N. *Antimicrobial Constitutional Immunity. [in Russian]* (Meditsina, Moscow. 1985).

[110] Samson M, Soularue P, Vassart G, and Parmentier M The genes encoding the human CC-chemokine receptors CC-CKR1 to CC-CKR5 (CMKBR1-CMKBR5) are clustered in the p21.3-p.24 regions of chromosome 3. *Genomics 36*, 522-526 (1996).

[111] Carrington M, Dean M, Martin MP, and O'Brien S Genetics of HIV-1infection: chemokine receptor CCR5 polymorphism and its consequences . *Human Molecular Genetics 8*, 1939-1945 (1999).

[112] Kaslow RA, Dorak T, and Tang J Influence of Host Genetic Variation on Susceptibility to HIV Type 1 Infection. *The Journal of Infectious Diseases 191*, S68-S77 (2005).

[113] Swann A.I. The relationship of erythrocyte age and parasitisation with Plasmodium gallinaceum in chickens. *Can. J. comp. Med. 38*, 391-397 (1974).

[114] Thornton R *American Indian holocaust and survival*(University of Oklahoma Press, Norman, 1987).

[115] Chang KC, Goldspink G, and Lida J Studies in the in vivo expression of the influenza resistance gene Mx by in-situ hybridisation. *Arch Virol 110*, 151-164 (1990).

[116] Rumyantsev SN The best defence against bioweapons has already been invented by evolution. *Infection, Genetics and Evolution 4*, 159-166 (2004).

[117] Soloukhin VZ *Tsirculatsiia virusov grippa A v biosfere:gipotezi i fakti*(Belarus, Minsk, 1987).

[118] Webster RG, Yakhno MA, Hinshaw VS, Bean WJ, and Murti KG Intestinal influenza: replication and characterization of influenza viruses in ducks. *Virology 84*, 268-278 (1978).

[119] Murphy BR *et al.* Evaluation of the A/Seal/Mass/1/80 virus in squirrel monkeys. *Infect Immun 42*, 426 (1983).

[120] Porter DD, Prince GA, and Hemming VG et al Pathogenesis of human Parainfluenza virus 3 infection in two species of cotton rats: Sigmodon hispidus develops bronchiolitis, while Sigmodon fulviventer develops interstitial pneumonia. *J Virology 65*, 103-111 (1991).

[121] Koehler TM *Anthrax*(Springer, Berlin, 2002).

[122] Cox J.H., Dietzschold B., and Weiland F et al. Preparation and characterization of rabies virus hemagglutinin. *Infect. Immun.* . *30*, 572-577 (1980).

[123] Scott FT *et al.* The burden of Herpes Zoster: A prospective population based study. *Vaccine 24*, 1308-1314 (2006).

[124] Hibberd ML, Sumiya M, Summerfield JA, Booy R, and Levin M Association of variants of the gene for mannose-binding lectin with susceptibility to meningococcal disease. *Lancet 353*, 1049-1053 (1999).

[125] Bell DM Occupational risk of human immunodeficiency virus infection in healthcare workers: an overview. *Am. J. Med. 102*, 9-15 (1997).

[126] Leynaert B, Downs AM, and de Vincenzi I Heterosexual transmission of human immunodeficiency virus: variability of infectivity throughout the course of infection. *Am J Epidemiol 148*, 88-96 (1998).

[127] Piatak M Jr *et al.* High levels of HIV-1 in plasma during all stages of infection determined by competitive PCR. *Science 259*, 1749-1754 (1993).

[128] Gibbons RJ and van Houte J Bacterial adherence and the formation of dental plaques in *Bacterial adherence* (ed. Beachey EH) 62-104 (Chapman and Hall, New York, 1980).

[129] van Riel D *et al.* H5N1 Virus Attachment to Lower Respiratory Tract . *Science ???*, ?? (2006).

[130] Ricketts T.F. *The diagnosis of small pox*1966).

[131] Christie A.B. *Infectious disease*(Churchill Livingstone, Edinburgh, 1987).

[132] Vavilov NI *Immunity of plants to infectious diseases (In Russian)* Moscow, 1918).

[133] Bodian D and Horstman D.M. Poliovirus in *Viral and ricketsial infections of man.* (eds. Horsfall F.L. and Tamm I.) (Pitman, London, 1965).

[134] Sweany H.C., Cook C.E., and Kegerreis R A study of the position of primary cavities in pulmonary tuberculosis. *Am. Rev. Tuberc. 24*, 558-582 (1931).

[135] Raoult D. Host factors in the severity of Q fever. Annals of the NYAS 590, 33-38. 1990. New York, NYAS. Ref Type: Magazine Article

[136] Neiburger E.J. Syphilis in a Pleistocene bear? *Nature*603 (1988).

[137] Russel CE *Plant breeding for pest and disease resistance*(Butterworth, London, 1978).

[138] Bridges CB, Kuehnert MJ, and Hall CB Transmission of influenza: implications for control in health care settings. *Clin Infect Dis 37*, 1094-1101 (2003).

[139] Salgado CD, Farr BM, Hall KK, and Hayden FG Influenza in the acute hospital setting. *Lancet Infect Dis 2*, 145-155 (2002).

[140] Rimmelzwaan GF *et al.* Influenza A virus (H5N1) infection in cats causes systemic disease with potential novel routes of virus spread within and between hosts. *Am J Pathol 168*, 176-183 (2006).

[141] Heatley R.V. *The Helicobacter pylory handbook*(Blackwell Science, Oxford, 1998).

[142] Rotterdam H and Enterline HT *Pathology of the stomach and duodenum.* (Springer Verlag Inc. New York: 1989).

[143] Genta G. Helicobacter pylori in a gastric pit. *N. Engl. J. Med. 335*, 250 (1996).

[144] *Gastroduodenal disease and Helicobacter pylori*(Springer, Berlin, Heidelberg, 1999).

[145] Heggie A.D. Pathogenesis of the rubella exanthem: distribution of rubella virus in the skin during rubella with and without rash. *J Inf Dis 137*, 74-76 (1978).

[146] Moehring J.M.,M.T.J. Characterization of the diphteria toxin resistance system in Chinese hamster ovary cells. Somat Cell Genet , 453-468. 1979. Peltola H. 1983. Meningococcal disease: still with us. *Reviews of Infectious Diseases* 5:71-91.

[147] Rumyantsev S.N., Shabalow N.P., Pyasetskaya M.F., Rogacheva N.M., and Bardakova L.I. Species, population and age diversity in cell resistance to adhesion of Neisseria meningitidis serogroups A, B and C. *Microbes and Infection 2*, 447-453 (2000).

[148] Rumyantsev SN Genetic immunity and influenza pandemics. *FEMS immunology and medical microbiology*(2006).

[149] Moehring J.M. and Moehring T.J. Comparison of diphtheria intoxication in human and non-human cell lines ant their resistant variants. *Infection and Immunity 13*, 221-228 (1976).

[150] Black-Schaffer B., Hiebert T.G., and Kerby G.P. Experimental study of purpuric meningococcaemia in relation to the Schwartzman phenomenon, with discussion of meningococcic purpura, the Waterhouse-Fridrichsen syndrome and bilateral renal cortical necrosis. *Arch. Path. 43*, 28-37 (1943).

[151] Arnheiter H.,S.S.N.M.C.S.a.M.E. Transgenic mice with intracellular immunity to influenza virus. *Cell 1990; 62*, 51-61 (1990).

[152] Thomas G.A., Williams D., and Williams E.D. The demonstration of tissue clonality by X-linked enzyme histochemistry. *J Pathol 155*, 101-108 (1988).

[153] Carbone KM, Luftig RB, and Buckley MR *Microbial Triggers of Chronic Human Illness.*(American Academy of Microbiology, Washington, DC, USA, 2005).

[154] Watkins H and Farrall M Genetic susceptibility to coronary artery disease: from promise to progress. *Nature Reviews Genetics 7*, 163-173 (2006).

[155] Hsu CY Recent Advances in the Development of Therapeutics to Reduce Brain Damage after Stroke in *7th International Conference on Health Problems Related to the Chinese* (The Chinese American Medical Society, New York, 1994).

[156] Chu J Epidemiology of Cerebrovascular Disease Among Chinese-Canadians in *7th International Conference on Health Problems Related to the Chinese* (The Chinese American Medical Society, New York, 1994).

[157] Fabricant CG *Herpesvirus-induced atherosclerosis*(Springer Verlag, New York, 1984).

[158] Sing S.F.,M.P.P. Genetics of atherosclerosis. Annu Rev Genet [24:], 171-187. 1990. Lusis AJ, Fogelman AM, and Fonarow GC Genetic Basis of Atherosclerosis: Part II. *Circulation 110*, 2066-2071 (2004).

[159] Flegal KM, Carroll MD, Kuczmarski RJ, and Johnson CL Overweight and obesity in the United States: prevalence and trends, 1960-1994. *Int J Obes Relat Metab Disord 22*, 39-47 (1998).

[160] Mann CC Provocative study says obesity may reduce U.S. life expectancy. *Science 307*, 1716-1717 (2005).

[161] Fernandez JR and Shiver MD Using genetic admixture to study the biology of obesity traits and to map genes in admixed populations. *Nutr Rev 62*, S69-S74 (2004).

[162] Flegal KM, Carroll MD, Ogden CL, and Johnson CL Prevalence and trends in obesity among US adults, 1999-2000. *JAMA 288*, 1723-1727 (2002).

[163] Mokdad AH, Ford ES, ,V.F., Marks JS, and Koplan JP . The continuing epidemics of obesity and diabetes in the United States. *JAMA 286*, 1195-1200 (2001).

[164] Pietrobelli A and Steinbeck K S Paediatric obesity: what do we know and are we doing the right thing? *International Journal of Obesity 28*, 2-3 (2004).

[165] Allison DB, Edlen-Nezin L, and Clay-Williams G Obesity among African American women: prevalence, consequences, causes, and developing research Womans Health. *Nature 33*, 243-274 (1997).

[166] Fernandez JR and Allison DB. Understanding racial differences in obesity and metabolic syndrome traits. Nutr Rev 61[9], 316-319. 2003.

[167] Allison D *et al.* The heritability *of* body mass index among an intemational sample of monozygotic twins reared apart. *Int. J Obes Relat Metab Disord 20*, 501-506 (1996).

[168] Stunkard AJ, Foch TT, and Hrubec Z . A twin study of human obesity. *J Am Med Assoc 256*, 51-54 (1986).

[169] Stunkard AJ, Harris JR, Pedersen NL, and McClean GE The body-mass index of twins who have been reared apart . *N Engl J Med 322*, 1483-1487 (1990).

[170] Barsh GS, ,F.I., and O'Rahilly S Genetics of body-weight regulation. *Nature 404*, 644-645 (2000).

[171] Roth,J., Qiang,X., Marban,S.L., Redelt,H., and Lowell,B.C. The Obesity Pandemic: Where Have We Been and Where Are We Going? *Obesity Research 12*, 88S-101 (2004).

[172] Friedman JM Modern science versus the stigma of obesity. *Nature Medicine 10*, 563-569 (2004).

[173] Gibson WT *et al.* Melanin-concentrating hormone receptor mutations and human obesity: functional analysis. *Obes Res 12*, 743-749 (2004).

[174] Friedman JM and Halaas JL. Leptin and the regulation of body weight in mammals. 396, 763-770. 1998. *Nature 396*, 763-770 (1998).

[175] Montague CT and Farooqi IS Congenital leptin deficiency is associated with severe early-onset obesity in humans. *Nature 387*, 903-908 (1997).

[176] Wan W, Farboud B, and Privalsky ML Pituitary resistance to thyroid hormone (RTH)-syndrome is associated with T3 receptor mutants that selectively impair{beta}2 isoform function . *Mol Endocrinol*(2005).

[177] Ledley FD *et al.* Sequence similarity between cholera toxin and glycoprotein hormones: implications for structure activity relationship and mechanism of action. *Biochem Biophys Res Commun 69*, 852-859 (1976).

[178] Katz MS and Greenough WB Selective inhibition of cholera toxin- and catecholamine stimulated lipolysis by blocking agents. *Infect Immun 12*, 964-968 (1975).

[179] Lee G *et al.* Structure:function studies of receptor for thyrotropin and tetanus toxin: Lipid modulation of effector binding to the glycoprotein receptor component. *Biochem Biophys Res Commun 83*, 313-320 (1978).

[180] Jondal M, Klein G, Oldstone MBA, and et al Association between complement and Epstein-Barr virus receptors on human lymphoid cells . *Scand J Immunol 5*, 401-410 (1976).

[181] Popovic M, Sarin PS, Robert-Gurroff M, and et al. Isolation and transmission of human retrovirus (human T-cell leukemia virus) . *Science 219*, 856-859 (1983).

[182] Enegren BJ and Burness ATH Chemical structure of attachment sites for viruses on human erythrocytes. *Nature 268*, 536-537 (1977).

[183] Nagashima K, Zabriskie JB, and Lyons MJ. Virus-induced obesity in mice: association with a hypothalamic lesion. *J Neuropathol Exp Neurol 51*, 101-109 (1992).

[184] Atkinson RL *et al.* . Evaluation of human adenoviruses as an etiology of obesity in chickens. *Int J Obes 25 (Suppl. 2)*, S12 (2001).

[185] So PW, Herlihy AH, and Bell JD Adiposity induced by adenovirus 5 inoculation. *Int JObes (Lond) 29*, 603-606 (2005).

[186] Whigham LD, Israel BA, and Atkinson RL Adipogenic potential of multiple human adenoviruses *in vivo* and *in vitro* in animals. *Am J Physiol Regul Integr Comp Physiol 290*, R190-R194 (2006).

[187] Atkinson RL *et al.* Human adenovirus-36 is associated with increased body weight and paradoxical reduction of serum lipids. *Int J Obes (Lond) 29*, 281-286 (2005).

[188] Vangipuram SD, Sheel J, Atkinson RL, Holland TC, and Dhurandhar NH A human adenovirus enhances preadipocyte differentiation. *Obes Res 12*, 770-777 (2004).

[189] Rumyantsev SN The intra-individual diversity in senescence. *Biogerontology 4*, 171-178 (2003).

[190] Rumyantsev,S.N. Chemical ecology and biomolecular evolution. *Acta. Biotheor. 45*, 65-80 (1997).

[191] Rumyantsev SN The mosaic of organism: physiological, pathological and biotechnological aspects (in Russian). in *The state and the development's perspectives of agricultural biotechnology* 210-215 Moscow, 1986).

[192] Farooqi IS Partial leptin deficiency and human adiposity. *Nature 414*, 34-35 (2001).

[193] Fernandez JR, Shriver MD, Beasley TM, and et al Association of African Genetic Admixture with Resting Metabolic Rate and Obesity Among Women . *Obesity Research 11*, 904-911 (2003).

[194] Genesis,24, 3 33 1984).

[195] Genesis 5-57 1984).

[196] Christensen K The quest for genetic determinants of human longevity: challenges and insights. *Nature Reviews Genetics 7*, 436-448 (2006).

[197] Allard M, Lebre V, and Calment J *Jeanne Calment: From Van Gogh's time to ours:122 extraordinary years*(WH Freeman and Co, New York., 1998).

[198] Finch CE *Senescence, Longevity, and the Genome* Chicago, IL, 1990).

[199] Friedman D and Johnson T A mutation in the *age-1* gene in *Caenorhabditis elegans* lengthens life and reduces hermaphrodite fertility. *Genetics 118*, 75-86 (1988).

[200] Huang S, Gray MD, Oshima J, Mian S, and Campisi J The premature aging syndrome protein WRN is a 3' to 5' exonuclease. *Nature Genetics 20*, 114-116 (1998).

[201] Scaffidi P and Misteli T Lamin A-Dependent Nuclear Defects in Human Aging. *Science 1*, 1-2 (2006).

[202] Semsei I On the nature of aging. *Mech. Ageing Dev 117*, 93-108 (2000).

[203] Rattan SIS Biogerontology: the next step. in *Molecular and cellular gerontology* 290 New York, 2006).

[204] Kligman AM, Grove GL, and Balin AK Aging of human skin in *Handbook of the biology of aging* (eds. Finch CE and Schneider EL) (Van Nostrand, New York, 1985).

[205] Giacometty L. Hair growth and aging in *Advances in the biology of the skin.* (ed. Montagna W) 97-118 (Pergamon Press, Oxford, 1965).

[206] Shock NW *Longitudinal studies of aging in humans.* (Van Nostrand Reinold, New York, 1985).

[207] Dimri GP *et al.* A novel biomarker identifies senescent human cells in culture and aging skin in vivo. Proc.Nat'l Acad.Sci.USA 92, 9363-9367. 1995. Ref Type: Magazine Article

[208] Strehler B *Time, Cells, and Aging*(Cyprus, 1999).

[209] Finch CE and Kirkwood T *Chance, development and aging*(Oxford University Press, Oxford, 2001).

[210] Robson H.G. and Vas S.I. Resistance of inbred mice to Salmonella typhimurium. *J Infect Dis 126*, 378-386 (1972).

[211] *Diagnostic and Statistical Manual*(American Psychiatric Association, Washington, DC, 1994).

[212] Andreasen NC *The Broken Brain: The Biological Revolution in Psychiatry*(Harper and Row, New York, 1984).

[213] Eaton WW Epidemiology of schizophrenia. *Epidemiol Rev 7*, 105-126 (1985).

[214] Allen JS and Sarich VM. Schizophrenia in an evolutionary perspective. *Perspectives in Biology andMedicine 32*, 132-153 (1988).

[215] Rin H and Lin T-Y Mental illness among Formosa aborigines as compared with the Chinese in Taiwan. *J Ment Sci 108*, 134-146 (1962).

[216] Dale PW Prevalence of schizophrenia in the Pacific Island populations of Micronesia. *J Psychiatr Res 16*, 103-111 (1981).

[217] Waldo MC Schizophrenia in Kosrae, Micronesia: prevalence, gender ratios, and clinical symptomatology. *Schizophr Res*(1999).

[218] Weiner BP and Marvit RC Schizophrenia in Hawaii: analysis of cohort mortality risk in a multi-ethnic population. *Br J Psychiatry 131*, 497-503 (1977).

[219] Lieberman JA, Perkins D, Belger A, and et al The early stages ofschizophrenia: speculations on pathogenesis, pathophysiology. *Biol Psychiatry 50*, 884-897 (2001).

[220] Sartorius N, Jablensky A, and Shapiro R Cross-cultural differences in the short-term prognosis of schizophrenic psychoses. *Schizophr Bull 4*, 102-103 (1978).

[221] Janowsky DS, el-Yousel MK, Davis JM, and Sekerke HJ Provocation of schizophrenic symptoms by intravenous administration of methylphenidate. *Arch Gen Psychiatry 28*, 185-191 (1973).

[222] Wong DF *et al.* Positron emission tomography reveals elevated D2 dopamine receptors in drug-naive schizophrenics. *Science 234*, 1558-1563 (1986).

[223] van Kammen DP *et al.* Amphetamine induced heterogeneous changes in psychotic behavior in schizophrenia. *Am J Psychiatry 139*, 991-997 (1982).

[224] Kornetov NA. [Interrelations between the main forms of the course of schizophrenia and the morphological phenotype of patients' constitution (clinico- anthropometric data][Article in Russian]. *Zh Nevropatol Psikhiatr Im S S Korsakova 91*, 104-108 (1991).

[225] Dvirskii AE [Problem of the etiology of schizophrenic pathomorphosis][Article in Russian]. *Zh Nevropatol Psikhiatr Im S S Korsakova 81*, 1863-1866 (1981).

[226] Waddington JL *et al.* Early cerebro-craniofacial dysmorphogenesis in schizophrenia: a lifetime trajectory model from neurodevelopmental basis to 'neuroprogressive' process. *J Psychiatr Res 33*, 477-489 (1999).

[227] Glinda S, Cooper JE, and Stroehlab BC The epidemiology of autoimmune diseases. *Autoimmunity Reviews 2*, 119-125 (2003).

[228] Agarwal SS and Naik S Current trends in autoimmune diseases. *Immunology Today 19*, 436 (1998).

[229] Lamberta N and Nelson L Microchimerism in autoimmune disease: more questions than answers? *Autoimmunity Reviews 2*, 133-139 (2003).

[230] Stacya S, Krolick KA, Infante AJ, and Kraig E Immunological memory and late onset autoimmunity . *Mechanisms of Ageing and Development 123*, 975-985 (2002).

[231] Howard JF. Myasthenia gravis – a summary. www.myasthenia.org/information/summary.htm . 1997. Myasthenia Gravis Foundation of America.

[232] Ankylosing Spondylitis. http://www.spondylitis.org/about/as.aspx . 2005. Spondylitis Association of America.

[233] Nesse R.M. and Williams G.C. Research desing that address evolutionary questions about medical disorders in *Evolution in health and disease* (ed. Stearns S.C) 16-23 (Oxford University Press, .Oxford, 1999).

[234] UNAIDS, *AIDS Epidemic Update,* December 2005; www.unaids.org/epi/2005/.

INDEX

B

C

D

E

F

G

H

I

J

K

L

M

N

O

Q

R

S

T

U

V

W

Y

Z